158415

CHARACTERISTIC RAMAN FREQUENCIES OF ORGANIC COMPOUNDS

CHARACTERISTIC RAMAN FREQUENCIES OF ORGANIC COMPOUNDS

FRANCIS R. DOLLISH
Carnegie-Mellon University
Pittsburgh, Pennsylvania

WILLIAM G. FATELEY
Kansas State University
Manhattan, Kansas

FREEMAN F. BENTLEY
Wright-Patterson Air Force Base
Ohio

A WILEY-INTERSCIENCE PUBLICATION

JOHN WILEY & SONS, New York • London • Sydney Toronto

PREFACE

In the decade that followed the announcement of the discovery of the Raman effect by Sir C. V. Raman in March 1928, over 2000 publications reporting the spectra of more than 4000 compounds appeared. The classic work of Kohlrausch summarizes the theoretical and experimental work on the correlation of Raman spectra with molecular structure during this time. In the late 1940's and early 1950's advances in the field of infrared instrumentation led to the introduction of infrared spectrometers which offered the vibrational spectroscopist an easier, quicker, and less expensive method for routine structural analysis. The rapid development of infrared spectroscopy soon relegated Raman spectroscopy to the province of the physicist and the research molecular spectroscopist. However, the introduction of the laser as an excitation source and other instrumental and sampling improvements in the mid 1960's have overcome many of the experimental difficulties heretofore associated with Raman spectroscopy, and spectra of most compounds may now be recorded in a routine manner.

Our first effort in the application of Raman spectroscopy to molecular structure identification was in obtaining and interpreting the laser Raman spectra of about 90 alkyl- and halogenated alkyl-benzenes. By combining these data, along with the older work using mercury arc excitation and the more recent investigations involving normal coordinate analysis of aromatic compounds, we were able to demonstrate the utility of laser Raman spectroscopy as a practical analytical tool in the analysis of benzene derivatives. This material was presented by one of us (W. G. F.) at the 1970 Laser Raman Institute and Workshop (sponsored by the Center for Materials Research, University of Maryland, Ellis R. Lippincott, Director) and the interest and encouragement engendered there persuaded us to extend this approach to other classes of chemical compounds. Approximately three years, eight hundred laser Raman spectra, and fourteen hundred references later we have completed this survey of the application of Raman spectroscopy to the identification of organic compounds.

We express our appreciation to Dr. Gerald L. Carlson, Dr. Ernesto C. Tuazon, Howard J. Sloane, and Professor F. A. Miller for many helpful conversations. We extend our thanks to David F. Pensenstadler, Francis Emory Kurka, and Ms. Peggy Johnston for recording the many Raman spectra and to Betty Ely for her assistance in the literature survey. We offer sincere thanks to Carolyn Keller who with patience and good cheer typed the first draft with its seemingly endless tables and references and who with the assistance of Marsha Melzer typed the final manuscript. One of us (W. G. F.) gratefully acknowledges partial support from an unrestricted grant provided by the Gulf Research and Development Foundation. We also appreciate the partial financial support received from the Mellon Institute of Science of Carnegie-Mellon University and from Air Force Contracts F 33 615-70-C-1382 and F 33 615-71-C-1157.

Francis R. Dollish
William G. Fateley
Freeman F. Bentley

Pittsburgh, Pennsylvania
Manhattan, Kansas
Wright-Patterson Air Force Base, Ohio
July 1973

CONTENTS

INTRODUCTION

The recent resurgence of interest in Raman spectroscopy has occasioned the publication of numerous review articles and books that deal with various aspects of the theory, experimental methods, and applications of this technique. Therefore, we refrain from giving a repetitious account of the basic theory and instrumentation. An elementary introduction to the Raman effect with special emphasis on inorganic chemistry has been provided by Tobias (1). General surveys of the application of laser Raman spectroscopy have been made by Hendra et al. (2,3). Reviews of recent developments in the field appear annually in the *Annual Reports on the Progress of Chemistry,* published by the Chemical Society (London) and biennially in the review issues of *Analytical Chemistry.* RAMAN NEWSLETTER, the monthly publication of the Raman Technical Group of the Optical Society of America (2100 Pennsylvania Avenue, Washington, D. C. 20037), affords a means of rapid exchange of information between active workers in the field of Raman spectroscopy and includes a current bibliography of the Raman literature.

The classic work in the theoretical interpretation of the Raman effect is that of Placzek (4). This topic is also well covered in the basic texts of Herzberg (5,6) and in a recent book by Koningstein (7). The experimental aspects of laser Raman spectroscopy are still in a state of flux, but the state-of-the-art expertise is reviewed in the books of Gilson and Hendra (8) and Tobin (9).

A detailed and comprehensive survey of the Raman studies during the 1930's and early 1940's is contained in the works of Kohlrausch (10-12). A similar task reviewing the Raman literature to 1938 was undertaken by Hibben (13) and extended by Glockler (14) to the years 1939-1942. Although parts of these works are dated, they still offer valid insights into the application of Raman spectroscopy to molecular structure elucidation.

Several attempts have been made to integrate the use of both infrared and Raman group frequencies in molecular structure determination (9,15-17). However, considering the relative volumes of the IR and Raman literature, these approaches have been, quite understandably, more infrared oriented. Such an integrated approach was also contemplated by us but the excellence of the texts available in the field of infrared group frequencies (15,16,18, and 19) dissuaded us from this formidable task. We have, however, discussed the infrared results together with the Raman in those cases where the information derived is highly complementary, for example, the vibrational spectra of carboxylic acid dimers.

We have also limited our discussion of Raman group frequencies to organic compounds. Extensive bibliographies have been published on the vibrational spectra of inorganic and organometallic compounds covering the periods 1935–1966 (20) and 1967–1969 (21). The infrared and Raman spectral data for inorganic and coordination compounds have been discussed by Siebert (22) and by Nakamoto (23). The application of Raman spectroscopy in the field of coordination chemistry has been reviewed by Hester (24) and extensive compilations of useful spectral data relating to metal-ligand vibrations can be found in the book of Adams (25). Laser Raman spectroscopy has also become an important technique in the characterization of polymers. This subject has been treated in detail by Gilson and Hendra (8) and in several comprehensive reviews (26–29).

Tabulations of the band positions and relative intensities in the mercury arc excited Raman spectra of many organic compounds are found in Kohlrausch (12) and in the Landolt-Börnstein tables (30) together with references to the Raman spectra of specific compounds in the literature up to 1950. The American Petroleum Institute Research Project 44, Raman Spectral Data, (Thermodynamics Research Center Data Project, Texas A&M University) includes over 400 Raman spectra and plans to issue laser Raman spectral data are in progress. It has also been announced that laser Raman spectra will be added to the Sadtler Standard Spectra Collection (Sadtler Research Laboratories, 3316 Spring Garden Street, Philadelphia, Pa.). Plans to publish a Raman Spectral Data Compendium, each data sheet of which contains two laser Raman spectra (parallel and perpendicular polarization) and one infrared spectrum for each compound, have been announced by Heyden & Son, Ltd. (Spectrum House, Alderton Cresent, London NW4, England).

REFERENCES

1. R. S. Tobias, *J. Chem. Ed., 44,* 2, 70 (1967).
2. P. J. Hendra and P. M. Stratton, *Chem. Rev., 69,* 325 (1969).
3. P. J. Hendra and C. J. Vear, *Analyst, 95,* 321 (1970).
4. G. Placzek, *Handbuch der Radiologie,* Vol. VI, Part 2, E. Mark, Ed., Akademische Verlagsgesellschaft, Leipzig, 1934, pp. 209-374 [Translation U. S. Atomic Energy Commission, UCRL-256 (L), (1962)].

5. G. Herzberg, *Molecular Spectra and Molecular Structure I. Spectra of Diatomic Molecules,* 2nd ed., Van Nostrand, New York, 1950.
6. G. Herzberg, *Molecular Spectra and Molecular Structure II. Infrared and Raman Spectra of Polyatomic Molecules,* Van Nostrand, Princeton, 1945.
7. J. A. Koningstein, *Introduction to the Theory of the Raman Effect,* D. Reidel, Dordrecht-Holland, 1972.
8. T. R. Gilson and P. J. Hendra, *Laser Raman Spectroscopy,* Wiley-Interscience, London, 1970.
9. M. C. Tobin, *Laser Raman Spectroscopy,* Wiley-Interscience, New York, 1971.
10. K. W. F. Kohlrausch, *Der Smekal-Raman-Effect,* Springer, Berlin, 1931.
11. K. W. F. Kohlrausch, *Der Smekal-Raman-Effect, Ergänzungsband 1931–1937,* Springer, Berlin, 1938; reprinted by Edwards, Ann Arbor, Michigan, 1944.
12. K. W. F. Kohlrausch, *Ramanspektren,* Band 9, Abschnitt VI, *Hand-und Jahrbuch der Chemischen Physik,* A. Euken and K. L. Wolf, Eds., Akademische Verlagsgellschaft Becker & Erler, Leipzig, 1943; reprinted by J. W. Edwards, Ann Arbor, Michigan, 1945; reprinted by Heyden & Son, Ltd., London, 1972.
13. J. H. Hibben, *The Raman Effect and Its Chemical Applications,* Reinhold, New York, 1939.
14. G. Glockler, *Rev. Mod. Phys., 15,* 112 (1943).
15. R. N. Jones and C. Sandorfy, in *Chemical Applications of Spectroscopy,* Techniques of Organic Chemistry, W. West, Ed., Interscience, New York, 1956.
16. N. B. Colthup, L. H. Daly, and S. E. Wiberley, *Introduction to Infrared and Raman Spectroscopy,* Academic Press, New York, 1964.
17. H. A. Szymanski, *Correlation of Infrared and Raman Spectra of Organic Compounds,* Hertillon Press, Cambridge Springs, Pa., 1969.
18. L. J. Bellamy, *The Infra-red Spectra of Complex Molecules,* 2nd ed., Methuen, London, 1958.
19. L. J. Bellamy, *Advances in Infrared Group Frequencies,* Methuen, London, 1968.
20. N. N. Greenwood, E. J. F. Ross, and B. P. Straughan, *Index of Vibration Spectra of Inorganic and Organometallic Compounds,* Vol. 1 (1935–1960), CRC Press, Cleveland, Ohio, 1972; Vol. 2 (1961–1966) in preparation.
21. N. N. Greenwood, Ed., *Spectroscopic Properties of Inorganic and Organometallic Compounds,* Vols. 1–3 (a review of the literature published during 1967–1969, respectively), The Chemical Society, London, 1968–1970.

22. N. Siebert, *Anwendungen der Schwengungspektroskopie in der Anorganischen Chemie,* Springer, Berlin, 1966.
23. K. Nakamoto, *Infrared Spectra of Inorganic and Coordination Compounds,* 2nd ed., Wiley, New York, 1970.
24. R. E. Hester, *Coord. Chem. Rev., 2,* 319 (1967).
25. D. M. Adams, *Metal-Ligand and Related Vibrations,* Arnold, London, 1967.
26. R. F. Schaufele, *Trans. N. Y. Acad. Sci., 30,* 69 (1967).
27. P. J. Hendra, *Adv. Polym. Sci., 6,* 151 (1969).
28. J. L. Koenig, *Appl. Spectrosc. Rev., 4,* 233 (1971).
29. F. J. Boerio and J. L. Koenig, *J. Macromol. Sci. – Rev. Macromol. Chem., C7,* 209 (1972).
30. H. Pajenkamp, *Landolt-Börnstein Zahlenwerte und Funktionen aus Physik, Chemie, Astronomie, Geophysik and Technik,* Sechste Auflage, A. Euken, Ed., I Band, Atom - und Molekularphysik, 2 Teil, Molekelen I and 3 Teil, Molekelen II, Springer, Berlin, 1951.

CHARACTERISTIC RAMAN FREQUENCIES OF ORGANIC COMPOUNDS

CHAPTER ONE

ALKANES

1.1 INTRODUCTION

An understanding of the spectra-structure correlations in alkanes and cycloalkanes is basic to the interpretation of the vibrational spectrum of organic molecules. Correlations in the infrared (IR) spectra of these molecules have been extensively cataloged (1–3). The far IR spectra between 700 and 300 cm^{-1} have been discussed by F. F. Bentley, L. D. Smithson, and A. L. Rozek (4). Normal coordinate analysis of saturated hydrocarbons has been performed by R. G. Snyder and J. H. Schachtschneider (5–7).

The early work in the Raman spectra of hydrocarbons has been summarized by K. W. F. Kohlrausch (8). M. R. Fenske et al. (9,10) have published Raman spectra of 291 compounds, including paraffins, naphthenes, olefins, and aromatics. Characteristic frequencies in the Raman spectra of *n*-alkanes have been discussed by S. Mizushima and T. Shimanouchi (11,12) and those of saturated aliphatic hydrocarbons by N. Sheppard (13,14).

1.2 *n*-ALKANES

In the liquid state, many normal alkanes exist as mixtures of two or more rotational isomers. For example, *n*-butane can occur in two distinct conformations; I is the trans form corresponding to the planar zigzag configuration and IIa and IIb are the two equivalent gauche forms. At room temperature about 80% of the *n*-butane molecules are in the trans conformation and about 20% are in the gauche conformation. When *n*-butane is cooled and crystallized into the solid state, the molecules all assume the trans or zigzag conformation. Each rotational isomer has its own distinct set of vibrational frequencies and selection rules due

I 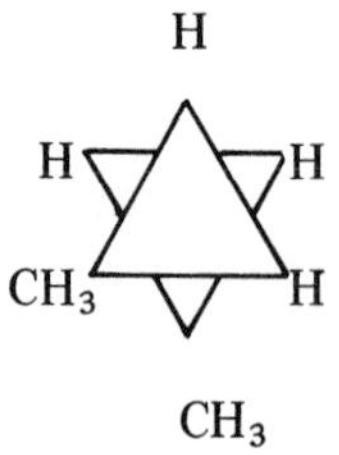 IIa 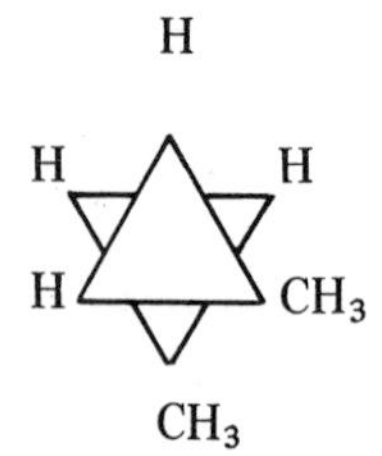 IIb

to the difference in symmetry of each conformation (e.g., the trans form of *n*-butane possesses C_{2h} symmetry and the gauche form C_2). In spite of this complication of rotational isomerism, characteristic group frequencies can be found in the alkanes, since certain similar vibrations of different isomers occur at nearly the same frequency. Also, the vibrational frequencies characteristic of such moieties as the *tert*-butyl and isopropyl groups are relatively unaffected by the conformation of the rest of the molecule.

J. H. Schachtschneider and R. G. Snyder (6) have carried out extensive normal coordinate calculations on a series of planar zigzag *n*-alkanes from C_2H_6 to *n*-$C_{14}H_{30}$. The general character of the normal modes and their frequency ranges are listed in Table 1.1. The carbon-hydrogen stretching vibrations appear in the range 3000-2700 cm^{-1}. These bands are generally intense in the Raman and the vibrations are more or less completely localized within the methyl and methylene groups and give rise to correlatable frequencies. Besides those bands given in Table 1.1, there are bands at 2912 and 2800-2700 cm^{-1}. These are overtone or combination bands of the deformation modes of CH_3 and CH_2 groups whose intensity is enhanced by interaction with carbon-hydrogen stretching fundamentals of the same symmetry classes. In the narrow range 1475–1450 cm^{-1} in the Raman spectra of *n*-alkanes, there is generally a broad band caused by the overlapping of the CH_3 antisymmetric deformation and the CH_2 scissors. The methylene twisting modes occur in the region 1310–1175 cm^{-1}. Actually, these modes are a mixture of rocking and twisting coordinates. The higher frequency corresponds to an in-phase twisting mode of the CH_2 groups and the lower frequency is an out-of-phase rocking mode. Intermediate frequencies are a combination of these modes. For *n*-alkanes, a band appears at 1305–1295 cm^{-1}, the intensity of which is proportional to the number of methylene groups in the molecule. This band is assigned to an in-phase twisting mode.

Carbon-carbon stretching vibrations fall in the range 1150–950 cm^{-1}. In the higher frequencies, there is coupling with the methyl rocking mode. Frequencies that are correlated with the *n*-alkane vibrations occur at 1150–1135 and

Table 1.1 Vibrational Analysis of *n*-Alkanes, C_nH_{2n+2}

Vibration	Frequency Range (cm^{-1})	
Methyl antisymmetric C-H stretching	For $n = 3$	2969–2965
	For $n > 3$	2967
Methyl symmetric C-H stretching	For $n = 3$	2884 and 2883
	For $n > 3$	2884
Methylene antisymmetric C-H stretching		2929–2912
Methylene symmetric C-H stretching		2861–2849
Methyl out-of-plane HCH deformation	For $n = 3$	1466 and 1465
	For $n > 3$	1465
Methyl and methylene in-plane HCH deformation		1473–1446
Methyl symmetric HCH deformation	For $n = 3,4$	1385–1368
	For $n > 4$	1376
Methylene wagging		1411–1174
C-C Stretching		1132–885
Methyl terminal rocking	For $n < 10$	975–835
	For $n > 10$	895
CCC Deformation		535–0
Methylene twisting-rocking		1310–1175
Methylene rocking-twisting		1060–719
CH_3-CH_2 Torsion		280–220
CH_2-CH_2 Torsion		153–0

1060–1056 cm^{-1}. In polyethylene, these Raman bands are observed at 1133 and 1061 cm^{-1}.

There are two series of strong bands that appear in the Raman spectra of *n*-alkanes, the frequencies of which depend upon the number of carbon atoms in the chain. The frequencies that are given in Table 1.2 satisfy the Born-Karmann periodicity condition $\lambda' = \pi/N$, where N is the number of carbon atoms. The Raman band that appears in the range 888-837 cm^{-1} is due to a mode that consists of in-plane methyl rocking, C-C stretching, and CCC deformation and is localized at the ends of the *n*-alkane chains. The second series (425–150 cm^{-1}) is

Table 1.2 Observed Skeletal Frequencies of *n*-Alkanes

n-Alkane	Skeletal Carbon Stretching (cm^{-1})	CCC Deformation Frequency (cm^{-1})
n-Butane	837	425
n-Pentane	869	406
n-Hexane	898	373
n-Heptane	905	311
n-Octane	899	283
n-Nonane	888	249
n-Decane	886	231
n-Dodecane	892	194
n-Hexadecane	888	150

attributed to a CCC bending mode with some C-C stretching, which tends to expand or contract the chain ("chain expansion" mode).

1.3 BRANCHED ALKANES

Characteristic frequencies in the Raman spectra of branched alkanes are shown in Figure 1.1. For the isopropyl group, the vibration in the range 1350–1330 cm^{-1} can be associated with the CH deformation. Those characteristic vibrations that lie between 1300–900 cm^{-1} can be described as varying mixtures of carbon-carbon stretching and methyl rocking which are specific for a given moiety. Those vibrations lying between 900 and 650 cm^{-1} can be ascribed to various carbon-carbon stretchings. Those vibrations located below 500 cm^{-1} are CCC deformations. Often a Raman band at 530 cm^{-1} indicates the presence of adjacent tertiary and quaternary carbon atoms. If an ethyl group is attached to the point of branching, bands at 1020–1000 cm^{-1} appear in the spectrum.

As an example of the procedure using the characteristic frequencies given in Figure 1.1 for identifying branched alkanes, the spectra of the nine C_7 isomers are analyzed. The spectra appear in the collection of representative Raman spectra in Appendix 2 (Spectra 1-9). These compounds have been the subject of a Raman investigation by Rosenbaum, Grosse, and Jacobson (15). The most easily identified vibration is the symmetric C_5 skeletal stretch present in the *tert*-butyl and internal quaternary carbon atom moieties. In the spectra of 2,2- and 3,3-dimethylpentane and in 2,2,3-trimethylbutane this vibration appears as an intense polarized band in the region 750–650 cm^{-1}. The Raman band at 524 cm^{-1} is characteristic of adjacent tertiary and quaternary carbon atoms and identifies

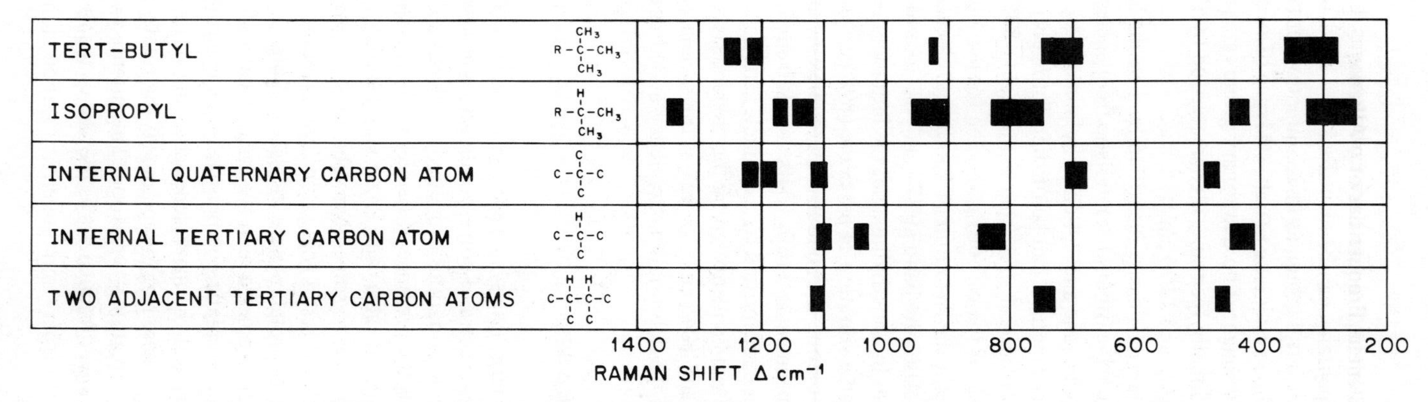

Figure 1.1 Characteristic frequencies in the Raman spectra of branched alkanes.

2,2,3-trimethylbutane. Internal quaternary carbon atoms, for example, those found in 3,3-dimethylpentane, can be identified by a band at 1190 cm^{-1}, whereas the presence of a *tert*-butyl group, as in 2,2-dimethylpentane, is confirmed by a band at 1250 cm^{-1}. In a systematic search, evidence of two adjacent tertiary carbon atoms is usually sought if no quaternary groups are found. This moiety, which is found in 2,3-dimethylpentane, is characterized by Raman bands at 750–720, 930–920, 1160, and 1190 cm^{-1}.

The isopropyl group can be easily identified by a strong Raman band near 1345 cm^{-1} together with the other characteristic frequencies given in Figure 1.1. The spectra of 2-methylhexane and 2,4-dimethylpentane show these specific vibrations. These two compounds can be further distinguished by the intensity of the symmetric carbon-carbon stretching in the region 835–749 cm^{-1} and the skeletal stretching mode of the carbon chain. The former vibration is approximately twice as intense in the spectrum of 2,4-dimethylpentane, which contains two isopropyl groups. Also, the band at 870 cm^{-1} is close to that of the skeletal frequency of *n*-pentane. In 2-methylhexane, the latter vibration is at 893 cm^{-1}, which is close to the same vibration of *n*-hexane (898 cm^{-1}). The presence of an internal tertiary carbon atom, as found in 3-methylhexane and 3-ethylpentane, is confirmed by Raman bands near 1160, 1040, and 450 cm^{-1}. The isomer containing the ethyl group can be identified by a characteristic band of this group at 1000 cm^{-1}. *n*-Heptane can be identified by the characteristic bands for *n*-alkanes found at 1303, 1139, and 1045 cm^{-1} as well as the skeletal stretch of the carbon chain at 905 cm^{-1} and the deformation frequency at 311 cm^{-1}.

1.4 CYCLOALKANES

The Raman spectra of large ring systems (C_{12} and higher) do not differ significantly from the Raman spectra of corresponding alicylic systems since the same strainless low energy conformations of methylene chains are adopted. However, for smaller ring systems the presence of Baeyer strain (deviation of valence angle from tetrahedral or trigonal value) and Pitzer strain (repulsion between neighboring nonbonded atoms) leads to some observable differences, as listed in Table 1.3. As the ring strain increases in going from cyclopentane down to cyclopropane, the methylene C-H stretchings are displaced to higher frequencies. In fact, the frequencies for cyclopropane are in the range usually identified with the C-H stretching of olefins and aromatic compounds. Also, cyclization produces a small decrease in the frequency of the methylene scissoring vibration. However, the most significant band in these ring compounds is the totally symmetric carbon-carbon stretch or ring "breathing" vibration. This intense polarized Raman band shifts progressively to lower frequencies as the size of the ring increases.

Table 1.3 Comparison of the Raman Spectra of Alkanes and Cycloalkanes

	Vibration Frequency (cm^{-1})			
Compound	Methylene CH_2 Antisymmetric Stretch	Methylene CH_2 Symmetric Stretch	Methylene Scissors	Ring "Breathing"
Cyclopropane	3101-3090	3038-3019	1443	1188
Cyclobutane	2987-2975	2895-2887	1443	1001
Cyclopentane	2959-2952	2866-2853	1455	886
Cyclohexane (chair isomer)	2933-2915	2897-2852	1452	802
Cycloheptane	2935-2917	2862-2851	1450	733
Cyclooctane	2925	2855	1467	703
n-Alkanes	2929-2912	2861-2849	1468	–

1.5 ALKYL-SUBSTITUTED CYCLOPROPANES

The ring "breathing" vibration of cyclopropane located at 1188 cm^{-1} in the Raman shifts to the region of 1220–1200 cm^{-1} in monosubstituted alkyl cyclopropanes and 1,2-disubstituted alkyl derivatives. In *gem*-1,1-disubstituted compounds, this vibration can be found at 1320 cm^{-1}. Other characteristic vibrations for these alkyl-substituted cyclopropanes have been reported (16-18). In other polysubstituted alkyl cyclopropanes, many of the ring vibrations are delocalized and strongly coupled with other ring vibrations and with vibrations of the alkyl substituents. In 1,1,2-trimethylcyclopropane and 1,1,2,2-tetramethylcyclopropane there are strong Raman bands at 1350 and 1308 cm^{-1}, respectively. In these compounds, there are also intense bands in the region of 700-650 cm^{-1} characteristic of the quaternary carbon atom.

1.6 ALKYL-SUBSTITUTED CYCLOBUTANES

In cyclobutane derivatives there are characteristic vibrations of the methylene groups at 1450–1420 cm^{-1} (CH_2 scissors vibration), 1260–1200 cm^{-1} (CH_2 wag), and near 740 cm^{-1} (CH_2 rock) (19). The intense Raman band from the carbon-carbon symmetric stretch is found at 933 cm^{-1} in alkyl-monosubstituted compounds. In *cis*-1,3-dimethylcyclobutane, this vibration is shifted to 887 cm^{-1} and in the *trans*-1,3-isomer it appears as a doublet at 891 and 855 cm^{-1} (20).

1.7 ALKYL-SUBSTITUTED CYCLOPENTANES

Table 1.4 contains the characteristic frequencies in the Raman spectra of monosubstituted alkyl cyclopentanes. The symmetric ring stretch of cyclopentane at 890 cm^{-1} is not greatly shifted upon alkyl substitution and is quite characteristic of the presence of the cyclopentyl group. The specific vibrations of di- and trisubstituted alkyl cyclopentanes are also given in Table 1.4. In these derivatives, the intensity of the Raman band that results from the symmetric ring stretch decreases with the amount of substitution. In 1,1 and 1,2-disubstituted derivatives the position of the symmetric ring "breathing" vibration is close to that of the monoalkyl compounds. However, for the 1,3-disubstitution and the trisubstituted alkyl cyclopentanes, this band shifts to lower frequencies depending on the number and positions of the substituents. The in-plane ring deformation and the methylene frequencies are in the same range as those of the monosubstituted compounds.

1.8 ALKYL-SUBSTITUTED CYCLOHEXANES

The specific vibrations for alkyl-substituted cyclohexanes are contained in Table 1.5. The prominent Raman band from the ring "breathing" vibration located at 802 cm^{-1} for the chair form of cyclohexane (Spectrum 10) is shifted to lower frequencies in the range 784–703 cm^{-1}. The Raman spectra of various dialkylcyclohexanes have been studied by G. N. Zhizhin et al. (21), who found certain frequencies could be used to distinguish the cis and trans isomers of these compounds. These correlations are:

Position of Substituents	Characteristic Frequencies (cm^{-1})
cis-1,2-	601–591
trans-1,2-	625–589
cis-1,3-	None
trans-1,3-	632–621
cis-1,4-	637–623
trans-1,4-	633–620

Table 1.4. Characteristic Frequencies in the Raman Spectra of Alkyl-Substituted Cyclopentanes

	Vibration Frequency (cm^{-1})					
Substituent	CH_2 Scissors	CH_2 Wag	CH_2 Twist	In-Plane Ring Bend	Ring "Breathing"	Ring Stretch
Monoalkyl-	1460–1446	1319–1293	1209–1187	1037–1012	899–889	861–837
1,1-Dimethyl-	1458	1309	1162	1034	888	838
cis-1,2-Dimethyl-	1459	1308	1198	1020	884	836
trans-1,2-Dimethyl-	1460	1292	1203	1021	896	862
cis-1-Methyl-2-ethyl-	1455	1292	1196	1026	889	838
cis-1,3-Dimethyl-	1464	1306	1190	1035	878	801
trans-1,3-Dimethyl-	1463	1317	1210	1028	823	775
cis-1-Methyl-3-ethyl-	1461	1316	1213	1045	838	781
1,1,3-Trimethyl-	1463	1311	1184	1050	841	790
1,2,2-Trimethyl-	1458	1286	1182	1060	890	832
cis, cis, cis-1,2,3-Trimethyl-	1457	1308	1200	1039	867	812
cis, trans, cis-1,2,3-Trimethyl-	1468	1277	1214	1035	876	814
cis, cis, trans-1,2,4-Trimethyl-	1464	1313	1173	1026	851	755
cis, trans, cis-1,2,4-Trimethyl-	1465	1314	1212	1031	813	768

Table 1.5 Characteristic Frequencies in the Raman Spectra of Alkyl-Substituted Cyclohexanes

	Vibration Frequency (cm^{-1})						
Substituent	CH_2 Scissors	CH_2 Wag	CH_2 Twist	CH_2 Rock	C-C Stretch	Ring "Breathing"	Ring Deformation
Monosubstituted							
Methyl-	1449	1350	1264	1168	1037	771	444
Ethyl-	1452	1352	1263	1165	1034	752,790	447
n-Propyl-	1453	1357	1268	1164	1035	784	443
Isopropyl-	1455	1348	1271	1166	1038	770	436
n-Butyl-	1450	1351	1269	1160	1033	773	443
Isobutyl-	1452	1352	1263	1162	1034	778	451
Disubstituted							
1,1-Dimethyl-	1446	1351	1268	1153	1028	703	456
cis-1,2-Dimethyl-	1453	1345	1259	1163	1058	729	472
trans-1,2-Dimethyl-	1459	1359	1256	1169	1081	749	499
cis-1,3-Dimethyl-	1465	1348	1271	1170	1058	769	446
trans-1,3-Dimethyl-	1451	1339	1269	1165	1056	753	454
cis-1,4-Dimethyl-	1450	1350	1268	1168	1057	761	469
trans-1,4-Dimethyl-	1467	1358	1255	1173	1067	762	475
Trisubstituted							
1,1,3-Trimethyl-	1452	1359	1290	1195	1080	728	465
cis,trans,cis-1,2,4-Trimethyl-	1459	1359	1280	1148	1079	769	490
cis,cis,trans-1,2,4-Trimethyl-	1459	1347	1313	1148	1066	748	498
cis,cis,trans-1,4-Dimethyl-2-ethyl-	1462	1348	1311	1149	1055	772	483
cis,trans,cis-1,4-Dimethyl-2-ethyl-	1460	1361	1313	1150	1042	761	474

REFERENCES

1. H. L. McMurry and V. Thornton, *Anal. Chem., 24,* 318 (1952).
2. L. J. Bellamy, *The Infrared Spectra of Complex Molecules,* Methuen, London, 1954; *Advances in Infrared Group Frequencies,* Methuen, 1968.
3. N. B. Colthup, L. H. Daly, and S. E. Wiberley, *Introduction to Infrared and Raman Spectroscopy,* Academic Press, New York, 1964.
4. F. F. Bentley, L. D. Smithson, and A. L. Rozek, *Infrared Spectra and Characteristic Frequencies 700-300 cm^{-1}*, Interscience, New York, 1968.
5. R. G. Snyder and J. H. Schachtschneider, *Spectrochim. Acta, 19,* 85 (1963).
6. J. H. Schachtschneider and R. G. Snyder, *Spectrochim. Acta, 19,* 117 (1963).
7. R. G. Snyder and J. H. Schachtschneider, *Spectrochim. Acta, 21,* 169 (1965).
8. K. W. F. Kohlrausch, *Ramanspektren,* Akademische Verlagsgesellschaft, Leipzig, 1943.
9. M. R. Fenske, W. G. Braun, R. V. Wiegand, Dorothy Quiggle, R. H. McCormick, and D. H. Rank, *Anal. Chem., 19,* 700 (1947).
10. W. G. Braun, D. F. Spooner, and M. R. Fenske, *Anal. Chem., 22,* 1074 (1950).
11. S. Mizushima and T. Shimanouchi, *J. Am. Chem. Soc., 71,* 1320 (1949).
12. T. Shimanouchi and S. Mizushima, *J. Chem. Phys., 17,* 1102 (1949).
13. N. Sheppard, *J. Chem. Phys., 16,* 690 (1948).
14. N. Sheppard and D. M. Simpson, *Quart. Rev., 7,* 19 (1953).
15. E. J. Rosenbaum, V. Grosse, and H. F. Jacobson, *J. Am. Chem. Soc., 61,* 689 (1939).
16. L. M. Sverdlov and E. P. Krainov, *Opt. Spectrosc., 7,* 296 (1959).
17. E. P. Krainov and L. M. Sverdlov, *Vysch. Ucheb. Zaved. Fiz., 10,* 31 (1967); *Chem. Abstr. 68,* 73628z.
18. V. T. Aleksanyan, M. R. Aliev, M. Yu. Lukina, O. A. Nesmeyanova, and G. A. Khotimiskaya, *Izv. Akad. Nauk SSSR, Ser. Khim., 1968,* 807; *Chem. Abstr. 69,* 31640 n.
19. H. E. Ulery and J. R. McClenon, *Tetrahedron, 19,* 749 (1963).
20. V. T. Aleksanyan, G. M. Kuz'yants, and M. Yu Lukina, *J. Struct. Chem. (USSR), 8,* 569 (1967).
21. G. N. Zhizhin, Kh. E. Sterin, V. T. Aleksanyan, and A. L. Liberman, *J. Struct. Chem. (USSR), 6,* 651 (1965).

CHAPTER TWO

HALOALKANES

2.1 HALOGENATED METHANES

The characteristic frequencies in the Raman spectra of some halogenated methanes are summarized in Table 2.1 Normal coordinate calculations for these molecules have been reviewed by Shimanouchi (1) and by Aldous and Mills (2,3). A study of the assignments in the vibrational spectra of eighteen completely halogenated methanes of the type $CX_nY_mZ_{4-n-m}$ where X = F, Y = C1 and Z = Br has been reported by Ngai and Mann (4).

2.2 MONOSUBSTITUTED HALOALKANES

Data on the correlation frequencies in the IR spectra for the carbon-halogen stretching vibrations of haloalkanes and a bibliography of the literature on this subject up to 1966 are contained in Chapter 4 of the book of Bentley, Smithson, and Rozek (16). A discussion of the Raman spectra of the isomeric monosubstituted haloalkanes, $C_nH_{2n+1}X$, where n = 1–5 and X = C1, Br, and I, is given in a series of five papers by Kohlrausch et al. (17–21). Further work on the Raman spectra of iodoalkanes was reported by Kahovec and Wagner (22). More recently, the structure and vibrational spectra of 3-chloro-3-ethylpentane were studied by Stokr, Caraculacu, and Schneider (23). This compound, as well as 3-chloropentane, was also investigated by Gasanov (24). A normal coordinate analysis for a series of *n*-alkyl chlorides was performed by Snyder and Schachtschneider (25) and one for secondary chlorides by Opaskar and Krimm (26). IR and Raman studies on the 2-halopropanes was published by Klaboe (27), on the 2-halobutanes by Keresturi, Ulyanova, and Pentin (28), and on the 2-halo-2-methylbutanes by Park and Wyn-Jones (29). The IR study of the rotational isomers in halopentanes and halooctanes was carried out by Gates, Mooney, and Willis (30).

Table 2.1 Characteristic Raman Frequencies of Halogenated Methanes[a]

	X = Chlorine		X = Bromine		X = Iodine	
	Frequency (cm^{-1})	Reference	Frequency (cm^{-1})	Reference	Frequency (cm^{-1})	Reference
1. CH_3X						
CX stretch	709 (vs)[f]	5	609 (s)	6	523 (vs)	7
2. CH_2X_2						
CX_2 symmetric stretch	703 (vs)	8	577 (s)	9	483 (vs)	9
CX_2 scissors	285 (s)		174 (s)		121 (vs)	
CX_2 antisymmetric stretch	742 (wk)		639 (wk)		566 (vs)	
3. CHX_3						
CX_3 symmetric stretch	668 (s)	10	539 (s)	11	437 (s)[b]	12
CX_3 symmetric deformation	366 (s)		222 (vs)		153 (vs)	
CX_3 degenerate stretch	761 (s)		656 (s)		578 (s)	
CX_3 degenerate deformation	262 (vs)		154 (s)		105 (vs)	
4. CX_4						
CX_4 symmetric stretch	459 (vs)	13	267 (s)[d]	14	178 (vs)[e]	15
CX_4 degenerate deformation (e)	217 (s)		122 (vs)		90 (m)	
CX_4 degenerate stretch	790,762 (m)[c]		671 (wk)		—	
CX_4 degenerate deformation (f_2)	314 (vs)		182 (m)		123 (m)	

[a]Unless otherwise specified, all values refer to the liquid state.
[b]In solution.
[c]Fermi resonance.
[d]In benzene solution.
[e]In solid state.
[f]The following notation is used in this table and occurs throughout the book: (wk) = weak, (m) = medium, (s) = strong, and (vs) = very strong.

The spectra-structure correlations for alkyl and cycloalkyl halogenides have been reviewed by Altona (31).

The data collected on the carbon-halogen stretching frequencies in the laser Raman spectra of primary haloalkanes are given in Tables 2.2 to 2.4. The frequencies of these vibrations are sensitive to: (1) the mass of the halogen atom, (2) the type of substitution (primary, secondary, or tertiary), and (3) the geometrical arrangement of the molecule in the vicinity of the carbon-halogen bond. To describe the various conformers of the haloalkanes the *P, S, T* nomenclature will be used. This system of notation was first suggested by Mizushima et al. (32), then later revised by Krimm et al. (26,33, 34). The type of substitution is characterized by *P* (primary), *S* (secondary), or *T* (tertiary). The atom or atoms in the antiperiplanar position are indicated by the subscript H if a hydrogen atom is trans to the halogen atom or C if a carbon atom is in the trans position. Primes are used on the subscripts to designate conformers in which, for the trans hydrogen, the next-nearest-neighbor carbon atoms have been rotated away from the planar zigzag position.

The haloethanes have only one possible rotational isomer, which is designated P_H and is illustrated in Figure 2.1 for chloroethane. The C-Cl stretch occurs at

Table 2.2 C-Cl Stretching Frequencies in the Raman Spectra of Primary Chloroalkanes

	Frequency (cm^{-1})		
Conformation	P_C	P_H	$P_{H'}$
Chloroethane	—	658	—
1-Chloropropane	725	651	—
1-Chlorobutane	722	650	—
1-Chloropentane	722	653	—
1-Chlorohexane	724	651	—
1-Chloroheptane	720	655	—
1-Chlorooctane	721	651	—
1-Chlorononane	729	651	—
1-Chlorodecane	725	651	—
1-Chloro-3-methylbutane	722	656	—
1-Chloro-2-methylpropane	730	—	688
1-Chloro-2-methylbutane	726	—	680
1-Chloro-2,2-dimethylpropane	722	—	—
Relative Raman intensity	Medium-strong	Very strong	Strong

Table 2.3 C-Br Stretching Frequencies in the Raman Spectra of Primary Bromoalkanes

Conformation	Frequency (cm^{-1})		
	P_C	P_H	P_H'
Bromoethane	–	559	–
1-Bromopropane	650	566	–
1-Bromobutane	645	564	–
1-Bromopentane	642	564	–
1-Bromohexane	644	564	–
1-Bromoheptane	645	563	–
1-Bromooctane	643	561	–
1-Bromodecane	646	564	–
1-Bromododecane	647	563	–
1-Bromo-3-methylbutane	646	563	–
1-Bromo-2-methylpropane	655	–	625
Relative Raman intensity	Very strong–strong	Very strong	Strong

Table 2.4 C-I Stretching Frequencies in the Raman Spectra of Primary Iodoalkanes

Conformation	Frequency (cm^{-1})		
	P_C	P_H	P_H'
Iodoethane	–	499	–
1-Iodopropane	600	510	–
1-Iodobutane	595	508	–
1-Iodopentane	591	504	–
1-Iodohexane	597	505	–
1-Iodoheptane	596	502	–
1-Iodooctane	597	504	–
1-Iodo-3-methylbutane	593	509	–
1-Iodo-2-methylpropane	608	–	588
Relative Raman intensity	Very strong	Very strong	Strong

CHLOROETHANE
3 Equivalent Forms

H
H H
H H
Cl
P_H

1 - CHLOROPROPANE
1 Trans, 2 Equivalent Gauche Forms

CH_3 H H
H H H H H H
H H H CH_3 CH_3 H
Cl Cl Cl
P_C P_H P_H

1-CHLORO-2-METHYLPROPANE
1 Gauche, 2 Equivalent Trans Forms

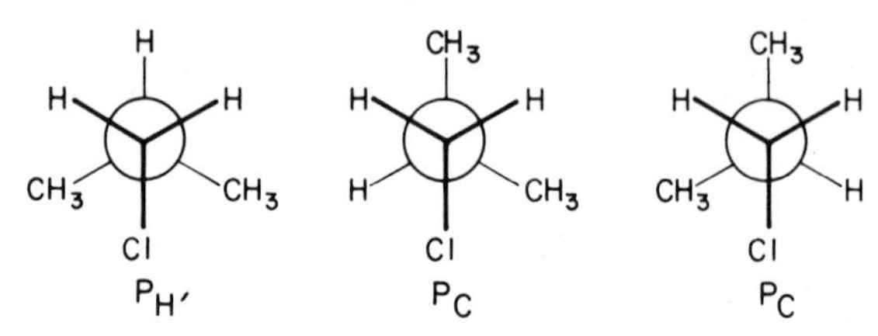

Figure 2.1 Rotational isomers of primary chloroalkanes.

658 cm^{-1}, the C-Br stretch at 559 cm^{-1}, and the C-I stretch at 499 cm^{-1}. These bands are the most intense ones in the Raman spectra below 2000 cm^{-1} and reflect the covalent character of the C-halogen bonds and the large change in polarizability that occurs during these vibrations. In the 1-halopropanes and the corresponding higher homologues (Spectra 11 and 13, Appendix 2), the P_H conformation appears as well as the P_C conformation, in which the halogen is trans to a carbon atom (see Figure 2.1). The carbon-chlorine stretching vibration for the P_C conformation occurs approximately 70 cm^{-1} higher in frequency than the corresponding P_H stretch. The C-Br stretch of the P_C isomer occurs 82 cm^{-1}

higher in frequency than in the P_H conformation and for the C-I vibration it is 90 cm^{-1} higher. This difference in stretching frequencies has been attributed by Colthup (35) to changes in the X-C-CH_2-C bending interaction during the vibration. In the P_H form, this angle is only slightly altered during the C-halogen stretch whereas in the P_C form this angle is sharply bent during the vibration. One other rotational isomer is possible for primary haloalkanes according to the Mizushima nomenclature. It is designated P_H' and describes a conformation in which the halogen atom is trans to a hydrogen atom as in the P_H isomer; however, the third carbon atom in the chain has been rotated away from the planar zigzag position. This rotational isomer occurs in compounds such as 1-chloro-2-methylpropane (Spectrum 12) together with the P_C conformation (see Figure 2.1). The C-X stretch of the latter vibration is only slightly higher in frequency (1-2%) than in other primary haloalkanes. Characteristic carbon-halogen stretching frequencies due to the P_H' conformation occur near 685 cm^{-1} for C-Cl, 614 cm^{-1} for C-Br, and 580 cm^{-1} for C-I compounds.

There are six possible rotational isomers for secondary haloalkanes in the Mizushima notation. These are illustrated in Figure 2.2 for the case of 3-chloro-

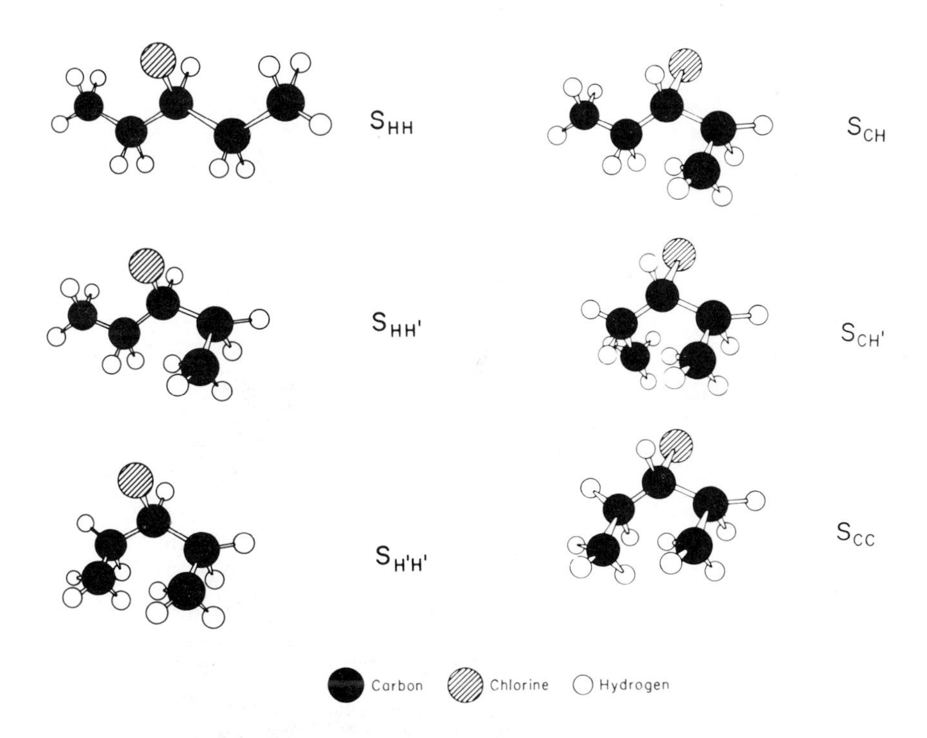

Figure 2.2 Rotational isomers of 3-chloropentane.

pentane. Data on the characteristic frequencies in the laser Raman spectra for the carbon-halogen stretching vibrations of secondary haloalkanes are given in Table 2.5. Assignments for the various conformations can be made with the aid of several model compounds. This method is illustrated for the case of the chloroalkanes. 2-Chloropropane can exist only in the S_{HH} form and a very strong band is found at 614 cm^{-1} in the Raman spectra. 3-Chloro-2,2-dimethylbutane exists only in the S_{CH} conformation with a C-Cl stretch at 667 cm^{-1}. 3-Chloro-2,2,4,4-tetramethylpentane has a Raman band at 758 cm^{-1}

Table 2.5 Carbon-Halogen Stretching Frequencies in the Raman Spectra of Secondary Haloalkanes

	Frequency (cm^{-1})				
Conformation	S_{CC}	$S_{CH'}$	S_{CH}	$S_{HH'}$	S_{HH}
A. Chloroalkanes					
2-Chloropropane	—	—	—	—	614
2-Chlorobutane	—	—	672	630	610
2-Chloropentane	—	—	670	—	614
3-Chloropentane	—	668	657	633	606
3-Chloro-2,2,4,4-tetramethylpentane	758	—	—	—	—
Relative Raman intensity	Variable	Weak	Medium-strong	Medium-strong	Very strong
B. Bromoalkanes					
2-Bromopropane	—	—	—	—	540
2-Bromobutane	—	—	615	587	538
Relative Raman intensity	—	—	Medium	Medium	Very strong
C. Iodoalkanes					
2-Iodopropane	—	—	—	—	495
2-Iodobutane	—	—	580	556	493
2-Iodooctane	—	—	586	—	488
Relative Raman intensity	—	—	Medium	Medium	Very strong

which can only be attributed to the C-Cl stretching of the S_{CC} rotational isomer. The assignment for the $S_{HH'}$ isomer can be made from the Raman spectra of 2-chlorobutane (Spectrum 14). It has three rotational isomers with bands occurring at 610 and 672 cm^{-1} which can be attributed to S_{HH} and S_{CH} forms, respectively, and a third band at 630 cm^{-1} that is due to the $S_{HH'}$ isomer. The assignments for this molecule have been confirmed by a normal coordinate analysis of the three conformations by Opaskar and Krimm (26). All six of the possible rotational isomers can exist in 3-chloropentane. Caraculacu, Stokr, and Schneider (36) found that the S_{CC} and $S_{H'H'}$ isomers of this molecule were improbable because of the proximity of the methyl groups. The frequencies of the C-Cl stretches for the other isomers of 3-chloropentane are given in Table 2.5.

Information on the characteristic frequencies in the Raman spectra of tertiary haloalkanes is contained in Table 2.6. In compounds such as 2-halo-2-methylpropane (*tert*-butyl halide) (Spectrum 15) only one rotational isomer (T_{HHH}) can

Table 2.6 Carbon-Halogen Stretching Frequencies in the Raman Spectra of Tertiary Haloalkanes

	Frequency (cm^{-1})			
Conformation	$T_{CHH'}$	T_{CHH}	T_{HHH}	$T_{HHH'}$
A. Chloroalkanes				
2-Chloro-2-methylpropane	–	–	570	–
2-Chloro-2-methylbutane	–	620	562	–
3-Chloro-3-ethylpentane	592	592	–	538
Relative Raman intensity	Medium-strong	Medium-strong	Very strong	Medium-strong
B. Bromoalkanes				
2-Bromo-2-methylpropane	–	–	520	–
2-Bromo-2-methylbutane	–	580	511	–
Relative Raman intensity	–	Medium	Very strong	–
C. Iodoalkanes				
2-Iodo-2-methylpropane	–	–	492	–
2-Iodo-2-methylbutane	–	571	496	–
Relative Raman intensity	–	Strong	Strong	–

exist. Two conformations can occur in tertiary haloalkanes in which the rotational isomers are generated by rotation about only one C-C bond, such as in the 2-halo-2-methylalkanes. These two forms are illustrated in Figure 2.3 for 2-chloro-2-methylbutane. The frequency of the C-Cl stretch for the T_{HHH} conformation occurs at 560 cm^{-1} and that for the T_{CHH} form at 620 cm^{-1}. The former configuration is often termed the gauche form and the latter the trans form of the molecule. For molecules in which the rotational isomers are generated by using two or three rotational axes, up to ten conformations are possible according to the modified Mizushima notation. Stokr, Caraculacu, and Schneider (23) have shown that although 3-chloro-3-ethylpentane could theoretically exist in seven different conformations, there are only three that are possible based on steric considerations. The band at 538 cm^{-1} was assigned to the C-Cl stretch from the molecule in the $T_{HHH'}$ conformation. The spectral differences between the two remaining isomers, T_{CHH} and $T_{CHH'}$, are very small and it was assumed that the stretching vibration for both was close to 592 cm^{-1}.

A summary of the characteristic frequencies in the Raman spectra of haloal-

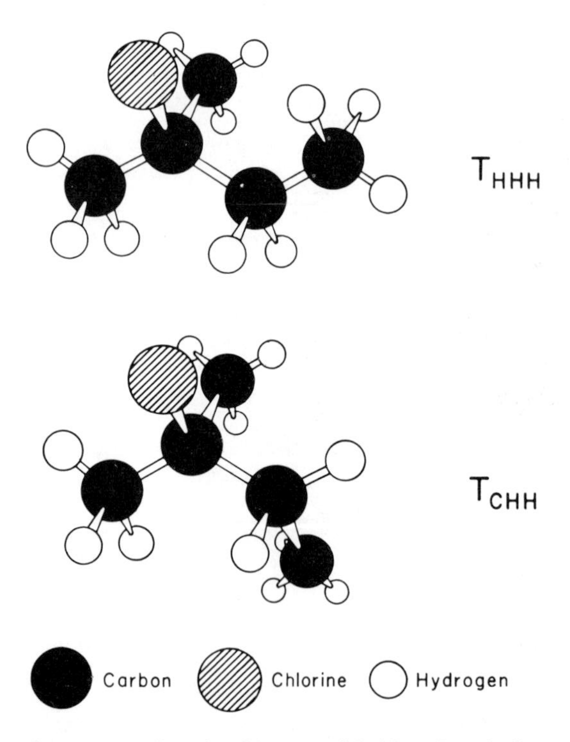

Figure 2.3 Rotational isomers of 2-chloro-2-methylbutane.

kanes is contained in Table 2.7. In the primary haloalkanes, the P_C band is present in most spectra and the presence of either the P_H or P_H' band gives further evidence on the structure of the molecule. In the secondary haloalkanes, a very strong Raman band from the S_{HH} conformation is found, except for molecules such as 2-halo-2,2,4,4-tetramethylpentane in which this conformation cannot occur. In this instance, the S_{CC} band is quite prominent. In the tertiary substituted compounds, most spectra contain bands due to the T_{HHH} and T_{CHH} rotational isomers. Using these correlations, it is possible to distinguish the halogen atom present within any type of substitution and the type of substitution present at the carbon attached to the halogen for any haloalkane. However, it may not be possible to distinguish between a secondary and a tertiary iodoalkane.

Table 2.7 Characteristic Carbon-Halogen Stretching Frequencies in the Raman Spectra of Haloalkanes

	Frequency (cm^{-1})		
Conformation	Cl	Br	I
A. Primary			
P_H	660–650	565–560	510–500
P_H'	690–680	625–615	590–580
P_C	730–720	650–640	600–590
B. Secondary			
S_{HH}	615–605	540–535	495–485
$S_{HH'}$	635–630	590–575	560–545
$S_{H'H'}$	690–680	670–650	–
S_{CH}	675–655	620–605	585–575
$S_{CH'}$	670	–	–
S_{CC}	760–740	700–680	–
C. Tertiary			
T_{HHH}	570–560	520–510	495–485
$T_{HHH'}$	540	–	–
T_{CHH}	620–590	590–580	580–570
$T_{CHH'}$	590	–	–

2.3 DI- AND POLYSUBSTITUTED HALOALKANES

The Raman spectra of the 1,1-dihaloethanes (CH_3CHX_2) have been reported by Kahovec and Wagner (37). The two carbon-halogen stretching frequencies, ν(antisymmetric CX_2) and ν(symmetric CX_2), occur at 689 and 641 cm^{-1} for X = Cl, at 606 and 546 cm^{-1} for X = Br, and at 552 and 466 cm^{-1} for X = I. The 2,2-dihalopropanes, $(CH_3)_2CX_2$, also exhibit two C-X stretching vibrations: at 653 and 557 cm^{-1} for X = C1, at 585 and 481 cm^{-1} for X = Br, and at 551 and 448 cm^{-1} for X = I. The Raman spectra of *gem*-dihalo derivatives of higher alkanes (21,37) generally exhibit more than two bands in the C-X stretching region.

Rotational isomerism in the vibrational spectra of 1,2-dihaloethane has been thoroughly discussed by Mizushima (38). A recent review of the carbon-halogen stretching frequencies and conformation of these and other vicinal dihaloalkanes has been reported by Altona and Hageman (39). In the liquid state these compounds exist as equilibrium mixtures of trans and gauche conformations with the trans form the predominant species in most cases. In the solid state almost all the molecules are in the trans form. The C-X stretching frequencies in the Raman spectra of the 1,2-dihaloethanes in the liquid state are:

C-X Stretching Frequencies (cm^{-1})

	Trans	Gauche
CH_2ClCH_2Cl	745 (vs); inactive (IR at 709)	645 (s); 677 (m)
CH_2BrCH_2Br	660 (vs); inactive (IR at 587)	551 (s); 583 (m)
CH_2ClCH_2Br	726 (vs); 630 (vs)	568 (s); 665 (m)

The Raman spectra of a large number of polychlorinated alkanes have been reported by Gerding and Rijnders (40). Detailed vibrational assignments in the spectra of the perhalogenated ethanes CX_3CX_3 have been discussed by Carney et al. (41). In the Raman spectra of the solid state the most intense band arises from the CX_3 symmetric stretch and occurs at 431 cm^{-1} for X = C1 and at 255 cm^{-1} for X = Br.

2.4 HALOGENATED CYCLOALKANES

A review of the IR and Raman studies of conformational isomerism in monohalogenated cycloalkanes has been made by Ekejiuba and Hallam (42). Carbon-halogen stretching frequencies for these compounds are given in Table 2.8. Raman special studies have also been conducted for 1-alkyl-1-halogenocyclopen-

Table 2.8 Characteristic Raman Carbon-Halogen Stretching Frequencies for Halocycloalkanes[a]

	X = Chlorine		X = Bromine	
	Frequency ν(CCl) (cm^{-1})	Reference	Frequency ν(CBr) (cm^{-1})	Reference
1. cyclopropyl–X	632[b]	46	545 (s)	47
2. cyclobutyl–X	618 (m)–Equatorial 528 (s)–Axial	48	534 (w)–Equatorial 481 (s)–Axial	49
3. cyclopentyl–X	624 (sh)–Equatorial 588 (m)–Axial	50	515 (vs)	50
4. cyclohexyl–X	733 (s)–Equatorial 688 (m)–Axial	51	689 (s)–Equatorial 660 (m)–Axial	51

[a]All frequencies and intensities refer to the liquid state at ambient temperature.
[b]IR frequency.

tanes (43), 1-alkyl-1-halogenocyclohexanes (44), and for some *trans*-1,2-dihalogeno derivatives of cyclopentane and cyclohexane (45).

2.5 FLUOROALKANES

An extensive review of the IR and Raman spectral data on organic fluorine compounds has been made by Brown and Morgan (52). In the vibrational spectra of fluoroalkanes the C-F stretching region has been assigned to the range 1400-1000 cm^{-1} (53). However, Tuazon et al. (54) have pointed out that in molecules in which the C-F bond is adjacent to a C-C single bond, there is interaction between these two stretching modes and often the frequency of the mode possessing the greater C-F stretching character lies below 1000 cm^{-1}. For example, in liquid fluoroethane the Raman band at 873 cm^{-1} is assigned to the C-F stretch while that at 1041 cm^{-1} has a greater contribution from the C-C stretch. Because of this vibrational coupling spectra structure correlations for these types of compounds are of limited applicability.

REFERENCES

1. T. Shimanouchi, *Pure Appl. Chem.*, *7*, 131 (1963).
2. J. Aldous and I. M. Mills, *Spectrochim. Acta, 18*, 1073 (1962).
3. J. Aldous and I. M. Mills, *Spectrochim. Acta,* 19, 1567 (1963).
4. L. H. Ngai and R. H. Mann, *J. Mol. Spectrosc., 38*, 322 (1971).
5. J. Wagner, *Z. Phys. Chem.,40B,* 439, (1938).
6. H. L. Welsh, M. F. Crawford, T. R. Thomas, and G. R. Love, *Can. J. Phys., 30*, 577 (1952).
7. D. F. Fenlon, F. F. Cleveland, and A. G. Meister, *J. Chem. Phys., 19,* 1561 (1951).
8. F. E. Palma, E. A. Piotrowski, S. Sundaram, and F F. Cleveland, *J. Mol. Spectrosc., 13*, 119 (1964).
9. J. Wagner, *Z. Phys. Chem., 45B,* 69 (1939).
10. R. W. Wood and D. H. Rank, *Phys. Rev., 48, 63* (1935).
11. O. Redlich and W. Stricks, *Monatsh. Chem., 67,* 328 (1936).
12. H.Stammreich and R. Forneris, *Spectrochim. Acta, 8,* 52 (1956).
13. J. P. Zietlow, F. F. Cleveland, and A. G. Meister, *J. Chem. Phys., 18,* 1076 (1950).
14. A. G. Meister, S. E. Rosser, and F. F. Cleveland, *J. Chem. Phys., 18,* 346 (1950).
15. H. Stammreich, Y. Tavares, and D. Bassi, *Spectrochim. Acta, 17*, 661 (1961).
16. F. F. Bentley, L. D. Smithson, and A. L. Rozek, *Infrared Spectra and Characteristic Frequencies 700–300 cm*$^{-1}$, Interscience, New York, 1968.
17. A. Dadieu, A. Pongratz, and K. W. F. Kohlrausch, *Monatsh. Chem., 61,* 369 (1932).
18. H. Kopper, R. Seka, and K. W. F. Kohlrausch, *Monatsh. Chem., 61,* 379 (1932).
19. A. Dadieu, A. Pongratz, and K. W. F. Kohlrausch, *Monatsh. Chem., 61,* 409 (1932).
20. K. W. F. Kohlrausch and F. Köppl, *Monatsh. Chem., 63,* 255 (1933).
21. K. W. F. Kohlrausch and F. Köppl, *Monatsh. Chem., 65,* 185 (1935).
22. L. Kahovec and J. Wagner, *Z. Phys. Chem., 42B,* 123 (1939).
23. J. Stokr, A. Caraculacu, and B. Schneider, *Coll. Czech. Chem. Comm., 30,* 683 (1965).

24. R. G. Gasanov, *J. Struct. Chem. (USSR), 10*, 54 (1969).
25. R. G. Snyder and J. H. Schachtschneider, *J. Mol. Spectrosc., 30*, 290 (1969).
26. C. G. Opaskar and S. Krimm, *Spectrochim. Acta, 23A*, 2261 (1967).
27. P. Klaboe, *Spectrochim. Acta, 26A*, 87 (1970).
28. G. Keresturi, O. D. Ulyanova, and Yu. A. Pentin, *Opt. Spectrosc., 26*, 390 (1969).
29. P. J. D. Park and E. Wyn-Jones, *J. Chem. Soc. (A)*, 1968, 2944.
30. P. N. Gates, E. F. Mooney, and H. A. Willis, *Spectrochim. Acta, 23A*, 2043 (1967).
31. C. Altona, *Tetrahedron Lett., 19*, 2325 (1968).
32. S. Mizushima, T. Shimanouchi, K. Nakamura, M. Hayashi, and T. Tsuchiya, *J. Chem. Phys., 26*, 970 (1957).
33. J. J. Shipman, V. L. Folt, and S. Krimm, *Spectrochim. Acta, 18*, 1603 (1962).
34. J. J. Shipman, V. L. Folt, and S. Krimm, *Spectrochim. Acta, 24A*, 437 (1968).
35. N. B. Colthup, *Spectrochim. Acta, 20*, 1843 (1964).
36. A. Caraculacu, J. Stokr, and B. Schneider, *Coll. Czech. Chem. Comm., 29*, 2783 (1964).
37. L. Kahovec and J. Wagner, *Z. Phys. Chem., 47B*, 48 (1940).
38. S. Mizushima, *Structure of Molecules and Internal Rotation*, Academic Press, New York, 1954.
39. C. Altona and H. J. Hageman, *Rec. Trav. Chim., 88*, 33 (1969).
40. H. Gerding and G. W. A. Rijnders, *Rec. Trav. Chim., 65*, 143 (1946).
41. R. A. Carney, E. A. Piotrowski, A. G. Meister, J. H. Braun, and F. F. Cleveland, *J. Mol. Spectrosc., 7*, 209 (1961).
42. I. O. C. Ekejiuba and H. E. Hallam, *J. Mol. Struct., 6*, 341 (1970).
43. H. R. Buys, C. Altona, and E. Havinga, *Rec. Trav. Chim., 87*, 53 (1968).
44. C. Altona, H. J. Hageman, and E. Havinga, *Rec. Trav. Chim., 87*, 353 (1968).
45. C. Altona, H. J. Hageman, and E. Havinga, *Spectrochim. Acta, 24A*, 633 (1968).
46. M. I. Kay, *Diss. Abstr., 25*, 130 (1964).
47. W. G. Rothschild, *J. Chem. Phys., 44*, 3875 (1966).
48. J. R. Durig and A. C. Morrissey, *J. Chem. Phys., 46*, 4854 (1967).
49. J. R. Durig and W. H. Green, *J. Chem. Phys., 47*, 673 (1967).
50. J. R. Durig, J. M. Karriker, and D. W. Wertz, *J. Mol. Spectrosc., 31*, 237 (1969).
51. M. Rey-Lafon, C. Rouffi, M. Camiade, and M. Forel, *J. Chim. Phys., 67*, 2030 (1970).

52. J. K. Brown and K. J. Morgan, in *Advances in Fluorine Chemistry,* Vol. 4, M. Stacy, J. C. Tatlow, A. G. Sharpe, Eds. Butterworths, London, 1965.
53. L. J. Bellamy, *Infrared Spectra of Complex Molecules,* Methuen, London, 1958.
54. E. C. Tuazon, W. G. Fateley, and F. F. Bentley, *Appl. Spectrosc., 25,* 374 (1971).

CHAPTER THREE

ALIPHATIC ALCOHOLS, ETHERS AND RELATED COMPOUNDS

3.1 ALIPHATIC ALCOHOLS

The early work on the Raman spectra of aliphatic alcohols and ethers is reviewed by Hibben (1) and Kohlrausch (2). More recently, the Raman spectra of methanol and its deuterated species were investigated by Falk and Whalley (3). A normal coordinate analysis of CH_3OH was carried out by Tanaka, Kurantani, and Mizushima (4). Krishnan (5) examined the Raman spectra of the normal C_1 through C_4 alcohols. These compounds, along with eight other secondary and tertiary alcohols, were studied by Venkateswarlu and Mariam (6). The vibrational spectra of ethyl alcohol and isopropyl alcohol and some of their deuterated derivatives were analyzed in a series of articles by Tanaka (7). A normal coordinate analysis of *n*-propanol was carried out by Fukushima and Zwolinski (8). The IR and Raman spectra of twelve isotopic species of ethanol in the gaseous, liquid, and solid states were examined by Perchard and Josien (9).

When an oxygen atom is substituted for a carbon atom in the carbon-carbon aliphatic chain, the frequencies of the skeletal stretching vibrations are altered only slightly, since the force constant for the C-O bond is nearly the same as that of the C-C bond and the relative mass of the hydroxyl group differs from that of the methyl group by only two units. In the IR, however, this substitution leads to a great change in intensity because of the polar nature of the C-O bond; but in the Raman spectrum, the intensity of the C-O vibration is weaker and of the same order of magnitude as the C-C vibrations. Therefore, it is difficult to distinguish an aliphatic alcohol from its corresponding hydrocarbon using only the Raman spectral region below 2000 cm^{-1}. Also, there is little difference in the Raman spectra of primary alcohols for molecules having nine carbon atoms or greater. Tasumi, Shimanouchi, Watanabe, and Goto (10), in their normal coordinate treatment of the normal higher alcohols $C_{12}H_{25}OH$ through $C_{37}H_{75}$ OH, demonstrated that the frequencies of the alcohols do not deviate much from those of the *n*-paraffins, although the presence of the carbon-oxygen bond causes great changes in the IR absorption intensities.

Another difficulty encountered in the study of the spectra of hydroxylic compounds arises from the highly polar nature of the OH bond, which leads to association of molecules through hydrogen bonding. The free OH stretch can only be observed in the vapor state or in dilute solution in nonpolar solvents. In the liquid and solid state, or in concentrated solutions, the alcohols form dimers and higher polymeric forms which may be either linear or cyclic in character. This self-association is sensitive to temperature, solvent, and molecular structure. In general, hydrogen bonding decreases the frequency of the OH stretch and increases the frequency of the OH deformation vibrations. For example, in gaseous methanol the OH stretch is at 3681 cm^{-1} and in the liquid is at 3328 cm^{-1}; the OH bend found at 1345 cm^{-1} in the gas is displaced to 1418 cm^{-1} in the liquid.

The intermolecularly hydrogen-bonded OH stretching vibration is evident in the argon ion laser Raman spectra of liquid alcohols as a broad weak absorption centered about 3400-3300 cm^{-1} (see Spectrum 16, Appendix 2). The half width of the band is about 200 cm^{-1}. Its intensity is only 1/20 of the CH_3 and CH_2 stretchings. In the corresponding Raman spectra taken using the He-Ne laser, this absorption is very weak or not visible at all. This phenomenon is due to the efficiency of the photomultiplier tube having an S-20 type spectral response. For the He-Ne laser, the spectral response is nearly flat out to 1500 cm^{-1}, whereupon it decreases as much as one hundredfold in going to 3000 cm^{-1}; the intensities of OH and CH stretchings in the uncorrected spectra are quite small compared to the Raman bands at lower frequencies. The spectral response of the spectrometer using the argon ion laser at 4880 Å is nearly flat over the entire range and is a more accurate representation of the relative intensities throughout the entire spectral region.

Some frequencies that are characteristic of aliphatic alcohols are listed in Table 3.1. In the primary alcohols, ν (C-O) is coupled with the C-C skeletal stretchings and the intense band in the IR, which is usually attributed to the C-O stretch, is assigned to an antisymmetric C-C-O stretch. The symmetric C-C-O

Table 3.1 C-OH Frequencies in the Raman Spectra of Liquid Aliphatic Alcohols

	Vibration Frequency (cm^{-1})					
Compound	OH Stretch	Antisymmetric CCO Stretch	Symmetric CCO Stretch	CH_2 Twist $+\delta$ (COH)	CO Stretch + CH_3 Rock $+\delta$ (COH)	δ(CCO)
A. Primary						
Methanol	3330	–	1033	1363	–	480
Ethanol	3360	1052	884	1276	1093	430
1-Propanol	3340	1058	971	1275	1100	460
1-Butanol	3340	1070	965	1250	1100	450
1-Pentanol	3340	1055	962	1208	1100	440
2-Methyl-1-propanol	3360	1050	965	1250	1127	435
B. Secondary						
2-Propanol	3360	1135	819	–	1125	490
2-Butanol	3360	1120	822	1260	1115	500
C. Tertiary						
2-Methyl-2-propanol	3380	1210	753	–	–	350
2-Methyl-2-butanol	3380	1200	730	–	–	360
Relative Raman intensity	Weak-broad	Strong-medium	Very strong-strong	Weak-medium	Weak-medium	Weak-medium

stretch located lower in frequency involves more C-C stretch than C-O stretch. In the secondary alcohols an intense band is observed around 820 cm^{-1} resulting from the symmetric stretching of the C_3O skeleton. In the analogous hydrocarbon compounds, the C_4 symmetric stretch is at 795 cm^{-1}. For tertiary alcohols, an intense band that arises from the symmetric C_4O skeletal stretch is observed in the region 760–730 cm^{-1}. In neopentane, $(CH_3)_4C$, this vibration is at 733 cm^{-1} and for 2,2-dimethylbutane at 714 cm^{-1}. The in-plane deformation of the C-OH group is not a good characteristic frequency since it is coupled with both the methyl twisting and rocking vibrations and some CO stretch. The out-of-plane CCO deformation, although of weak to medium intensity in the Raman, is quite characteristic of the type of alcohol. Primary alcohols above ethanol absorb in the region 460–430 cm^{-1}, secondary alcohols around 495 cm^{-1}, and tertiary alcohols at 360–350 cm^{-1}

3.2 ALIPHATIC ETHERS

Raman spectra of the simple ethyl, propyl, butyl, and amyl ethers were reported by Cleveland et al. (11). Kanazawa and Nukada have reviewed the assignment of the vibrational spectra of dimethyl ether and its d_6 isotope (12). Wieser et al. (13) have assigned the vibrational spectra for the conformers of diethyl ether and its deuterated analogues, as well as calculating a valence force field for these molecules. Perchard (14) has investigated the IR and Raman spectra of the eight isotopic species of methyl ethyl ether. The spectra, assignments, valence force field, and molecular conformations of ten aliphatic ethers were reported by Snyder and Zerbi (15). Mashiko et al. have reported on the vibrational assignments of aliphatic ethers including acetals and ketals (16).

As we noted in the section on alcohols, since the force constant of the C-O bond is close to that of the C-C bond and the relative mass of the oxygen differs slightly from that of the CH_3 group, the Raman spectra of alcohols, ethers, and hydrocarbons are very similar. Except in complex molecules, however, there are enough differences to distinguish these compounds from one another. The specific vibrations of the C-O-C linkage include: (1) an antisymmetric COC stretch, (2) a symmetric COC stretch, and (3) a COC symmetric deformation. The position of these vibrations in the Raman spectra of aliphatic ethers is listed in Table 3.2 (see Spectrum 17). In the liquid state, many of these ethers exhibit rotational isomerism. In those cases in which complete assignments have been made the frequencies for each rotamer are given; otherwise, the position of the most intense band for each type vibration in the Raman spectra is listed. Of these three vibrations, the COC symmetric stretch has the greatest intensity in the Raman. For *n*-alkyl ethers, it is is located in the range 930–830 cm^{-1}. Branching on the α-carbon decreases the frequency of the symmetric stretch to 798 cm^{-1} for the isopropyl group and to 700 cm^{-1} for the *tert*-butyl group. These latter vibrations involve the symmetric stretching of the C_3O and C_4O skeletons, respectively. The antisymmetric stretch is generally found between 1070 and 1150 cm^{-1} except for di-*tert*-butyl ether (1205 cm^{-1}). The position of the COC symmetric deformation lies between 500 and 400 cm^{-1} and is not a specific vibration since it also mixes with CCC and OCC deformations and to some extent with C-C stretchings, as shown in the potential energy distributions for aliphatic ethers in ref. 15. Compounds containing more than one ether linkage

Table 3.2 Characteristic Frequencies in the Raman Spectra of Liquid Aliphatic Ethers

Compound	Frequency (cm^{-1}) Anti-symmetric COC Stretch	Symmetric COC Stretch	COC Symmetric Deformation
A. Symmetric substitution			
Dimethyl ether 1 rotamer-C_{2v} symmetry	1102	924	428
Diethyl ether 2 rotamers-trans-trans	1120	845	440
trans-gauche	1140	835	499
Di-*n*-propyl ether	1120	890	480
Di-*n*-butyl ether	1115	840	460
Di-*n*-amyl ether	1120	850	498
Di-isobutyl ether	1125	840	420
Di-isoamyl ether	1140	870	420
Di-isopropyl ether	1125	798	492
Di-*tert*-butyl ether	1205	700	425
B. Asymmetric substitution			
Methyl ethyl ether 2 rotamers-trans	1120	853	475
gauche	1068	843	—
Ethyl isopropyl ether 1 rotamer-C_1 symmetry	1125	830	492
Ethyl *n*-butyl ether	1125	860	465
Ethyl isobutyl ether	1140	845	480
Ethyl *tert*-butyl ether	1115	780	490

such as bis-(2-methoxyethyl) ether, $CH_3(OCH_2CH_2)_2OCH_3$, usually exhibit three broad bands at 300, 440, and 530 cm^{-1} in most compounds.

Other specific frequencies can be found in the Raman spectra that are indicative of the presence of CH_3-O or C-CH_2-O groups. The symmetric HCH deformation in both the methyl and methylene groups that are attached to oxygen atoms occurs at 1480–1470 cm^{-1}. In ethers, this band appears in the Raman superimposed on the high frequency side of the broad band due to the usual aliphatic methyl and methylene deformations that occur at 1460–1450 cm^{-1}. The carbon-hydrogen stretching vibrations of methyl, methylene, and isopropyl groups attached to oxygen are sufficiently different from those attached to carbon that they can be readily identified in the Raman spectra obtained using the argon ion laser. These specific frequencies are presented in Table 3.3.

Table 3.3 Carbon-Hydrogen Stretching Vibrations of Methyl, Methylene, and Isopropyl Groups Attached to Oxygen

	Frequency (cm^{-1})				
Compound	$-OCH_3$ Anti-symmetric C-H Stretch	$-OCH_3$ Symmetric C-H Stretch	$-OCH_2-$ Anti-symmetric C-H Stretch	$-OCH_2-$ Symmetric C-H Stretch	Isopropyl C-H Stretch
Dimethyl ether	2989	2815	–	–	–
Methyl ethyl ether	2981	2817	2981	2869	–
Methyl isopropyl ether	–	–	–	–	2935
Diethyl ether	–	–	2977	2869	–
Di-*n*-propyl ether	–	–	2970	2860	–
Di-*n*-butyl ether	–	–	2970	2870	–
Di-*n*-amyl ether	–	–	2965	2871	–
Di-isobutyl ether	–	–	2963	2866	–
Di-isoamyl ether	–	–	2965	2866	–
Di-isopropyl ether	–	–	–	–	2937
Ethyl isopropyl ether	–	–	2970	2861	2935

3.3 ACETALS

Detailed assignments in the IR and Raman spectra have been made for methylal [dimethoxymethane-$H_2C(OCH_3)_2$] (17,18), diethylformal [diethoxymethane-$H_2C(OC_2H_5)_2$] (19) and methoxyethoxymethane (20). Little and Martell (21) have reported the Raman spectra of seven acetals and Nukada (22) has given a detailed assignment for acetal [(ethylidene diethyl ether or 1,1-diethoxy-ethane)-$C_2H_5OCH(CH_3)OC_2H_5$] (Spectrum 18) and the dimethyl and methyl ethyl derivatives. The characteristic IR and Raman frequencies for these compounds have been discussed by Mashiko et al. (16). The totally symmetric skeletal stretching of the COCOC moiety appears as an intense highly polarized Raman band in the range 900–800 cm^{-1}. The antisymmetric stretch has been assigned to a medium to strong Raman band at 1145–1129 cm^{-1}. The three skeletal deformations of the COCOC group have been assigned to bands at 654–556 cm^{-1}, 537–370 cm^{-1}, and 396–295 cm^{-1}.

3.4 PEROXIDES

Vibrational assignments for hydrogen peroxide in the vapor and liquid states have been made by Giguere (23). The O-O stretching vibration gave rise to the most intense Raman band in the liquid at 877 cm^{-1}. In the IR spectra of both liquid and vapor ν(O-O) was assigned to a weak to very weak band near 882 cm^{-1}. Leadbeater (24), in a Raman study of several organic peroxides, assigned ν(O-O) to a strong band near 882 cm^{-1}. More recent Raman data on alkyl- and fluoro-substituted alkyl-peroxides include the following:

Compound	ν(O-O)(cm^{-1})	Reference
CF_3OOCl	943	25
CF_3OOCF_3	886	26
CF_3OOH	870	25
$(CF_3)_3COOC(CF_3)_3$	781	25
CH_3OOCH_3	779	27
$CH_3(CF_3)_2COOC(CF_3)_2CH_3$	774	25
$(CH_3)_3COOC(CH_3)_3$	771	25,27

In these Raman spectra ν (O-O) is usually the most intense band and is almost completely polarized. The presence of electron withdrawing groups adjacent to the peroxide bond leads to an increase in ν(O-O).

REFERENCES

1. J. H. Hibben, *The Raman Effect and Its Chemical Applications,* Reinhold Publishers, New York, 1939.
2. K. W. F. Kohlrausch, *Ramanspektren,* Akademische Verlagsges., Leipzig, 1943.
3. M. Falk and E. Whalley, *J. Chem. Phys., 34,* 1554 (1961).
4. C. Tanaka, K. Kuratani, and S. Mizushima, *Spectrochim. Acta, 9,* 265 (1957).
5. K. Krishnan, *Proc. Indian Acad. Sci. Sect. A, 53,* 151 (1961).
6. K. Venkateswarlu and S. Mariam, *Acta Phys. Austriaca, 15,* 362 (1962).

7. C. Tanaka, *Nippon Kagaku Zasshi, 83,* 521, 655, 661,792 (1962).
8. K. Fukushima and B. J. Zwolinski, *J. Mol. Spectrosc., 26*, 368 (1968).
9. J. Perchard and M. Josien, *J. Chim. Phys., 64,* 1834, 1856 (1968).
10. M. Tasumi, T. Shimanouchi, A. Watanabe, and R. Goto, *Spectrochim. Acta, 20,* 629 (1964).
11. F. F. Cleveland, M. J. Murray, H. H. Haney, and J. Shackelford, *J. Chem. Phys., 8,* 153 (1940).
12. Y. Kanazawa and K. Nukada, *Bull. Chem. Soc. (Jap.), 35,* 612 (1962).
13. H. Wieser, W. G. Laidlaw, P. J. Krueger, and H. Fuhrer, *Spectrochim. Acta, 24A,* 1055 (1968).
14. J. P. Perchard, *Spectrochim. Acta, 26A,* 707 (1970).
15. R. G. Snyder and G. Zerbi, *Spectrochim. Acta, 23A*, 391 (1967).
16. Y. Mashiko, S. Saeki, K. Nukada, Y. Kanazawa, and T. Suzuki, *Proc. Int. Symp. Mol. Struct. Spectrosc.,* Tokyo, *A220,* (1962).
17. J. K. Wilmhurst, *Can. J. Chem., 36*, 285 (1958).
18. K. Nukada, *Spectrochim. Acta, 18,* 745 (1962).
19. K. Nukada, *Bull. Chem. Soc. (Jap.), 34,* 1624 (1961).
20. K. Nukada, *Bull. Chem. Soc. (Jap.), 34,* 1615 (1961).
21. M. H. Little and A. E. Martell, *J. Phys. Colloid Chem., 53,* 472 (1949).
22. K. Nukada, *Bull. Chem. Soc. (Jap.), 35,* 3 (1962).
23. P.A. Giguere, *J. Chem. Phys., 18,* 88 (1950).
24. R. Leadbeater, *Compt. Rend., 230,* 829 (1950).
25. A. J. Melveger, L. R. Anderson, C. T. Ratcliffe, and W. B. Fox, *Appl. Spectrosc., 26* 381 (1972).
26. J. R. Durig and D. W. Wertz, *J. Mol. Spectrosc., 25*, 467 (1968).
27. K. O. Christie, *Spectrochim. Acta, 27A,* 463 (1971).

CHAPTER FOUR

ALIPHATIC AMINES, NITROALKANES,

AND THEIR DERIVATIVES

4.1 ALIPHATIC AMINES

The early work in the Raman spectra of the simple aliphatic amines containing one to five carbon atoms is summarized in Chapter 6 of Kohlrausch's book (1). In the IR, a series of 20 primary and 13 secondary amines have been studied by Stewart (2). Also, the IR solution spectra of the aliphatic normal primary monoamines from ethyl to decyl and the alpha-omega diamines from ethylene to octamethylene have been presented in a paper by Segal and Eggerton (3). Extensive work has been done on methyl, dimethyl, and trimethyl amines and their deuterated analogues. The experimental and theoretical data on the vibrational assignments of these molecules have been reviewed by Dellepiane and Zerbi (4) who also report their own normal coordinate calculations based on both a Urey-Bradley force field and a valence force field. Similar calculations have been carried out by Popov et al. (5,6). More recent work includes that of Durig et al. (7) on methyl amine and four of its deuterated species and that of Clippard and Taylor (8) on $(CH_3)_3N$ and $(CD_3)_3N$. The torsional bands in the far IR spectra of *n*-propyl, isopropyl, and *t*-butyl amines have been examined by Scott and Crowder (9). Characteristic frequencies in the vibrational spectra of primary amines have been discussed by Hirakawa and Tsuboi (10) and by Cain, Freeman, and Henshall (11) who used the group frequency factorization procedure developed by

Table 4.1 Characteristic Frequencies in the Raman Spectra of Liquid Primary Aliphatic Amines

	Vibration Frequency (cm^{-1})			
Compound	ν_{asym} (NH_2)	ν_{sym} (NH_2)	δ_{sym} (NH_2)	ν (CN)
A. Primary α-carbon				
Methyl amine	3372	3320	1600	1040
Ethyl amine	3371	3318	1619	1085
n-Propyl amine	3378	3325	1624	1072
n-Butyl amine	3379	3320	1610	1083
n-Hexyl amine	3375	3320	1620	1075
Isobutyl amine	3380	3330	1616	1061
B. Secondary α-carbon				
Isopropyl amine	3363	3308	1622	1034, 1135
sec-Butyl amine	3364	3321	1605	1044, 1145
C. Tertiary α-carbon				
tert-Butyl amine	3360	3310	1623	1040, 1240
Relative Raman Intensity	Strong	Very strong	Weak	Medium

King and Crawford (12). This procedure was also used by Finch, Hyams, and Steele (13) for interpreting the vibrational spectra of compounds containing the dimethylamino group.

The characteristic frequencies in the laser Raman spectra of primary aliphatic amines are listed in Table 4.1 (see Spectrum 19, Appendix 2). For the C-NH_2 group the following vibrations are possible: (1) an antisymmetric NH stretch, (2) a symmetric NH stretch, (3) a NH_2 symmetric deformation (scissors), (4) a carbon-nitrogen stretch, (5) a NH_2 wag, and (6) a NH_2 twist. The nitrogen-hydrogen stretches are quite prominent in the Raman spectra of the primary amines. The strongest N-H stretching band in the Raman is due to the symmetric (in-phase) stretch and is located at 3450–3250 cm^{-1} in the vapor state or in dilute solution. In the pure liquid state, this vibration is shifted through hydrogen bonding down to 3330–3250 cm^{-1}. The less intense band from the antisymmetric (out-of-phase) nitrogen-hydrogen stretch is found at 3550–3300 cm^{-1} in the unassociated state and at 3400–3330 cm^{-1} in the liquid or other associated states. The symmetric NH_2 deformation that occurs in the region 1650–1590 cm^{-1} leads to a strong band in the IR. However, in the laser Raman spectra it occurs as a weak to very weak broad band. This absorption escaped detection in the early Raman work on the amines using the mercury arc source since its frequency was in close proximity to a strong mercury line.

The position of the carbon-nitrogen stretching vibration is affected by the branching on the α-carbon atom. Those with a primary carbon atom (RCH_2-NH_2) have the C-N stretch at 1090–1060 cm^{-1} in the Raman. In this region, the carbon-carbon stretches from the rest of the molecule also occur and ν(CN) can often be obscured. Those amines with a secondary α-carbon atom(R_2CH-NH_2) have Raman bands at 1045–1035 cm^{-1} and at 1140–1080 cm^{-1}. However, the strongest band occurs from the symmetric skeletal stretch of the C_3N moiety around 800 cm^{-1}. Primary amines with tertiary alpha carbon atoms (R_3C-NH_2) possess Raman bands in the 1040–1020 cm^{-1} and 1240–1170 cm^{-1} regions, but the most intense band arises from the symmetric stretch of the C_4N skeleton that is found at about 745 cm^{-1}. These skeletal stretches for the secondary and tertiary α-carbon substituted amines are quite intense in the Raman and easily identifiable.

The two remaining vibrations of the C-NH_2 group, the NH_2 twist and wag, do not lead to correlatable frequencies in the Raman. In the liquid primary aliphatic amines, the NH_2 wag in the IR gives rise to a broad band of strong intensity in the range 850–750 cm^{-1}. However, in the Raman spectrum of methyl amine it occurs as a weak broad band at 875 cm^{-1}. In the higher primary amines, it is obscured by methylene twisting and rocking vibrations which are stronger in intensity and occur in the same region. The position of the NH_2 twisting mode has long been a subject of controversy (4). Durig et al. (7) have assigned this vibration in methyl amine to a band at 1353 cm^{-1} which they observed only in the IR of the solid.

The characteristic frequencies in the Raman spectra of secondary (e.g., Spectrum 20) and tertiary (e.g., Spectrum 21) aliphatic amines are given in Table 4.2. For the secondary amines in the liquid state there is only one N-H stretching frequency that occurs in the region of 3350–3300 cm^{-1}. There are two carbon-nitrogen stretches for the secondary amines. The antisymmetric stretch where the two C-N bonds are out-of-phase gives rise to a strong band in the IR and a moderate intensity band in the Raman in the range 1145–1130 cm^{-1}. On the other hand, the symmetric CNC stretch where the C-N vibrations are in-phase produces a strong band in the Raman and a weak to very weak band in the IR in the region 900–850 cm^{-1}. Buttler and McKean (14) have assigned the two N-H bending modes in dimethyl amine to frequencies near 1500 and 730 cm^{-1}. In the Raman spectra of secondary amines

Table 4.2 Characteristic Frequencies in the Raman Spectra of Liquid Secondary and Tertiary Amines

	Vibration Frequency (cm^{-1})			
A. Secondary Amines	ν(NH)	ν_{asym} (CNC)	ν_{sym} (CNC)	δ(CNC)
Dimethyl amine	3343	1154	931	397
Diethyl amine	3315	1137	871	427
Dipropyl amine	3320	1140	865	409
Dibutyl amine	3320	1130	895	415
Relative Raman Intensity	Medium-strong	Medium	Medium-strong	Medium
B. Tertiary Amines	ν_{asym} (CNC)	ν_{sym} (CNC)	δ_{asym} (CNC)	δ_{sym} (CNC)
Trimethyl amine	1048	833	429	375
Triethyl amine	1070	740	470	340
Tributyl amine	1055	830	470	350
Relative Raman Intensity	Medium	Medium-strong	Medium-weak	Medium-weak

taken with the argon ion laser, some very weak broad bands in these two regions can be detected in some of the spectra. In the IR spectra of these compounds, there is a strong broad band in the 750–700 cm^{-1} range assigned to the a' N-H deformation. The CNC skeletal deformation in secondary amines appears at 427 ± 14 cm^{-1} in the Raman spectra.

The tertiary amines are more difficult to characterize. In the Raman, the carbon-nitrogen antisymmetric stretch is found at 1070–1050 cm^{-1} and the symmetric stretch near 830 cm^{-1} in most compounds. The CNC skeletal deformations are located near 350 and 470 cm^{-1}. In secondary and tertiary amines, the symmetric stretch of the methylene groups attached to the nitrogen atom have a band at 2820–2760 cm^{-1}.

In summary, the type of substitution of aliphatic amines can best be distinguished in the Raman from an examination of the N-H stretching region (3380–3300 cm^{-1}). Primary amines have two bands in this region, secondary amines one, and tertiary amines none. Further structural information is obtained from the position of the C-N stretches and skeletal deformation frequencies, as listed in Tables 4.1 and 4.2.

4.2 AMINE SALTS

Raman spectra have been reported for aqueous solutions of alkyl-ammonium chlorides, $R\text{-}NH_3^+Cl^-$ [R = CH_3 (15,16); C_2H_5 (16); $n\text{-}C_3H_7$ (17)]. Several aliphatic diammonium chlorides, $NH_3^+(CH_2)_nNH_3^+$ (n = 1–5), have also been studied (18). A detailed assignment of the methylammonium ion has been reported by Waldron (19). A strong Raman band was observed for all these compounds in the range 2986–2974 cm^{-1} and was attributed to the symmetric stretching of the NH_3^+ group. The aqueous Raman spectra of the polymethyl ammonium chlorides reported by Edsall (17) have been interpreted by Ebsworth and Sheppard (20). The most prominent Raman bands are for $(CH_3)_2NH_2^+$ – 3048 cm^{-1} [$\nu_{asym}(CH_3)$], 2990 cm^{-1} [$\nu_{sym}(CH_3) + \nu(\overset{+}{N}H_2)$], 895 cm^{-1} [$\nu(C\text{-}\overset{+}{N})$], and 412 cm^{-1} [$\delta\ \overset{+}{N}C_2)$]; for $(CH_3)_3\overset{+}{N}H$ – 3030 cm^{-1} [$\nu_{asym}(CH_3)$], 2969 cm^{-1} [$\nu_{sym}(CH_3)$], 821 cm^{-1} [$\nu(C\text{-}\overset{+}{N})$], and 468–406 cm^{-1} [$\delta(C_3\overset{+}{N})$]; and for $(CH_3)_4N^+$ – 3037 cm^{-1} [$\nu_{asym}(CH_3)$], 2930 cm^{-1} [$\nu_{sym}(CH_3)$], 752 cm^{-1} [$\nu(C\text{-}\overset{+}{N})$], and 455–372 cm^{-1} [$\delta(\overset{+}{N}C_4)$].

4.3 HYDROXYLAMINES

The most intense band in the Raman spectrum of hydroxylamine, $HO\text{-}NH_2$, located at 906 cm^{-1} (21) arises from the O-N stretching mode. In hydroxylamine hydrochloride, $[HO\text{-}NH_3]^+Cl^-$, this band shifts to 1006 cm^{-1} in aqueous solution (17). George et al. (22) have measured the liquid state Raman spectra of several *O*-alkylhydroxylamines, $n\text{-}C_nH_{2n+1}ONH_2$ (n = 1, 6, 8, 10). The following assignments were made for these compounds: 3251–3237 cm^{-1} $\nu_{sym}(NH_2)$, 1595–1589 cm^{-1} – $\delta(NH_2)$, and 851–840 cm^{-1} – $\nu_{sym}(CON)$.

4.4 HYDRAZINES

The vibrational data for hydrazine, $H_2N\text{-}NH_2$, have been reviewed by Durig et al. (23). In the liquid state Raman spectrum (24), aside from a series of bands in the N-H stretching region (3336–3190 cm^{-1}), a very strong polarized band due to the N-N stretch was observed at 1111 cm^{-1}. Other strong bands occurred at 1628 cm^{-1} (NH_2 deformation) and 882 cm^{-1} (NH_2 rocking). For liquid tetrafluorohydrazine, $F_2N\text{-}NF_2$, a Raman band at 600 cm^{-1} was assigned to ν(N-N) for the trans isomer and one at 588 cm^{-1} to ν (N-N) for the gauche form. Complete assignments have been reported for the IR and Raman spectra of methylhydrazine (H_2NNHCH_3) (25) and unsymmetrical dimethylhydrazine [$H_2NN(CH_3)_2$] (26).

4.5 NITROALKANES

The assignments of the fundamental frequencies in the IR and Raman spectra of nitromethane were made by Wells and Wilson (27). Early work in the Raman spectra of nitroalkanes was done by Mathieu and Massignon (28) [nitromethane, nitroethane, 1-nitropropane, $CH(NO_2)_3$, $C(NO_2)_4$, $CCl_3(NO_2)$, $CBr_3(NO_2)$, $CBr(NO_2)_3$, 1-chloro-1-nitroethane, and 1-bromo-1-nitroethane]; by Wittek (29) [C_1 through C_5 normal alkyl, isopropyl, and isobutyl nitro derivatives; CCl_3NO_2, CBr_3NO_2, $C(NO_2)_4$]; and by Smith, Pan, and Nielson (30) (nitroethane, 1-nitropropane, and 2-nitropropane). More recently, Geiseler et al. (31, 32) have investigated the IR and Raman spectra of a large number of straight chain primary and isomeric secondary nitroalkanes.

The six fundamental vibrations of the C-NO_2 group are listed in Table 4.3. The frequencies and approximate intensities of these vibrations in the laser Raman spectra of some primary and secondary nitroalkanes are given in Table 4.4. In the IR the two most useful bands for identifying the presence of the nitro group are due to the symmetric and antisymmetric stretches. This is also true in the Raman. However, in the Raman the intensity of the symmetric stretch is generally stronger than that of the antisymmetric stretch whereas the converse is true in the IR. Popov and Shlyapochnikov (33) in their study of polynitroalkanes determined that of all the C-NO_2 vibrations, the antisymmetric NO_2 stretch is localized to a greater degree within the limits of the C-NO_2 group and the frequency of this vibration and its intensity per nitro group can be associated with the parameters of the nitro group alone. The symmetric stretch is more variable in frequency and is very sensitive to the structure of the groups attached to the C-NO_2 group. Slovetskii et al. (34) have found that primary and secondary nitroalkanes can be distinguished from tertiary ones by the bands of the symmetric and antisymmetric vibrations of the nitro group, whereas the primary compounds can be distinguished from the secondary ones by the position of the band resulting from the symmetric vibration. Also *gem*-disubstituted and 1,1,1-trinitroalkanes could be differentiated using these two vibrational bands. Their data are summarized in Table 4.5

The C-N stretch in CH_3NO_2 occurs as a strong band in the Raman at 920 cm^{-1} and at 878 cm^{-1} in $C_2H_5NO_2$. However, for 1-nitropropane (Spectrum 22), two lines of equal intensity are found at 876 and 900 cm^{-1}. Geiseler and

Table 4.3 Fundamental Vibrations of the C-NO_2 Group

Vibration	Frequency (cm^{-1}) in Nitromethane	Name
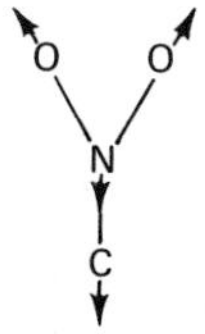	1376	NO_2 Symmetric stretch
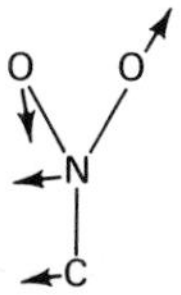	1561	NO_2 Antisymmetric stretch
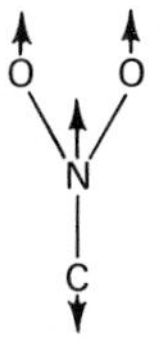	920	C-N Stretch
	656	NO_2 Symmetric deformation (scissors)
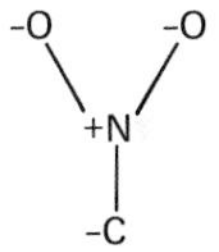 -O -O +N -C	609	C-NO_2 Out-of-plane deformation (− = below plane of paper; + = above plane of paper)
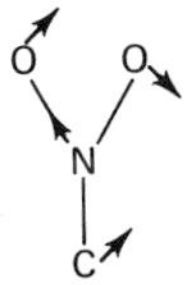	483	NO_2 In-plane deformation (rocking)

Table 4.4 C–NO_2 Frequencies in the Raman Spectra of Nitroalkanes

	Vibration Frequency (cm^{-1})						
Compound	NO_2 Anti-symmetric Stretch	NO_2 Sym-metric Stretch	C-N Stretch Gauche	C-N Stretch Trans	NO_2 Sym-metric Defor-mation	NO_2 Out-of-plane Defor-mation	NO_2 In-plane Defor-mation
1. Primary							
nitroethane	1555	1368, 1395	878	-	618	614	494
1-nitropropane	1553	1382	876	900	618	633	478
1-nitrobutane	1555	1382	883	914	615	633	481
1-nitropentane	1555	1382	885	911	613	637	480
1-nitrohexane	1555	1382	889	907	613	638	472
1-nitroheptane	1557	1386	878	898	615	637	483
1-nitrooctane	1557	1383	885	909	611	638	481
1-nitrononane	1557	1388	873	904	609	643	472
1-nitrodecane	1557	1384	894	909	615	639	481
Raman intensity	Medium–strong	Strong	Medium–strong	Medium–strong	Weak–medium	Weak (shoulder)	Weak–medium
2. Secondary							
2-nitropropane	1551	1360	851	-	620	-	525
2-nitrobutane	1550	1362	853	880	619	656	522
2-nitropentane	1553	1362	858	876	622	650	539
2-nitrooctane	1552	1361	863	891	627	664	551
3-nitropentane	1551	1370	847	868	613	658	516
3-nitrooctane	1554	1374	852	887	622	665	552
4-nitrooctane	1553	1368	863	908	625	670	557
Raman intensity	Medium	Strong	Strong	Medium	Medium	Very weak–weak	Strong

Kessler (31) attribute this splitting to the occurrence of rotational isomers analogous to the behavior of other homologous series of alkane derivatives. Their assignments for the trans and gauche isomers are given in Table 4.4. The assignment of these bands is more difficult for higher chain nitroalkanes since they lie in the region of C-C skeletal vibrations that are of comparable intensity. Therefore, these vibrations are of limited value for use as characteristic frequencies.

The symmetric deformation of the nitro group is a good characteristic frequency. It occurs with medium intensity in the region 630–610 cm^{-1} for all the primary and secondary nitroalkanes studied. The frequency of the out-of-plane deformation for this group is much more variable (670–615 cm^{-1}) and is generally very weak in the Raman and appears as a shoulder on the Raman band of the symmetric deformation. The frequency of the NO_2 in-plane

Table 4.5 Frequencies of the Antisymmetric and Symmetric Stretching Vibrations of the Nitro Group in Monosubstituted, *gem*-Disubstituted, and 1,1,1-Trisubstituted Nitroalkanes

		Frequency (cm^{-1})	
Type of Nitroalkane	Formula	Antisymmetric Stretch	Symmetric Stretch
Primary	$R_1CH_2NO_2$	1554 ± 6	1382 ± 6
Secondary	$R_1R_2CHNO_2$	1550.5 ± 2.5	1369.5 ± 3.5
Tertiary	$R_1R_2R_3CNO_2$	1538.5 ± 4.5	1348.5 ± 4.5
gem-Disubstituted	$R_1R_2C(NO_2)_2$	1575 ± 12	1332 ± 5
1,1,1-Trisubstituted	$R_1C(NO_3)_3$	1600 ± 3	1302 ± 5

deformation or rock can be used to distinguish primary and secondary nitroalkanes. For primary nitrosubstitution, a broad band occurs in the region 490–470 cm^{-1} whereas in the secondary nitroalkanes a strong sharp band occurs at 560–520 cm^{-1}.

4.6 ALKYL NITRATES

Raman spectra have been reported for a series of alkyl nitrates, RO-NO_2 (35–37). The symmetric NO_2 stretch gives rise to an intense polarized band at 1282–1275 cm^{-1} (Spectrum 23). The antisymmetric NO_2 stretch is assigned to a band of medium intensity at 1634–1622 cm^{-1}. The NO_2 in-plane deformation leads to a strong polarized band in the range 610–562 cm^{-1}.

4.7 ALKYL NITRITES

Dadieu et al. (35) have examined the Raman spectra of five alkyl nitrites, RO-N=O. The N=O stretching vibration was assigned to a strong Raman band in the region 1648–1640 cm^{-1}.

REFERENCES

1. K. W. F. Kohlrausch, *Ramanspektren,* Akademische Verlagsges., Leipzig, 1943.
2. J. E. Stewart, *J. Chem. Phys., 30,* 1259 (1959).
3. L. Segal and F. V. Eggerton, *Appl. Spectrosc., 15,* 112 (1961).
4. G. Dellepiane and G. Zerbi, *J. Chem. Phys., 48,* 3573 (1968).
5. M. R. Yagudaev, E. M. Popov, I. P. Yakovlev, and Yu. N. Sheinker,

Izv. Akad. Nauk SSSR, Ser. Khim., 1105 (1963).

6. E. M. Popov, V. N. Zheltova, and G. A. Kogan, *J. Struct. Chem. (USSR)*, *10*, 1002 (1969).
7. J. R. Durig, S. F. Bush, and F. G. Baglin, *J. Chem. Phys.*, *49*, 2106 (1968).
8. P. H. Clippard and R. C. Taylor, *J. Chem. Phys.*, *50*, 1472 (1969).
9. D. W. Scott and G. A. Crowder, *J. Mol. Spectrosc.*, *26*, 477 (1968).
10. A. Y. Hirakawa and M. Tsuboi, *Proc. Int. Symp. Mol. Struct. Spectrosc.*, *A107*, Tokyo (1962).
11. B. R. Cain, J. M. Freeman, and T. Henshall, *Can. J. Chem.*, *47*, 2947 (1969).
12. W. T. King and B. L. Crawford, *J. Mol. Spectrosc.*, *5*, 421 (1960); *8*, 58 (1962).
13. A. Finch, I. J. Hyams, and D. Steele, *J. Mol. Spectrosc.*, *16*, 103 (1965).
14. M. J. Buttler and D. C. McKean, *Spectrochim. Acta*, *21*, 465 (1965).
15. J. T. Edsall, *J. Chem. Phys.*, *4*, 1 (1936).
16. J. T. Edsall, and H. Scheinberg, *J. Chem. Phys.*, *8*, 520 (1940).
17. J. T. Edsall, *J. Chem. Phys.*, *5*, 225 (1937).
18. S. A. S. Ghazanfar, J. T. Edsall, and D. V. Myers, *J. Am. Chem. Soc.*, *86*, 559 (1964).
19. R. D. Waldron, *J. Chem. Phys.*, *21*, 734 (1953).
20. E. A. V. Ebsworth and N. Sheppard, *Spectrochim. Acta*, *13*, 261 (1959).
21. L. Medard, *Compt. Rend.*, *199*, 421 (1934).
22. W. O. George, J. H. S. Green, and M. J. Rix, *Spectrochim. Acta*, *26A*, *2007* 1970).
23. J. R. Durig, S. F. Bush, and E. E. Mercer, *J. Chem. Phys.*, *44*, 4238 (1966).
24. J. S. Ziomek and M. D. Zeidler, *J. Mol. Spectrosc.*, *11*, 163 (1963).
25. J. R. Durig, W. C. Harris, and D. W. Wertz, *J. Chem. Phys.*, *50*, 1449 (1969).
26. J. R. Durig and W. C. Harris, *J. Chem. Phys.*, *51*, 4457 (1969).
27. A. J. Wells and E. B. Wilson, Jr., *J. Chem. Phys.*, *9*, 314 (1941).
28. J. P. Mathieu and D. Massignon, *Ann. Phys.*, *16*, 5 (1941).
29. H. Wittek, *Z. Phys. Chem.*, *51B*, 103, 157 (1942).
30. D. C. Smith, Chi-Yuan Pan, and J. R. Nielsen, *J. Chem. Phys.*, *18*, 706 (1950).
31. G. Geiseler, and H. Kessler, *Ber. Bunsenges. Phys. Chem.*, *68*, 571 (1964).
32. G. Geiseler, H. Kessler, and J. Fruwert, *Ber. Bunsenges. Phys. Chem.*, *70*, 918 (1966).

33. E. M. Popov, and V. A. Shylapochnikov, *Opt. Spectrosc.*, *15*, 174 (1963).
34. V. I. Slovetskii, V. A. Shylapochnikov, S. A. Shevelev, A. A. Fainzil'berg, and S. S. Novikov, *Izv. Akad. Nauk SSSR., Otd. Khim. Nauk*, *1961*, 330.
35. A. Dadieu, F. Jele, and K. W. F. Kohlrausch, *Monatsh. Chem.*, *58*, 428 (1931).
36. J. Lecomte and J. Mathieu, *Compt. Rend.*, *213*, 721 (1941).
37. H. Wittek, *Z. Phys. Chem.*, *52B*, 153 (1942).

CHAPTER FIVE

ORGANOSULFUR COMPOUNDS

5.1 THIOLS (MERCAPTANS)

A detailed investigation of the vibrational assignments in the IR and Raman spectra of methanethiol (methyl mercaptan) and its deuterated analogue has been made by May and Pace (1). In the Raman spectrum of liquid CH_3SH strong polarized bands are found at 2575 cm^{-1} (S-H stretch), 806 cm^{-1} (C-SH deformation), and 704 cm^{-1} (C-S stretch). For ethanethiol two strong Raman bands are found in the C-S stretching region (710–590 cm^{-1}). These bands at 665 and 658 cm^{-1} have been assigned (2) to a Fermi resonance between the C-S stretching fundamental and an overtone of the CCS deformation (331 cm^{-1}). In higher alkanethiols the presence of rotational isomerism may lead to several Raman bands in this range, for example, in the spectrum of 1-propanethiol (3,4) a very strong band at 651 cm^{-1} is attributed to the C-S stretch of the isomer possessing the gauche configuration about the C_1-C_2 bond and a band of moderate intensity at 701 cm^{-1} to the rotamer with the trans configuration. Scott and El-Sabban (5) have derived a valence force field for nine alkanethiols, critically reexamined their vibrational assignments, and discussed the probable conformations of these molecules. Several reviews correlating characteristic vibrational frequencies in the IR and Raman spectra of organic sulfur compounds have also been reported (6–8).

The characteristic frequencies in the Raman spectra of thiols (e.g., Spectrum 24, Appendix 2) are presented in Table 5.1. The frequency range of the S-H stretching mode is 2590–2560 cm^{-1} for aliphatic and aryl thiols. In the IR this band is weak but in the Raman it appears as a strong, polarized band. In aliphatic thiols generally one or more highly polarized Raman bands

Table 5.1 Characteristic Frequencies in the Raman Spectra of Thiols (Liquid State)

Compound	ν(S-H) (cm^{-1})	ν(C-S) (cm^{-1})	References
1. Methanethiol	2575	704	1,9,10
2. Ethanethiol	2571	665,658	2,10,11
3. R-SH (R = *n*-alkyl)	2580–2570	660–650	4, 12–14
4. 2-Propanethiol	2564	623,616	2,14,15
5. 2-Methyl-2-propanethiol	2570	587	16,17
6. 2-Chloro-1-ethanethiol	2570	695,640	18
7. 2-Bromo-1-ethanethiol	2562	735,662	18
8. Cyclohexanethiol	2568	734,710	19
9. Benzenethiol	2566	–	20,21
10. 4-Fluorobenzenethiol	2589	–	22
11. 4-Chlorobenzenethiol	2569	–	23

of very strong to strong intensity arising from the C-S stretching vibration occur in the range 735–590 cm^{-1}. In aryl thiols the C-S stretch couples with the ring stretching modes and no distinct ν (C-S) can be assigned. However, the C-SH deformation in aryl thiols appears as a Raman band of medium intensity near 920 cm^{-1}.

5.2 SULFIDES (THIAALKANES)

The early Raman studies of dimethyl sulfide (2-thiapropane) have been reviewed by Fonteyne (24). Complete vibrational assignments in the IR and

Table 5.2 Characteristic Raman Frequencies for Liquid Organic Sulfides

Compound	ν(C–S) cm^{-1}	References
Dimethyl sulfide	742(s)-ν_{asym};691(vs)-ν_{sym}	25–27
Methyl ethyl sulfide	723(m)-trans;675(m)-gauche; 650(vs)-trans,gauche	31,32
Methyl isopropyl sulfide	723(s);636(vs);610(m)	33
Methyl *t*-butyl sulfide	722(m);589(vs)	34
Diethyl sulfide	693(m);657(s);639(s)	10,35,36
Di-isopropyl sulfide	651(vs)	19
Di-*t*-butyl sulfide	585(vs)	–
Di-*n*-propyl sulfide	715(m);652(s)-broad	–
Di-*n*-butyl sulfide	715(w);652(m)-broad	–
$H_3CSCH_2CH_2SCH_3$	735(m);690(s);651(m);644(w)	37
Diallyl sulfide	752(vs);731(vs)	38
Methyl vinyl sulfide	739(s);698(w);678(s)	39,40
Methyl aryl sulfides	732–710(s);–CH_3–S stretch	20,22,41–43

Raman have been reported for $(CH_3)_2S$ (25–27) and for $(CD_3)_2S$ (27,28). Several normal coordinate analyses have also been reported for these (5, 29, 30) and other thiaalkanes (5). In the Raman spectrum of liquid $(CH_3)_2S$ an intense polarized band at 691 cm^{-1} is assigned to ν_{sym} (CSC), the symmetric stretching of the C-S-C bonds, and a strong depolarized band at 742 cm^{-1} to ν_{asym} (CSC), the antisymmetric stretching of this group. The in-plane C-S-C deformation, δ (CSC), is assigned to a strong polarized band at 284 cm^{-1}.

The characteristic Raman frequencies for various classes of organic sulfides (e.g., Spectrum 25) are listed in Table 5.2 For alkyl sulfides generally one or more strong polarized Raman bands attributed to ν (C-S) occur in the range 740–585 cm^{-1}. Sulfides containing the *t*-butyl group exhibit an intense highly polarized band near 585 cm^{-1}. One or more polarized Raman bands of moderate-to-weak intensity occur in the range 380–320 cm^{-1} for alkyl sulfides of the type -C-C-S-R. These bands arise from the in-plane deformation of the C-C-S moiety.

5.3 DI- AND POLYSULFIDES

For the disulfides, detailed vibrational assignments have been reported for dimethyl disulfide (2,3-dithiabutane) (44), diethyldisulfide (3,4-dithiahexane) (45), and di-*t*-butyl disulfide (2,2,5,5-tetramethyl-3,4-dithiahexane) (46). Allum et al. (47) have examined the IR and Raman spectra of seventeen aliphatic, unsaturated, and aromatic disulfides and assigned the stretching frequencies for the C-S and S-S bonds. In straight chain dialkyl disulfides the C-S stretches occur in the Raman as one or more bands in the region 715–620 cm^{-1} and the S-S stretch is assigned to two bands that occur in the narrow range 525–510 cm^{-1} (see Spectrum 26). In the Raman spectrum of di-*t*-butyl disulfide, ν (C-S) is assigned to a strong polarized band at 576 cm^{-1} and ν (S-S) to a medium intensity polarized band at 543 cm^{-1}. The S-S stretching frequency for diaryl disulfides has been assigned to the range 540–520 cm^{-1}.

Raman spectra have been reported for dimethyl trisulfide (10) and diethyl trisulfide (10,35). The S-S stretching frequencies for these polysulfides occur in the Raman as strong bands in the range 510–480 cm^{-1}. In the Raman spectra of a series of sulfane derivatives of the type RS_nH and RS_nCl (n = 2–4) Feher and Kruse (48) assigned the S-S stretching frequency to a strong band in the range 500–450 cm^{-1}.

5.4 THIOCARBOXYLIC ACIDS AND ESTERS

The Raman spectra of the thiol acids, R-C(=O)-SH ($R=CH_3$, C_2H_5), have been reported by Kohlrausch and Pongratz (49). The vibrational assignment for thiolacetic acid was made by Sheppard (50) who assigned the Raman band at 2568 cm^{-1} to the S-H stretch, that at 1694 cm^{-1} to the C=O stretch, and that at 626 cm^{-1} to the C-S stretch. In the Raman spectrum of thiobenzoic acid (51) these vibrations are located at 2591 cm^{-1} – ν (S-H), 1676 cm^{-1} – ν (C=O), and 608 cm^{-1} – ν (C-S).

IR and Raman data for the thiol chloroformate, Cl-C(=O)-SCH_3, have been presented by Nyquist (52) who found that the three characteristic IR frequencies of the -S-C(=O)-Cl group at 1767 cm^{-1} – ν(C=O), 846 cm^{-1} – ν(antisymmetric S-C-Cl) and 581 cm^{-1} – ν(symmetric S-C-Cl) were also found in the liquid phase Raman spectrum. For methyl dithioacetate, CH_3-C(=S)-SCH_3, Herzog et al. (53) have assigned the antisymmetric and symmetric C–C–S stretchings to strong Raman bands at 857 and 572 cm^{-1}, respectively, and ν(C=S) to an intense band at 1203 cm^{-1}.

5.5 THIOAMIDES AND THIOUREAS

The difficulties encountered in attempting to assign the C=S stretching frequency in molecules in which the thiocarbonyl group is attached to one or two nitrogen atoms have been thoroughly reviewed by Rao and Venkataraghavan (54) and by Jensen and Nielsen (55). Frequencies between 1570 and 950 cm^{-1} have been proposed in the literature. Normal coordinate calculations have shown (56) that in thioformamide the 843 cm^{-1} band corresponds to an almost pure ν(C=S) mode. However, in thioacetamide (57) and *N*-mono and *N*-disubstituted thioamides (58,59) this vibration is coupled with several other vibrational modes. For example, in thioacetamide strong IR bands at 1306, 975, and 718 cm^{-1} have been attributed to modes that involve considerable contributions from both the C-C and C=S stretching vibrations and, in the case of the two higher frequencies, the NH_2 rocking vibration as well. In the Raman spectrum of crystalline thioacetamide (60) (Spectrum 27) a band of moderate intensity occurs at 1299 cm^{-1} and a band at 716 cm^{-1} is the most intense one observed.

In the Raman spectra of thiourea in the crystalline state (60) and in aqueous solution (61) the strongest band is located near 735 cm^{-1}. A normal coordinate analysis of the planar vibrations of thiourea was carried out by Aitken, Duncan, and McQuillan (62) who found that the carbon-sulfur stretching coordinate was distributed mainly among three vibrations, the major proportion to that at 733 cm^{-1} and approximately equal proportions to those at 1414 and 487 cm^{-1}. In the Raman spectra of *N,N'*-dimethyl- and *N,N'*-diphenyl-thiourea (63) bands at 722 and 688 cm^{-1}, respectively, were assigned to a vibration that contains 50% or more C=S character.

5.6 THIOCARBONYL HALIDES

The Raman spectrum of thiophosgene, $Cl_2C{=}S$, was first measured by Thompson (64) and more recently the vibrational spectra of thiophosgene and several of its polymeric species were investigated by Frenzel et al. (65). A strong polarized band at 1120 cm^{-1} was assigned to the C=S stretching mode. Raman spectral data have also been reported for thiocarbonyl chlorofluoride, FClC=S (66) and for thiocarbonyl chlorobromide, BrClC=S (67). For these compounds ν(C=S) occurs at 1242 and 1125 cm^{-1}, respectively.

5.7 THIOPHOSPHORYL COMPOUNDS

Delwaulle and Francois (68-70) have reported the Raman spectra of ten thiophosphoryl trihalides, $X_3P{=}S$ (X = F, Cl, Br). Complete vibrational assignments for the series $Cl_nF_{3-n}P{=}S$ ($n = 0$–3) have been made by Durig and Clark (71). In the Raman spectra of these compounds the P=S stretching vibration was assigned to a band in the range 750-695 cm^{-1}. In the liquid Raman spectrum of methylthiophosphonic dichloride, $CH_3P({=}S)Cl_2$, ν (P=S) occurs at 671 cm^{-1} (72) and in the dimethylphosphinothioc halides, $(CH_3)_2P({=}S)X$, ν(P=S) is located at 612 cm^{-1} for the chloride and at 600 cm^{-1} for the bromide (73). In the solid state Raman spectrum of triethylphosphine sulfide $(C_2H_5)_3P{=}S$, the most intense band in the spectrum at 539 cm^{-1} is assigned to the P=S stretching mode (74). A series of compounds containing the CH_3O-P(=S) moiety have been investigated by Raman spectroscopy (75-79). A doublet with a separation of 10-30 cm^{-1} is found in the liquid phase spectra in the range 720-600 cm^{-1} and is assigned to ν(P=S) of two rotational isomers. The intensity of one band of the doublet decreases significantly upon cooling to the solid state.

5.8 SULFINYL (THIONYL) COMPOUNDS

Raman spectral data for the frequency of the S=O stretching vibration in various compounds are collected in Table 5.3. The influence of various inter- and intramolecular factors on the position of the ν (S=O) band in both the IR and Raman spectra has been reviewed by Simon and Kriegsmann (80) and by Steudel (81). Detailed vibrational assignments have been reported for the thionyl halides: SOF_2 (82-84), $SOCl_2$ (84,85), and $SOBr_2$ (86). Normal coordinate analyses have also been attempted for this series of compounds (87,88). Dimethyl sulfoxide, $(CH_3)_2SO$, has been the subject of numerous vibrational studies (27,29,30,87-91).

In the spectra of symmetrically substituted aliphatic sulfoxides (e.g., Spectrum 28) one or two bands occur in the ν(S=O) stretching region 1070-1040 cm^{-1}. This splitting, which is usually less than 10 cm^{-1}, has been attributed to the presence of rotational isomerism about the C-S axes (96). Unsymmetrically substituted sulfoxides generally exhibit two bands in this range. Hydrogen bonding of the S=O group results in a decrease in ν(S=O) by 15-40 cm^{-1}. The exact dependence of ν(S=O) upon the electronegativity of the substituents has been discussed in detail by Steudel (81). In general, bonding to oxygen or halogens results in a shift to higher frequencies (see Table 5.3).

Table 5.3 Characteristic Raman Frequencies for the Sulfinyl Group (-SO-)

Structure	ν(S=O) cm^{-1}	Other Prominent Raman Bands cm^{-1}	References
SOF_2	1308	801-ν_{sym}(FSF); 721-ν_{asym}(FSF)	83,92
$SOCl_2$	1233	490-ν_{sym}(ClSCl); 442-ν_{asym}(ClSCl)	89
$SOBr_2$	1121	405-ν_{sym}(BrSBr); 379-ν_{asym}(BrSBr)	86
$(CH_3)_2SO$	1042	384-δ (CSO)	87,91
R_2SO	1050–1020	—	81,89
R–SO-aryl	1050–1040	—	42,81
$(aryl)_2SO$	1042–1035	—	43,81
$(CH_3O)_2SO$	1207	723-ν_{sym}(OSO); 579-δ (O_2SO); 444-δ (COS)	89,93,94
$(RO)_2SO$	1209–1198	740–707-ν_{sym}(OSO); 584–581-δ (O_2SO); 447–437-δ (COS)	80,89,93
RO–SO–Cl	1221–1214	447–437-ν (SCl)	89
$[(CH_3)_2N]_2SO$	1108	674,657-SN_2 stretches of rotamers	95
$(CH_3)_2N$-SO-Cl	1189	701-ν (SN); 394-ν (SCl)	95

5.9 SULFONYL (SULFURYL) COMPOUNDS

The characteristic frequencies in the vibrational spectra of molecules containing the -SO_2-group have been thoroughly reviewed (80,89,97,98). The Raman data available for this class of compounds are collected in Table 5.4. Robinson (97) has shown that the -SO_2- stretching frequencies are essentially free from mass and coupling effects and depend in a systematic manner upon the electronegativities of the attached groups. Both the antisymmetric and symmetric -SO_2-stretching vibrations lead to very strong bands in the IR spectra. In the Raman spectra $\nu_{asym}(SO_2)$ is quite variable in intensity and often is not easily detected. However, $\nu_{sym}(SO_2)$ occurs as an intense polarized band in the ranges specified in Table 5.4.

Table 5.4 Characteristic Frequencies in the Raman Spectra of Sulfonyl (Sulfuryl) Compounds

Compound		ν_{asym} (SO_2) cm^{-1}	ν_{sym} (SO_2) cm^{-1}	δ (SO_2) cm^{-1}	References
1. Sulfones					
a. CH_3SO_2-CH_3	aqueous solution	1289(m)	1138(vs)	502(s)	98
	liquid	1295	1128	498	99
	crystal	1269	1121	496	27,89,99
b. R-SO_2-R		1330–1305	1145–1125(vs)	512–485(m)	80,89
c. CH_3-SO_2-aryl		1325–1300(wk)	1155–1140(vs)	573–540(m)	42
2. Thiolsulfonates					
a. R-SO_2-SR		1334–1305(s-m)	1128–1126(s)	559–553 (s-m)	100,101
3. Sulfonates					
a. R-SO_2-OR		1358–1352(m)	1172–1165(vs)	555–517 (2 bands)	102,103
b. aryl-SO_2-OR		1363–1338(w)	1192–1185(vs)	589–562(wk-m)	103
c. Cl-SO_2-OR		1406–1401(s-m)	1191–1184(vs)	598–544 (2 bands)	89
4. Sulfonamides					
a. CH_3-SO_2-NH_2		1322(m)	1138(vs)	524(s)	104
b. $(CH_3$-$SO_2)_2NH$		—	1163(vs)	510(vs)	104
c. C_6H_5-SO_2-NH_2		—	1157(s)	—	105

5. Sulfates					
a. $RO\text{-}SO_2\text{-}OR$		1388–1372(s)	1196–1188(vs)	—	89
6. Organic sulfonyl halides					
a. $CH_3\text{-}SO_2\text{-}F$		1401(m)	1186(vs)	531(s)	106
b. aryl-SO_2-F		1412–1402(m–wk)	1197–1167(vs)	596–588(m)	107
c. $CH_3\text{-}SO_2\text{-}Cl$		1361(m)	1168(vs)	538(s)	80,106
d. aryl-SO_2-Cl		1384–1361(m–wk)	1184–1169(vs)	587–565(m)	107
7. Sulfuryl halides					
a. SO_2F_2	liquid	1497(w)	1263(vs)	547(s)	83,84
b. SO_2Cl_2		1414(s)	1182(s)	560(s)	84,89

REFERENCES

1. I. W. May and E. L. Pace, *Spectrochim. Acta, 24A,* 1605 (1968).
2. D. Smith, J. P. Devlin, and D. W. Scott, *J. Mol. Spectrosc., 25,* 174 (1968).
3. R. E. Pennington, D. W. Scott, H. L. Finke, J. P. McCullough, J. F. Messerly, I. A. Hossenlopp, and G. Waddington, *J. Am. Chem. Soc., 78,* 3266 (1956).
4. M. Hayashi, Y. Shiro, and H. Murata, *Bull. Chem. Soc. Jap., 39,* 112 (1966).
5. D. W. Scott and M. Z. El-Sabban, *J. Mol. Spectrosc., 30,* 317 (1969).
6. I. F. Trotter and H. W. Thompson, *J. Chem. Soc., 1946,* 481.
7. N. Sheppard, *Trans. Faraday Soc., 46,* 429 (1950).
8. D. W. Scott and J. P. McCullough, *J. Am. Chem. Soc., 80,* 3554 (1958).
9. J. Wagner, *Z. Phys. Chem., 40B,* 36 (1938).
10. R. Vogel-Högler, *Acta Phys. Austriaca, 1,* 311 (1948).
11. J. Wagner, *Z. Phys. Chem., 40B,* 439 (1938).
12. E. A. Crigler, *J. Am. Chem. Soc., 54,* 4199 (1932).
13. A. Dadieu, A. Pongratz, and K. W. F. Kohlrausch, *Monatsh. Chem., 61,* 409 (1932).
14. G. Radinger and H. Wittek, *Z. Phys. Chem., 45B,* 329 (1940).
15. J. P. McCullough, H. L. Finke, D. W. Scott, M. E. Gross, J. F. Messerly, R. E. Pennington, and G. Waddington, *J. Am. Chem. Soc., 76,* 4796 (1954).
16. N. Sheppard, *Trans. Faraday Soc., 46,* 527 (1950).
17. J. P. McCullough, D. W. Scott, H. L. Finke, W. N. Hubbard, M. E. Gross, C. Katz, R. E. Pennington, J. F. Messerly, and G. Waddington, *J. Am. Chem. Soc., 75,* 1818 (1953).
18. M. Hayashi, Y. Shiro, M. Murakami, and H. Murata, *Bull. Chem. Soc. Jap., 38,* 1740 (1965).
19. D. W. Scott and G. A. Crowder, *J. Chem. Phys., 46,* 1054 (1967).
20. L. Kahovec and A. W. Reitz, *Monatsh. Chem., 69,* 363 (1936).
21. D. W. Scott, J. P. McCullough, W. N. Hubbard, J. F. Messerly, I. A. Hossenlopp, F. R. Frow, and G. Waddington, *J. Am. Chem. Soc., 78,* 5463 (1956).
22. J. H. S. Green, D. J. Harrison, W. Kynaston, and D. W. Scott, *Spectrochim. Acta, 26A,* 1515 (1970).
23. R. A. Nyquist and J. C. Evans, *Spectrochim. Acta, 17,* 795 (1961).
24. R. Fonteyne, *J. Chem. Phys., 8,* 60 (1940).
25. J. P. McCullough, W. N. Hubbard, F. R. Frow, I. A. Hossenlopp, and G. Waddington, *J. Am. Chem.* Soc., *79,* 561 (1957).

26. J. R. Allkins and P. J. Hendra, *Spectrochim. Acta, 22,* 2075 (1966).
27. G. Geiseler and G. Hanschmann, *J. Mol. Struct., 11,* 283 (1972).
28. M. Tranquille, M. Fouassier, M. Lautie-Mouneyrac, P. Dizabo, and M. Forel, *Compt. Rend., 270C,* 1085 (1970).
29. M. Tranquille, P. Labarbe, M. Fouassier, and M. T. Forel, *J. Mol. Struct., 8,* 273 (1971).
30. G. Geiseler and G. Hanschmann, *J. Mol. Struct., 8,* 293 (1971).
31. D. W. Scott, H. L. Finke, J. P. McCullough, M. E. Gross, K. D. Williamson, G. Waddington, and H. M. Huffman, *J. Am. Chem. Soc., 73,* 261 (1951).
32. M. Hayashi, T. Shimanouchi, and S. Mizushima, *J. Chem. Phys., 26,* 608 (1957).
33. J. P. McCullough, H. L. Finke, J. F. Messerly, R. E. Pennington, I. A. Hossenlopp, and G. Waddington, *J. Am. Chem. Soc., 77,* 6119 (1955).
34. D. W. Scott, W. D. Good, S. S. Todd, J. F. Messerly, W. T. Berg, I. A. Hossenlopp, J. L. Lacina, A. Osborn, and J. P. McCullough, *J. Chem. Phys., 36,* 406 (1962).
35. P. Donzelot and M. Chaix, *Compt. Rend., 202,* 851 (1936).
36. D. W. Scott, H. L. Finke, W. N. Hubbard, J. P. McCullough, G. D. Oliver, M. E. Gross, C. Katz, K. D. Williamson, G. Waddington, and H. M. Huffman, *J. Am. Chem. Soc., 74,* 4656 (1952).
37. M. Hayashi, Y. Shiro, T. Oshima, and H. Murata, *Bull. Chem. Soc. Jap., 39,* 118 (1966).
38. K. W. F. Kohlrausch and W. Stockmair, *Z. Phys. Chem., 29B,* 292 (1934).
39. J. Fabian, H. Kröber, and R. Mayer, *Spectrochim. Acta., 24A,* 727 (1968).
40. E. M. Popov and G. I. Kagan, *Opt. Spectrosc., 11,* 394 (1961).
41. J. H. S. Green, *Spectrochim. Acta, 18,* 39 (1962).
42. G. Kresze, E. Ropte, and B. Schrader, *Spectrochim. Acta, 21,* 1633 (1965).
43. J. H. S. Green, *Spectrochim. Acta, 24A,* 1627 (1968).
44. D. W. Scott, H. L. Finke, M. E. Gross, G. B. Guthrie, and H. M. Huffman, *J. Am. Chem. Soc., 72,* 2424 (1950).
45. D. W. Scott, H. L. Finke, J. P. McCullough, M. E. Gross, R. E. Pennington, and G. Waddington, *J. Am. Chem. Soc., 74,* 2478 (1952).
46. J. H. S. Green, D. J. Harrison, W. Kynaston, and D. W. Scott, *Spectrochim. Acta, 25A,* 1313 (1969).
47. K. G. Allum, J. A. Creighton, J. H. S. Green, G. J. Minkoff, and L. J. S. Prince, *Spectrochim. Acta, 24A,* 927 (1968).
48. F. Feher and W. Kruse, *Chem. Ber., 91.* 2528 (1958);
49. K. W. F. Kohlrausch and A. Pongratz, *Z. Phys. Chem., 27B,* 176 (1934).

50. N. Sheppard, *Trans. Faraday Soc., 45,* 693 (1949).
51. K. W. F. Kohlrausch and A. Pongratz, *Monatsh. Chem., 64,* 374 (1934).
52. R. A. Nyquist, *J. Mol. Struct., 1,* 1 (1967).
53. K. Herzog, E. Steger, P. Rosmus, S. Scheithauer, and R. Mayer, *J. Mol. Struct., 3,* 339 (1969).
54. C. N. R. Rao and R. Venkataraghavan, *Spectrochim. Acta, 18,* 541 (1962).
55. K. A. Jensen and P. H. Nielsen, *Acta Chem. Scand., 20,* 597 (1966).
56. I. Suzuki, *Bull. Chem. Soc. Jap., 35,* 1286 (1962).
57. I. Suzuki, *Bull. Chem. Soc. Jap., 35,* 1449 (1962).
58. I. Suzuki, *Bull, Chem. Soc. Jap., 35,* 1456 (1962).
59. C. A. I. Chary and K. V. Ramiah, *Proc. Indian Acad. Sci., 69A,* 18 (1969).
60. K. W. F. Kohlrausch and J. Wagner, *Z. Phys. Chem., 45B,* 229 (1939).
61. J. T. Edsall, *J. Phys. Chem., 41,* 133 (1937).
62. G. B. Aitken, J. L. Duncan, and G. P. McQuillan, *J. Chem. Soc. (A), 1971,* 2695.
63. R. K. Ritchie, H. Spedding, and D. Steele, *Spectrochim. Acta, 27A,* 1597 (1971).
64. H. W. Thompson, *J. Chem. Phys., 6,* 748 (1938).
65. C. A. Frenzel, K. E. Blick, C. R. Bennett, and K. Niedenzu, *J. Chem. Phys., 53,* 198 (1970).
66. D. C. Moule and C. R. Subramaniam, *Can. J. Chem., 47,* 1011 (1969).
67. J. L. Brema and D. C. Moule, *Spectrochim. Acta, 28A,* 809 (1972).
68. M. Delwaulle and F. Francois, *Compt. Rend., 224,* 1422 (1947).
69. M. Delwaulle and F. Francois, *Compt. Rend., 225,* 1308 (1947).
70. M. Delwaulle and F. Francois, *Compt. Rend., 226,* 894 (1948).
71. J. R. Durig and J. W. Clark, *J. Chem. Phys., 46,* 3057 (1967).
72. J. R. Durig, F. Block, and I. W. Levin, *Spectrochim. Acta, 21,* 1105 (1965).
73. J. R. Durig, D. W. Wertz, B. R. Mitchell, F. Block, and J. M. Greene, *J. Phys. Chem., 71,* 3815 (1967).
74. J. R. Durig, J. S. Di Yorio, and D. W. Wertz, *J. Mol. Spectrosc., 28,* 444 (1968).
75. R. A. Nyquist, *Spectrochim. Acta, 23A,* 1499 (1967).
76. R. A. Nyquist and W. W. Muelder, *J. Mol. Struct., 2,* 465 (1968).
77. J. R. Durig and J. W. Clark, *J. Chem. Phys., 50,* 107 (1969).
78. J. R. Durig and J. S. Di Yorio, *J. Mol. Struct., 3,* 179 (1969).
79. J. R. Durig and J. W. Clark, *J. Cryst. Mol. Struct., 1,* 43 (1971).
80. A. Simon and H. Kriegsmann, *Z. Phys. Chem. (Leipzig), 204,* 369 (1955).

81. R. Steudel, *Z. Naturforsch.*, *25B*, 156 (1970).
82. J. K. O'Loane and M. K. Wilson, *J. Chem. Phys.*, *23*, 1313 (1955).
83. P. Bender and J. M. Wood, Jr., *J. Chem. Phys.*, *23*, 1316 (1955).
84. R. J. Gillespie and E. A. Robinson, *Can. J. Chem.*, *39*, 2171 (1961).
85. D. E. Martz and R. T. Lagemann, *J. Chem. Phys.*, *22*, 1193 (1954).
86. H. Stammreich, R. Forneris, and Y. Tavares, *J. Chem. Phys.*, *25*, 1277 (1956).
87. W. D. Horrocks, Jr. and F. A. Cotton, *Spectrochim. Acta*, *17*, 134 (1961).
88. D. A. Long and R. T. Bailey, *Trans. Faraday Soc.*, *59*, 792 (1963).
89. R. Vogel-Högler, *Acta Phys. Austriaca*, *1*, 323 (1948).
90. J. H. Carter, J. M. Freeman, and T. Henshall, *J. Mol. Spectrosc.*, *20*, 402 (1966).
91. M.-T. Forel and M. Tranquille, *Spectrochim. Acta*, *26A*, 1023 (1970).
92. D. M. Yost, *Proc. Indian Acad. Sci.*, *8*, 333 (1938).
93. A. Simon, H. Kriegsmann, and H. Dutz, *Chem. Ber.*, *89*, 2390 (1956).
94. P. Klaeboe, *Acta Chem. Scand.*, *22*, 2817 (1968).
95. R. Paetzold and E. Rönsch, *Spectrochim. Acta*, *26A*, 569 (1970).
96. M. Oki, I. Oka and K. Sakaguchi, *Bull. Chem. Soc. Jap.*, *42*, 2944 (1969).
97. E. A. Robinson, *Can. J. Chem.*, *39*, 247 (1961).
98. W. R. Feairheller, Jr. and J. E. Katon, *Spectrochim. Acta*, *20*, 1099 (1964).
99. R. D. McLachlan and V. B. Carter, *Spectrochim. Acta*, *26A*, 1121 (1970).
100. A. Simon and D. Kunath, *Z. Anorg. Allg. Chem.*, *311*, 203 (1961).
101. S. S. Block and J. P. Weidner, *Appl. Spectrosc.*, *20*, 73 (1966).
102. A. Simon, H. Kriegsmann, and H. Dutz, *Chem. Ber.*, *89*, 2378 (1956).
103. D. E. Freeman and A. N. Hambly, *Aust. J. Chem.*, *10*, 239 (1957).
104. A. Blaschette and H. Bürger, *Z. Anorg. Allg. Chem.*, *378*, 104 (1970).
105. W. R. Angus, A. H. Leckie, and T. I. Williams, *Trans. Faraday Soc.*, *34*, 793 (1938).
106. N. S. Ham and A. N. Hambly, *Aust. J. Chem.*, *6*, 33 (1953).
107. N. S. Ham and A. N. Hambly, *Aust. J. Chem.*, *6*, 135 (1953).

CHAPTER SIX

ALKENES

6.1 ETHYLENE

The data on the IR and Raman spectra of ethylene up to 1945 have been reviewed by Herzberg (1). Crawford et al. (2,3) proposed a new assignment for the fundamental vibrations of C_2H_4 and C_2D_4 based on gas phase IR studies of normal ethylene and its deuterated species. In 1964 Smith and Mills (4) reported the high-resolution IR spectra of the ν_7 and ν_{10} perpendicular bands and analyzed the rotational structure in terms of Coriolis perturbations. They also discussed and summarized the reported values of all the vibrational and rotational constants for ethylene. The IR spectra of carbon-13 enriched C_2H_4 and C_2D_4 have been examined by Becher and Adrian (5). IR studies have also been performed on the liquid (6) and the crystalline (6–8) states of C_2H_4 and C_2D_4. Work on the Raman spectrum of gaseous ethylene has been reviewed by Feldman, Romanko, and Welsh (9). A comparison of the Raman spectra of ethylene and the deuteroethylenes in the gaseous and liquid states has been presented by DeHemptinne and Charette (10). Assignments in the Raman spectra of both liquid and solid ethylene have been given by Blumenfeld, Reddy, and Welsh (11). The fundamental frequencies in the IR and Raman spectra of ethylene in various phases are given in Table 6.1.

Normal coordinate analyses of the vibrational frequencies of ethylene were performed by Kilpatrick and Pitzer (12) and by Crawford, Lancaster, and Inskeep (3). The latter proposed two equally plausible force fields [Set I ($k_{C=C} \sim 11$ mdyn/Å) and Set II ($k_{C=C} \sim 9$ mdyn/Å)] with Set I regarded as more probable. However, more recent work involving a study of ^{13}C frequency shifts (5,13), the hybrid orbital force field (14), and an *ab initio* calculation (15) indicate unambiguously that the Set II type force field is the correct one.

Table 6.1 Fundamental Frequencies in the Vibrational Spectra of Ethylene in Various Phases

D_{2h} Symmetry Class	Vibration Number	Approximate Description of Vibration	Frequency (cm^{-1})					
			IR[a]			Raman[b]		
			Gas	Liquid	Solid	Gas	Liquid	Solid
a_g	1	CH_2 Symmetric stretch	Inactive	3016 (vw)	—	3026 (vs)	3004 (vs)	2997 (vs); 3006 (sh)
	2	CC Stretch	Inactive	1620 (w)	—	1623 (s)	1619 (vs)	1602 (vw); 1616 (vs)
	3	CH_2 Scissors	Inactive	1339 (m)	—	1342 (vs)	1339 (vs)	1348 (vs); 1331 (vs)
a_u	4	CH_2 Twist	Inactive	—	—	Inactive	—	—
b_{1g}	5	CH_2 Antisymmetric stretch	Inactive	—	—	3103 (w)	3074 (s)	3067 (vs); 3069 (sh)
	6	CH_2 Rock	Inactive	1239 (vw)	—	—	1238 (vw)	1235 (vw)
b_{1u}	7	CH_2 Wag	949 (vs)	961 (vs)	970 (vs)	Inactive	—	—
b_{2g}	8	CH_2 Wag	Inactive	—	—	950 (vw)	942 (m)	950 (w); 941 (s)
b_{2u}	9	CH_2 Antisymmetric stretch	3106 (s)	3085 (vs)	3075 (vs)	Inactive	—	—
	10	CH_2 Rock	826 (w)	828 (vs)	828 (s); 826 (m)	Inactive	—	—
b_{3u}	11	CH_2 Symmetric stretch	2989 (s)	2983 (vs)	2973 (s)	Inactive	—	—
	12	CH_2 Scissors	1444 (s)	1437 (vs)	1440 (vs); 1436 (vs)	Inactive	—	—

[a] Assignments from refs. 2 and 6.
[b] Assignments from refs. 9 and 11.

6.2 ALKYL SUBSTITUTED ETHYLENES

Extensive investigations of the Raman spectra of alkyl substituted ethylenes have been conducted. The early work of Bourguel (16-21), Gredy (22-26), and their co-workers has been reviewed by Hibben (27) and Kohlrausch (28). Subsequent Raman spectral studies of these alkenes were reported by Cleveland (29-31), Gerding and van der Vet (32), Goubeau et al. (33-35), Fenske et al. (36, 37), and Landsberg et al. (38). Characteristic frequencies in the Raman Spectra of the olefinic hydrocarbons have been reviewed by Piaux (39), Sheppard and Simpson (40), Gruzdev (41), Sverdlov (42), and Rea (43).

The most characteristic frequencies in the Raman spectra of alkyl substituted ethylenes arise from the C=C stretching vibrations and occur as highly polarized bands in the range 1680–1630 cm^{-1}. These frequencies for the various types of alkyl substitution are summarized in Table 6.2 (see Spectra 29-34, Appendix 2). For primary straight chain alkyl substitutents, the C=C stretching frequency falls in a narrow range characteristic for each type of substitution. However, the presence of a methyl group attached to one of the double bond carbon atoms increases this average frequency for each substitution pattern by about 5 cm^{-1} because of interaction between the methyl deformation and C=C stretching modes. The presence of branching at the carbon atom α to the double bond in 1-alkenes decreases the C=C stretching frequency only 1-2 cm^{-1}, if at all. For other types of substitution, this frequency can be significantly lowered if there is steric hindrance between the substituent groups, for example, in 2-methyl-1-butene ν (C=C) occurs at 1651 cm^{-1}, whereas for 2,3,3-trimethyl-1-butene this frequency shifts to 1640 cm^{-1}. Studies of the variation of the intensity of the Raman ν(C=C) band (43,44) show that there is a significant increase in the intensity as the number of substituents on the double bond increases.

Other characteristic bands in the Raman spectra of alkyl substituted ethylenes are listed in Table 6.3. Alkenes containing the $=CH_2$ group exhibit a characteristic antisymmetric carbon-hydrogen stretch in the frequency range 3095–3070 cm^{-1}; those containing the =CHR group have a band located at 3040–3000 cm^{-1} due to the carbon-hydrogen stretch. The in-plane =C-H deformation vibrations occur in the range 1420–1250 cm^{-1}. The symmetric deformation of the $=CH_2$ group gives rise to a Raman band at 1420–1400 cm^{-1} whereas the range for the in-plane C-H deformation of the =CHR group occurs at 1360–1250 cm^{-1}. The out-of-plane deformation vibrations of these groups that lead to strong characteristic bands in the IR spectra between 1000–800 cm^{-1} appear in the Raman as very weak bands.

Some Raman bands that are useful in the identification of the type of alkyl substituent on the C=C bond are summarized in Table 6.4. The Raman intensity of the symmetric deformation band of a methyl group is greatly

Table 6.2 Frequency of the C=C Stretching Vibration on the Raman Spectra of Alkyl Substituted Ethylenes

Type of Alkene	ν(C=C) (cm^{-1})	Raman Intensity	IR Intensity	Type of Substituent	ν(C=C) (cm^{-1})
Ethylene (liquid)	1620	Very strong	Weak	—	—
Monosubstituted	1648–1638	Strong	Medium	1-Methyl (propene)	1648
				1-*n*-Alkyl	1642–1641
				1-Branched alkyl	1642–1640
trans-Disubstituted	1676–1665	Strong	Weak or absent	*trans*-2-Butene	1676
				trans CH_3CH=CHR	1673–1670
				trans CHR=CHR	1669–1665
cis-Disubstituted	1660–1654	Strong	Medium	*cis*-2-Butene	1660
				cis CH_3CH=CHR	1660–1658
				cis CHR=CHR	1654
Vinylidene	1658–1644	Strong	Medium	Isobutene	1658
				$CH_3CR=CH_2$	1652–1650
				$R_2C=CH_2$	1648
Trisubstituted	1678–1664	Strong	Medium or absent	2-Methyl-2-butene	1678
				$(CH_3)_2C$=CHR	1677–1675
				$CH_3CR=CHCH_3$	1673–1671
				CH_3CR=CHR	1669–1667
Tetrasubstituted	1680–1665	Strong	Weak or absent	2,3-Dimethyl-2-butene	1675
				$(CH_3)_2C=CRCH_3$	1670

Table 6.3 Other Characteristic Bands in the Raman Spectra of Alkyl Substituted Ethylenes[a]

Type of Substitution	Raman Frequency (cm^{-1})	Relative Raman Intensity	Description of Vibration
1. Monoalkyl	3086–3079	Medium	Antisymmetric carbon-hydrogen stretch of = CH_2 group
$H_2C{=}CHR$	3015–2993	Strong	Carbon-hydrogen stretch
	1419–1415	Medium	Symmetric deformation of =CH_2 group
	1309–1288	Medium	In-plane C-H deformation
	928–909	Weak	Out-of-plane=CH_2 deformation
	634–621	Weak	Out-of-plane C-H deformation
2. Asymmetric dialkyl	3092–3073	Weak	Antisymmetric carbon-hydrogen stretch of =CH_2 group
$H_2C{=}CR_1R_2$	2990–2983	Strong	Symmetric carbon-hydrogen stretch of =CH_2 group
	1413–1399	Medium	Symmetric deformation of =CH_2 group
	909–885	Weak	Out-of-plane =CH_2 deformation
	711–684	Weak	Out-of-plane =CH_2 deformation
	450–400	Very weak	Skeletal deformation of C=C-C group
3. *cis*-Dialkyl	3016–3001	Medium	Carbon-hydrogen stretch
$R_2HC{=}CHR_1$ (cis)	1270–1251	Strong	In-plane C-H deformation
	970–952	Medium	Antisymmetric carbon-carbon stretch
	720–700	Very weak	Out-of-plane C-H deformation
	592–545	Very weak	Skeletal deformation of C=C–C group

4. *trans*-Dialkyl	3007–2995	Weak	Carbon-hydrogen stretch
$H(R_2)C{=}C(R_1)H$	1314–1290	Strong	In-plane C-H deformation
	776–745	Very weak	In-plane C-H deformation
	492–455	Very weak	Skeletal deformation of C=C-C group
5. Trialkyl	3040–3020	Very weak	Carbon-Hydrogen stretch
$H(R_3)C{=}C(R_1)R_2$	1360–1322	Weak	In-plane C-H deformation
	830–800	Very weak	Out-of-plane C-H deformation
	522–488	Weak	Skeletal deformation of C=C-C group
6. Tetraalkyl	690–678	Strong–medium	Symmetric carbon-carbon stretch
$R_3(R_4)C{=}C(R_1)R_2$	510–485	Medium	In-plane skeletal deformation
	424–388	Weak	In-plane skeletal deformation

[a]See Table 6.2 for the characteristic C=C stretching frequencies.

Table 6.4 Substituent Bands in the Raman Spectra of Alkyl Ethylenes

Type of Substituent	Vibration	Type of Alkene	Frequency (cm^{-1})
Methyl attached to α-carbon	Symmetric deformation	$H_2C{=}CH{-}CH(CH_3){-}R$	1375–1372
		$H_2C{=}CH{-}C(CH_3)_2{-}R$	1380
		$H_2C{=}CH{-}C(CH_3)_2{-}CH_3$	1386
Methyl attached to double bond	Symmetric deformation	Asymmetric	1392–1377
		Cis	1372–1368
		Trans	1380–1379
		Tri	1385–1375
		Tetra	1392–1386
Isopropyl group attached to double bond	C_4 skeletal "breathing"	$C{=}C(H)(CH(CH_3)_2)$	870–800
		$C{=}C(CH_3)(CH(CH_3)_2)$	730–680
Tertiary butyl group attached to double bond	C_5 skeletal "breathing"	$C{=}C(H)(C(CH_3)_2)$	760–720
		$C{=}C(CH_3)(C(CH_3)_3)$	700–670

enhanced if it is located on a double-bond carbon atom or on the carbon α to the double bond because of a mixing of the C=C stretch and CH_3 deformation modes. Alkenes with an isopropyl or *t*-butyl group attached to the double bond can be detected by the presence of a strong, highly polarized band arising from the skeletal "breathing" vibration of each group.

Detailed vibrational assignments have been made for the following alkyl substituted ethylenes: propene (12,45), 1-butene (46), 1-hexene (47), 3,3-dimethyl-1-butene (48), 2-methylpropene (isobutene) (12,49–52), *cis* and *trans*-2-butene (12,53,54), *cis*-3-hexene (47), and 2,3-dimethyl-2-butene (tetramethylethylene) (54,55).

6.3 HALOGENATED ETHYLENES

The frequencies of the polarized Raman bands arising from the C=C stretching modes for various alkenes in which one or more fluorine atoms are attached to the double bond are listed in Table 6.5 The unusually high ν(C=C) frequencies for those alkenes containing two or more fluorines have been attributed to polarization of the C=C bond by the fluorine atoms (85); however, Scherer and Overend (86) have shown that in the vinylidene halide series frequency changes are not entirely associated with changes in the force field and that there can exist appreciable coupling between the C=C stretching and CH_2 bending coordinates. The carbon-fluorine bond stretching frequencies of these compounds generally lie in the region 1400–1000 cm^{-1} (87) but the exact band positions and Raman intensities for each substitution type are quite variable. Fluoroalkenes possessing the vinylidene type grouping, $=CF_2$, exhibit a strong Raman band in the range 580–560 cm^{-1}, which is assigned to the CF_2 wag.

Table 6.6 lists the strongest characteristic bands in the Raman spectra of the homologous series of haloalkenes in which the halogen substituents are either chlorine, bromine, or iodine. Kirrmann et al. (122–127) have investigated the Raman spectra of a series of halogenated isobutenes and di- and trihalogenated propenes. Gerding and Rijnders (128,129) have reported the Raman spectra of 25 various polychloro-substituted alkenes. Compounds of the type $CHX{=}CCl_2$ have been studied by Yarwood and Orville-Thomas (62).

Kirrmann's conclusions (126,127) as to the effect of halogen substitution on the frequency of ν(C=C) in halogenated ethylenes can be summarized as follows: (1) substitution of chlorine atoms on the double bond carbons progressively lowers the C=C stretching frequency; (2) substitution of bromine atoms produces a greater lowering of the frequency than does chlorine (the decrease is about 0–10 cm^{-1} greater in the bromo derivative compared with

Table 6.5 C=C Stretching Frequencies in the Raman Spectra of Fluoroalkenes

Fluorine Substitution	Frequency Range of ν(C=C) (cm^{-1})	Specific Examples: Compound	ν(C=C) (cm^{-1})[a]	References
Monofluoro-	1689–1644	$H_2C{=}CHF$	1654 (IR, gas)	56,57
		$Cl_2C{=}CClF$	1644	58,59
		$H_2C{=}CClF$	1654 (gas)	60
		$Cl_2C{=}CHF$	1654	61-63
		cis-ClHC=CClF	1648	64
		trans-ClHC=CClF	1645	64
		cis-ClHC=CHF	1660	65
		trans-ClHC=CHF	1643	65
		$H_2C{=}CFCH_3$	1689 (solid)	66
Symmetric difluoro-	1712–1694	*cis*-FHC=CHF	1712	67
		trans-FHC=CHF	1694	67
		cis-FClC=CClF	1704	68
		trans-FClC=CClF	1710	68,69
Asymmetric difluoro-	1739–1714	$H_2C{=}CF_2$	1728	70,71
		$Cl_2C{=}CF_2$	1739	56,72,73
		$ClHC{=}CF_2$	1742	74
		$BrHC{=}CF_2$	1728	75
		$Br_2C{=}CF_2$	1714	76,77
		$BrClC{=}CF_2$	1727	78
Trifluoro-	ca. 1793	$F_2C{=}CFCF_3$	1792	79,80
		$F_2C{=}CFCl$	1794	81,82
Tetrafluoro-	1872	$F_2C{=}CF_2$	1872 (gas)	83,84

[a]Unless otherwise specified, all values refer to the liquid state.

the chloro derivative); (3) substitution of a halogen in an allylic position lowers the frequency of ν(C=C) by 3–6 cm^{-1} in the case of C1 and by 6–11 cm^{-1} in the case of Br; (4) for polyhalogen derivatives, the decrease in frequency is roughly additive; and (5) a doublet due to the presence to rotational isomers is found in the ν(C=C) region for compounds of the type $CH_2{=}CC1CH_2X$ where X = halogen, OH, or OAc.

6.4 ALLYL AND VINYL DERIVATIVES

The characteristic frequencies in the Raman spectra of various allyl (e.g., Spectrum 35) and vinyl derivatives together with references to the spectra of specific compounds are listed in Tables 6.7 and 6.8, respectively. The frequencies of the carbon hydrogen stretches of the H_2C=CH group for the allyl derivatives fall in the same range as that for the monoalkyl substituted compounds (Table 6.3). Those of the vinyl derivatives are located in a wider range with those derivatives in which a multiple bond is conjugate to the C=C bond exhibiting these bands about 20–30 cm^{-1} higher than in the nonconjugated derivatives. The ν(C=C) frequency occurs at 1649–1625 cm^{-1} in the allyl compounds and in the vinyl compounds it is more substitutent dependent, decreasing in the following order: alkyl > -COOR > phenyl > -C(=O)-X (halogen), -C(=O)-R (alkyl) > -C≡N> -C≡C-R > -C≡C- ϕ > -S-R. The in-plane carbon-hydrogen deformations of the vinyl group generally give rise to Raman bands of medium to strong intensity near 1415 and 1295 cm^{-1}. The out-of-plane hydrogen wagging modes of the vinyl group that produce the strongest bands in the IR spectra of these alkenes usually appear as very weak bands in the Raman.

6.5 DISUBSTITUTED ETHYLENES

Table 6.9 contains some of the Raman frequencies that are characteristic of the vinylidene group. The frequency dependence of the C=C stretch on type of substituent parallels that found for the vinyl series. The in-plane symmetric deformation ("scissors") of the $=CH_2$ group leads to a strong Raman band near 1400 cm^{-1}.

The positions of the strong, polarized Raman bands arising from the C=C stretching mode for both cis and trans isomers of 1,2-disubstituted ethylenes are listed in Table 6.10. In the IR spectra of these compounds this band is often weak or forbidden by symmetry in the trans isomer. The frequency of the trans isomer generally occurs about 0–20 cm^{-1} higher than the cis isomer with an average shift close to 15 cm^{-1} (127). Vessiere (137) has found that in the series ClHC=$CHCH_2X$ (X = Cl, CN, COOH) the difference between ν(C=C) for the two isomers was only 0–8 cm^{-1}. The in-plane C–H deformations give rise to medium-to-strong Raman bands in the range 1420–1385 cm^{-1} for the cis-disubstituted ethylenes and at 1310–1280 cm^{-1} for the trans-derivatives.

6.6 TRI- AND TETRASUBSTITUTED ETHYLENES

The information concerning the Raman ν(C=C) frequencies for the tri- and tetrasubstituted ethylenes are given in Table 6.11. For trisubstituted

Table 6.6 Prominent Bands in the Raman Spectra of Haloalkenes[a]

Haloalkene	X=Chlorine Frequency (cm^{-1})	X=Chlorine References	X=Bromine Frequency (cm^{-1})	X=Bromine References	X = Iodine Frequency (cm^{-1})	X = Iodine References
1. $H_2C{=}CHX$		88–93		21,88,89,91,94		95,96
a. C=C Stretch	1603		1596		1581	
b. C-X Stretch	706		601		535	
c. C=C–X Rock	396		344		309	
2. *cis*-HXC=CHX		97–100		101–104		105
a. C=C stretch	1587		1583		1543	
b. Symmetric C-X stretch	711		589		496	
c. In-plane C-X deformation	173		114		85	
3. *trans*-HXC=CHX		97–100		101–104		105,106
a. C=C Stretch	1576		1581		1537	
b. Symmetric C-X stretch	844		746		663	
c. In-plane C-X deformation	349		216		154	
4. $H_2C{=}CX_2$		107–109		86,110		—
a. C=C Stretch	1616		1593		—	
b; CX_2 Symmetric stretch	601		467			
c. CX_2 Symmetric deformation	299		184			

5. $X_2C{=}CHX$		111–113		102–114		—
a. C=C Stretch	1589		1552		—	
b. C-X Stretch	932 (wk)		826			
c. C-X Stretch	840 (wk)		699			
d. C-X Stretch	628		500 (wk)			
e. In-plane skeletal deformation	381		238			
f. In-plane skeletal deformation	274		184 (wk)			
g. In-plane skeletal deformation	172		115 (wk)			
6. $X_2C{=}CX_2$		112,115–117		117,118		119–121
a. C=C Stretch	1571		1547, 1515[b]		1465 (solid)	
b. CX_2 Symmetric stretch	447		265		180 (solid)	
c. CX_2 Symmetric deformation	237		144		105 (solid)	

[a]Unless otherwise stated the frequencies refer to the liquid state.

[b]Fermi resonance.

Table 6.7 Characteristic Raman Frequencies for Allyl Compounds

$H_2C=CHX$	Vibrational Frequency (cm^{-1})					
	Asymmetric $=CH_2$ Stretch	$=CH$ Stretch	C=C Stretch	$=CH_2$ Deformation	=CH Deformation	Reference
Substituent X						
1. Allyl Compounds						
$-CH_2CH_2OH$	3079	3006	1641	1416	1292	88
$-CH_2CH_2Si(CH_3)_3$	3080	3002	1639	1415	1298	130
$-CH_2CH_2CO_2C_2H_5$	3050	3005	1642	1417	1295	131
$-CH_2CHBrCO_2C_2H_5$	3084	3014	1643	1418	1303	131
$-CH_2\text{-}C_5H_9$	3079	3000	1642	1416	1297	21,39
$-CH_2\text{-}C_6H_{11}$	3081	3001	1642	1415	1296	21,39
$-CH_2F$	3091	3023	1649	1409	1286	132
$-CH_2C1$	3091	3020	1643	1412	1293	88,95,132,133
$-CH_2Br$	3085	3017	1637	1407	1294	21,88,95,132,133
$-CH_2I$	NR[a]	NR	1625	1402	1284	132,134
$-CH_2NH_2$	3082	3005	1635	1416	1286	88,135
$-CH_2\text{-}C_6H_5$	3057	3011	1645	1416	1294	21,136
$-CH_2\text{-}OC(=O)\text{-}CH_3$	3090	3022	1649	1414	1295	39
$-CH_2OH$	3087	3011	1644	1406	1285	21,88
$-CH_2\text{-}COOH$	3090	3026	1647	1410	1296	137
$-CH_2\text{-}COOC_2H_5$	3091	3026	1647	1409	1296	137
$-CH_2\text{-}C{\equiv}C\text{-}CH_3$	3093	3018	1642	1425	1299	138
$-CH_2\text{-}C{\equiv}C\text{-}C_6H_5$	3060	3020	1642	1420	1295	21
$-CH_2\text{-}C{\equiv}N$	3090	3026	1641	1409	1290	88,139–141

$-CH_2-Si-Y_3$ (Y = F, C1, Br)	3098–3091	3025–3015	1643–1635	1398–1390	1307–1304	142
$-CH_2-GeY_3$ (Y = Cl,Br)	3090–3085	3020–3013	1635–1633	1393–1388	1301–1298	142
2. Diallyl Compounds						
$(H_2C{=}CHCH_2)_2S$	3080	3005	1633	1400	1293	88
$(H_2C{=}CHCH_2)_2NH$	3083	3023	1645	1420	1292	143
$(H_2C{=}CHCH_2)_2SiF_2$	3092	3016	1640	1423	1308	142
$(H_2C{=}CHCH_2)_2SiCl_2$	3089	3026	1636	1423	1307	142

[a] NR –Raman data in this region not reported.

Table 6.8 Characteristic Frequencies for Vinyl Compounds

$H_2C{=}C(H)X$	Vibrational Frequency (cm^{-1})					
	Carbon-Hydrogen Stretching Vibrations		C=C Stretch	$=CH_2$ Deformation	=CH Deformation	Reference
Substituent X						
$-C_6H_5$	3012	2983	1632	1414	1305	136,144,145
$-C_6H_4$-R (ortho)	3020–2986	2990–2981	1632–1626	1419–1412	1310–1305	144,145
$-C_6H_4$-R (meta)	3040–3010	3014–2981	1636–1627	1417–1413	1313–1304	144,145
$-C_6H_4$-R (para)	—	—	1634–1631	1423–1422	1313–1303	145
-C(=O)-H	—	—	1618	1426	1278	21
-C(=O)-F	3124	3004	1631	1411	1297	146
-C(=O)-C1	3124	2989	1620	1399	1290	147,148
-C(=O)-R	3036–3020	3008–2987	1619–1616	1425–1408	1291–1286	149,150
-C(=O)-OH	3112	3028	1637	1396	1282	21,147
-C(=O)-OR	3107–3102	3039–3034	1635–1632	1407–1400	1293–1279	147,151,152
$-C(=O)-NH_2$	3119	3050	1636	-	1289	153
-C≡N	3115	3036	1607	1406	1275	154
-C≡C-H	3102	3012	1595	1405	1288	46,155
-C≡C-R	3110–3101	3038–3014	1606	1414–1409	1291	155
$-C{\equiv}C-C_6H_5$	-	-	1589	1410	1289	156
$-C{\equiv}C-SiR_3$	3105–3102	3011–3008	1607–1605	1412–1408	1294–1292	157
-O-R (R = alkyl)	3098–3080	3024–3018	1654–1651[a] 1643–1637 1612–1610	1419–1410	1310–1304	158–162
$-O-C(=O)-CH_3$	3125	3050	1647	-	1295	151
-S-R (R = alkyl)	3097–3095	3014–3006	1587–1585	1431–1430	1280–1271	163,164
$-SiH_3$	3061	2986	1595	1404	1269	165
$-NO_2$	3135	3120	1639	-	1265	166

[a] Complex structure of band attributed to rotational isomerism.

Table 6.9 Characteristic Raman Frequencies for Asymmetrically Disubstituted Ethylenes (Vinylidene Compounds)

$H_2C{=}C(X_1)(X_2)$		Vibrational Frequency (cm^{-1})			
		$=CH_2$ Stretch	C=C Stretch	$=CH_2$ In-plane Deformation	Reference
Substituents					
X_1	X_2				
$-CH_2Cl$	$-CH_2Cl$	3097	1649	1416	167
$-CH_2Br$	$-CH_2Br$	3094	1638	1398	167
$-CH_3$	-Cl	3114	1640	1413	95,168
$-CH_3$	-Br	3107	1637	1400	95
$-CH_3$	-I	-	1625	1402	95
$-CH_3$	-COOH	-	1638	1402	169
$-CH_3$	$-COOCH_3$	3106	1639	1403	169-172
$-CH_3$	-COOR (R = alkyl)	-	1639-1637	1403-1402	169
$-CH_3$	-COCl	-	1634	1408	169
$-CH_3$	-CN	3114	1628	1404	173,174
$-CH_3$	$-C{\equiv}CH$	3103	1612	1387	155
$-CH_3$	-S-R (R = alkyl)	3100-3099	1609-1606	1404-1402	163
-Cl	$-CH_2CN$	3127	1647	1413	137
-Cl	$-CH_2COOC_2H_5$	3129	1644	1413	137
-Cl	$-CH_2Cl$	3120	1633	1389	175
-Cl	$-COOC_2H_5$	-	1611	1391	169
-Cl	-CN	3129	1605	1374	176
-Cl	$-Ge(CH_3)_3$	3080	1601	1409	177
-Br	$-CH_2CH_2COOC_2H_5$	-	1642	1417	131
-Br	$-CH_2CHBrCOOC_2H_5$	3105	1630	1405	131
-Br	$-CH_2-C_6H_5$	-	1631	1406	136
$-Ge(CH_3)_3$	$-Ge(CH_3)_3$	3001	1584	1399	177

Table 6.10 C=C Stretching Frequencies in the Raman Spectra of cis-trans Disubstituted Ethylenes

$X_1HC=CHX_2$		Frequency ν (C=C) (cm^{-1})		Reference
		Cis Isomer	Trans Isomer	
Substituents				
X_1	X_2			
$-CH_3$	$-CH_2OCOCH_3$	1665	1679	23
$-CH_3$	$-CH_2OH$	1658	1677	23
$-CH_3$	$-CHOH-C\equiv CH$	-	1676	23
$-CH_3$	$-CH_2-C_5H_9$	1658	1674	23
$-CH_3$	$-CH_2-C_6H_{11}$	1657	-	23
$-CH_3$	$-CH_2C1$	-	1671	23
$-CH_3$	$-CHC1_2$	-	1666	23
$-CH_3$	$-CH_2Br$	1651	1666	23
$-CH_3$	$-C_6H_5$	1642	1665	23
$-CH_3$	$-COOR$ ($R = CH_3, C_2H_5$)	-	1663	152
$-CH_3$	$-COOH$	1645	1652	23
$-CH_3$	$-CN$	1628	1641	178
$-CH_3$	$-CHO$	1625	1642	23
$-C_2H_5$	$-CH_2OH$	-	1674	26
$-C_2H_5$	$-CH_2OCOCH_3$	-	1673	26
$-C_2H_5$	$-CH_2C1$	-	1667	26
$-C_2H_5$	$-CH_2Br$	-	1662	26
$-C_2H_5$	$-C_6H_5$	1640	1652	26
$-C_2H_5$	$-CHO$	-	1638	26
$-C_5H_{11}$	$-CH_2OCH_3$	1650	1674	179
$-C_5H_{11}$	$-CH_2OH$	1657	1674	179
$-C_5H_{11}$	$-CH_2OCOCH_3$	1659	1673	179
$-C_6H_5$	$-CH_2OCOCH_3$	1645	1659	179
$-C_6H_5$	$-CH_2OCH_3$	1643	1657	179
$-C_6H_5$	$-CH_2OH$	1642	1657	179
$-C_6H_5$	$-CH_2Br$	—	1646	179
$-C_6H_5$	$-C_6H_5$	—	1632	179
$-C_6H_5$	$-COOH$	—	1628	179
$-C_6H_5$	$-CHO$	—	1624	179
$-C_6H_5$	$-COCH_3$	—	1624	179
$-C_6H_5$	$-COCl$	—	1620	179
$-(CH)_7CH_3$	$-(CH_2)_7COOC_2H_5$	1655	1669	180
$-CH_2Cl$	$-COOC_2H_5$	1648	1663	137

Table 6.10 C=C Stretching Frequencies in the Raman Spectra of cis-trans Disubstituted Ethylenes (Continued)

$X_1HC=CHX_2$		Frequency ν (C=C) (cm^{-1})		
		Cis Isomer	Trans Isomer	Reference
Substituents				
X_1	X_2			
$-CH_2Cl$	-CN	1633	1642	137
$-CH_2OH$	-CN	—	1643	137
-Cl	$-CH_2CN$	1634	1642	137
-Cl	$-CH_2COOH$	1639	1639	137
-Cl	$-CH_2OH$	—	1634	137
-Cl	$-CH_2Cl$	1629	1629	137
-CN	-CN	—	1612	181

derivatives of the type $X_1(X_2)C{=}CHCH_3$ the C=C stretching frequency falls in the range 1660–1640 cm^{-1}. A strong band in the range 1395–1375 cm^{-1} arising from the =CH in-plane deformation is also found in the trisubstituted ethylenes.

6.7 DIENES

Unconjugated dienes (i.e., those in which the double bonds are separated from each other by one or more saturated carbon atoms) display in their Raman spectra bands characteristic of the corresponding monoalkene (Tables 6.2 and 6.3) for each double bond. For example, bands characteristic of the terminal vinyl groups appear in the spectra of 1,4-pentadiene (3084, 3010, 1644, 1413, and 1295 cm^{-1}) (39) and 1,5-hexadiene (3081, 3004, 1641, 1416, and 1298 cm^{-1}) (39) (Spectrum 36). The Raman spectrum of $(CH_3)_2C{=}CHCH_2OC$-$(CH_3)_2CH{=}CH2$ (187) contains bands arising from the vinyl group (3138, 3046, 1640, 1413, and 1294 cm^{-1}) and the trisubstituted ethylene moiety (2993, 1675, 1326, and 822 cm^{-1}).

The simplest conjugated diene, 1,3-butadiene, has been the subject of numerous IR and Raman investigations. The early work has been reviewed by Aston et al. (188) and more recently, the Raman spectra of the gaseous and liquid states have been studied by Richards and Nielsen (189). The rotational Raman spectra of 1,3-butadiene and 1,3-butadiene-d_6 has been measured by Marais, Sheppard, and Stoicheff (190). IR and Raman data are also available for the following deuterated species of 1,3-butadiene: 2-d (191), -1,1,2-d_3 (192) and -d_6 (193). A normal coordinate analysis of 1,3-butadiene was carried out by Tarasova and Sverdlov (194,195). From these investigations,

Table 6.11 Raman Frequencies for the C=C Stretch in Tri- and Tetrasubstituted Ethylenes

Type of Substitution	Frequency ν (C=C) (cm^{-1})	Reference
1. Trisubstituted		
$CH_3(Br)C{=}CHCH_3$	1658	24
$CH_2Br(Br)C{=}CHCH_3$	1648	24
$CH_3(CHO)C{=}CHCH_3$	1648	24
$C_2H_5(CHO)C{=}CHCH_3$	1647	24
$CH_3(COCH_3)C{=}CHCH_3$	1646	182
$C_2H_5(COCH_3)C{=}CHCH_3$	1640	182
$Cl(COOC_2H_5)C{=}CHCH_3$	1637	137
$(CH_3)_2C{=}CHCOOCH_3$	1653	28
$CH_3(COOCH_3)C{=}CHOCH_3$	1648	169
$CH_3(COOC_2H_5)C{=}CHOC_2H_5$	1647	169
$(CH_3)_2C{=}CHCN$	1637	28
$CH_3(N(C_2H_5)_2)C{=}CHCOOC_2H_5$	1627	137
$(CH_3)_2C{=}CHCOCH_3$	1623	182
$CH_3(COOCH_3)C{=}CHCl$	1619	169
$CH_3(COOCH_3)C{=}CHBr$	1613	169
2. Tetrasubstituted		
$(CH_3)_2C{=}C(CH_3)Cl$	1672	25
$Cl_2C{=}C(COOH)Cl$	1658	183
$CH_3(C_2H_5)C{=}C(CH_3)COCH_3$-*trans*	1623	182
$(CH_3)_2C{=}C(CH_3)COCH_3$	1622	182
$NH_2(CN)C{=}C(CN)NH_2$-*cis*	1621	184
$Cl(CN)C{=}C(CN)Cl$-*trans*	1569	185
$Br(CN)C{=}C(CN)Br$-*trans*	1564	185
$Br_2C{=}C(CN)Br$	1541	186
$I(CN)C{=}C(CN)I$-*trans*	1534	185
$I_2C{=}C(CN)I$	1498	186

it has been shown that practically all of the frequencies observed in the IR and Raman spectra can be interpreted in terms of a planar trans configuration (I) rather than in terms of the cis configuration (II). The *trans*-1,3-butadiene molecules belong to the point group C_{2h} that possesses a center of symmetry.

Of the coupled vibrations of the two double bonds, the symmetric (in-phase) stretch appears only in the Raman and the antisymmetric (out-of-phase) stretch only in the IR. The observed frequencies (cm^{-1}) for *trans*-1,3-butadiene are:

	Symmetric ν(C=C) Raman Active	Antisymmetric ν(C=C) IR active
Gas	1643	1599
Liquid	1638	1592
Solid	1630	1596

In isoprene (2-methyl-1,3-butadiene), the center of symmetry is destroyed and both coupled double bond stretching modes are allowed in the IR and Raman. In the Raman spectrum of liquid isoprene (Spectrum 37), an intense band at 1637 cm^{-1} is assigned to the symmetric C=C stretch and a very weak band at 1604 cm^{-1} to the antisymmetric stretch. From the experimental (196–200) and theoretical (200, 201) studies it was concluded that isoprene exists predominantly in the trans configuration. Other 2-alkyl-1,3-butadienes (198) were also found to exist in the trans form with a strong Raman band in the range 1645–1636 cm^{-1} and a very weak band at 1595–1580 cm^{-1}. For chloroprene (2-chloro-1,3-butadiene) (202, 203) these bands appear at 1628 and 1581 cm^{-1} in the Raman spectrum of the liquid.

In monoalkyl substituted dienes of the type H_2C=CH–CH=CH–R and in more highly substituted dienes, there exists the possibility of cis-trans isomerism about the substituted double bond as well as the cis-trans isomerism of the C=C–C=C moiety. This results in a complex Raman band pattern in the 1700–1600 cm^{-1} region. For example, in 1,3-pentadiene (199) bands at 1654 and 1604 cm^{-1} are attributed to the trans isomer (III), while those at 1645 and 1594 cm^{-1} are assigned to the cis isomer (IV). In the Raman spectrum of 2,4-hexadiene (199) a band at 1655 cm^{-1} was assigned as the symmetric C=C stretch of the trans, trans isomer (V) and a band at 1668 cm^{-1} to that of the cis, trans isomer (VI). This type of isomerism has also been observed in the

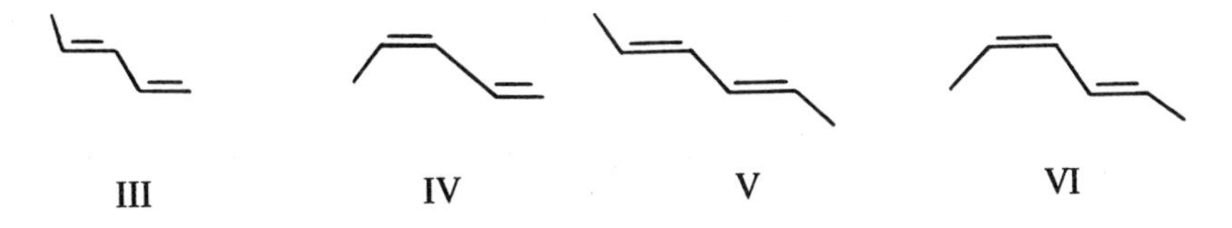

Raman spectra of 1-alkoxy-1,3-dienes (204), 1,2,3-trisubstituted-1,3-butadienes (205), and muconic acid derivatives (206).

The vibrational spectra of 1,1,4,4-tetrafluoro-1,3-butadiene has been discussed by Conrad and Dows (207). The $\nu_{sym}(C=C)$ appears at 1765 cm^{-1} in the Raman spectrum of the liquid and $\nu_{asym}(C=C)$ at 1715 cm^{-1} in the IR spectrum of the gas, which is consistent with a trans (C_{2h}) structure. For hexafluoro-1,3-butadiene (208) these vibrations appear in both the Raman and IR spectra near 1796 and 1768 cm^{-1}, respectively, which were interpreted on the basis of a cis configuration of symmetry C_{2v}. In the Raman spectrum of hexachloro-1,3-butadiene (209) two strong bands were observed at 1611 and 1566 cm^{-1}.

6.8 POLYENES

Detailed assignments of the bands in the Raman spectra have been reported for a trans isomer (210,211) and a cis isomer (211) of 1,3,5-hexatriene and for 1,3,5,7-octatetraene (212). Raman data have also been reported for $CH_3(CH=CH)_nCO_2H$ (n = 1–5) (213), β-carotene (214), and $C_2H_5OOC–(CH=CH)_n–COOC_2H_5$ (n = 1–8) (215,216). An analysis of the planar vibrations of the conjugated chain in polyenes has been carried out by Tric (217). Bands attributed to the polyene chain are observed near 1140 and 1600 cm^{-1} in the Raman spectra of polyenes. As the number of conjugated double bonds increases and the exciting frequency approaches that of the ultraviolet absorption band (resonance Raman conditions), these Raman bands exhibit a very rapid increase in intensity and various overtones and combination tones of the first and higher orders appear in the Raman spectra. Under certain excitation conditions, the intensities of these overtones exceed those of the fundamentals.

The decrease in the Raman frequency of the coupled double bond stretching band with the increase in the number of conjugated double bonds is illustrated below for two series of polyenes:

	Frequency (cm^{-1})	
n	$CH_3(CH=CH)_nCOOH$ (ref. 213)	$C_2H_5OOC–(CH=CH)_n–COOC_2H_5$ (ref. 215)
1	1655	1664
2	1644	1644
3	1618	1621
4	1559	1596
5	1576	1570
6	–	1562
7	–	1550
8	–	1542

6.9 METHYLENE AND ALKYLIDENE-CYCLOALKANES

Raman data for these molecules that contain a double bond linked to a cycloaliphatic ring are listed in Table 6.12. The double bond stretching frequency increases steadily from its almost normal open-chain value of 1651 cm^{-1} for methylene cyclohexane to 1736 cm^{-1} for the highly strained cyclopropane analogue. A further increase in this frequency is observed upon replacement of the hydrogens of the vinylidene group with alkyl substituents. A strong polarized band, which arises from the symmetric ring stretching ("breathing") vibration, is also found in the Raman spectra of these compounds. This band occurs at 1033 cm^{-1} for methylene cyclopropane, at 957 cm^{-1} for methylene cyclobutane, and at 897 cm^{-1} for methylene cyclopentane. It has been reported (221) that the characteristic Raman frequencies for alkylidene cyclopentanes occur near 430, 900, 1025, 1155, 1212, 1435, 1680, and 3040 cm^{-1}.

6.10 CYCLOPROPENES

The vibrational spectra of cyclopropene and cyclopropene-1,2-d_2 have been assigned by Mitchell, Dorko, and Merritt (223). In the Raman spectrum of the liquid, bands at 3158 and 3124 cm^{-1} are attributed to the =CH stretching vibrations and a strong polarized band at 1647 cm^{-1} to the C=C stretch. In 1-methylcyclopropene (224) the Raman frequency of the double bond stretch increases to 1782 cm^{-1}. For tetrahalocyclopropenes (225) this band was observed at 1811 and 1763 cm^{-1} for the chloro and bromo derivatives, respectively. The assignment of the IR and Raman bands in cyclopropenones has been discussed by Krebs, Schrader, and Höfler (226,227). Two very strong characteristic bands at 1865-1840 cm^{-1} and 1660-1600 cm^{-1} are found for the cyclopropenones. The first band can be assigned to an out-of-phase stretching vibration of the C=C and C=O bonds with a predominance of the C=O coordinate and the lower frequency band to the corresponding in-phase stretching vibrations with a predominance of the C=C coordinate. A band was also observed near 880 cm^{-1} for these derivatives and was assigned to a symmetric ring stretching vibration.

6.11 CYCLOBUTENES

The vibrational spectra of cyclobutene and cyclobutene-d_6 have been interpreted by Lord and Rea (228). The double bond stretching frequency in the Raman spectrum of liquid cyclobutene is located at 1566 cm^{-1}. In 1-methyl-

Table 6.12 Characteristic Frequencies in the Raman Spectra of Methylene- and Alkylidene-Cycloalkanes

Compound	Structure	Frequency ν (C=C) (cm^{-1})	Reference
1. Three-membered ring			
Methylene cyclopropane	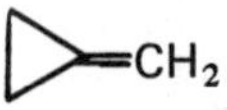 $=CH_2$	1736	218
2. Four-membered ring			
Methylene cyclobutane	 $=CH_2$	1679	218,220
Isopropylidene cyclobutane	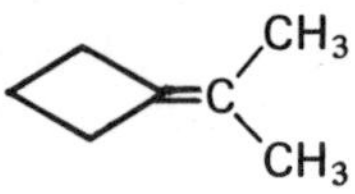 $=C$ (CH_3, CH_3)	1718	221
3. Five-membered ring			
Methylene cyclopentane	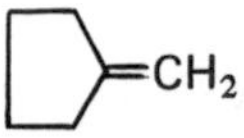 $=CH_2$	1657	220,221
Ethylidene cyclopentane	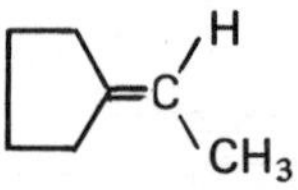 $=C$ (H, CH_3)	1682	220,221
n-Propylidene cyclopentane	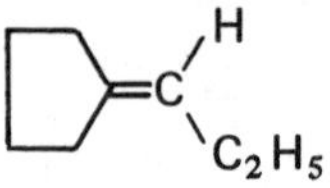 $=C$ (H, C_2H_5)	1679	221
n-Butylidene cyclopentane	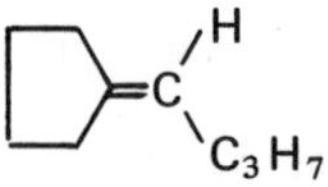 $=C$ (H, C_3H_7)	1678	221
Isopropylidene cyclopentane	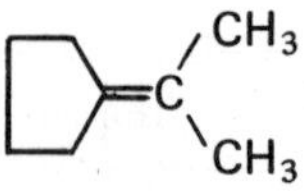 $=C$ (CH_3, CH_3)	1687	220
4. Six-membered ring			
Methylene cyclohexane	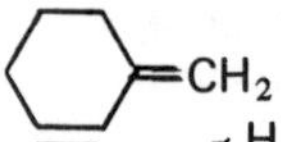 $=CH_2$	1651	220,222
Ethylidene cyclohexane	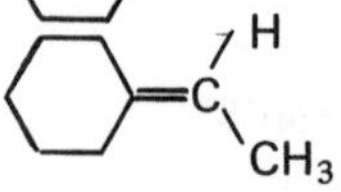 $=C$ (H, CH_3)	1676	220,222
Isopropylidene cyclohexane	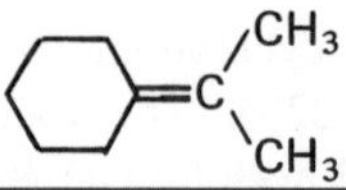 $=C$ (CH_3, CH_3)	1668	220,222

cyclobutene (219) ν(C=C) increases to 1640 cm^{-1}. In liquid hexafluorocyclobutene (229), a strong polarized band at 1794 cm^{-1} is assigned to the C=C stretching mode.

6.12 CYCLOPENTENES AND CYCLOPENTADIENES

The Raman spectra of cyclopentene has been extensively investigated (39, 230-234). The rotational Raman spectrum of cyclopentene has also been obtained (235). Normal coordinate analyses of cyclopentene have been performed by Sverdlov and Krainov (236) and by Petzuch (237). The most prominent bands in the Raman spectrum of the liquid occur at 3060 cm^{-1} (=C-H stretch), 1614 cm^{-1} (C=C stretch), and the ring stretches at 1108, 965, and 900 cm^{-1}.

The characteristic frequencies in the Raman spectra of substituted cyclopentenes are listed in Table 6.13. The early work on the Raman spectra of cyclopentene derivatives has been reviewed by Piaux (39) and more recently that of alkyl substituted cyclopentenes by Sverdlov and Krainov (236). For alkyl substituted cyclopentenes with no substituents on the C=C bond the frequency of ν(C=C) falls near 1618 cm^{-1}, for those with one alkyl substituent on the double bond it is located in the range 1666-1650 cm^{-1}, and for those derivatives with two substituents attached to the double bond it is found at 1693-1671 cm^{-1}. All the cyclopentene derivatives exhibit a Raman band near 900 cm^{-1}, which is attributed to the symmetric ring stretching vibration.

The Raman spectrum of cyclopentadiene has been reported by several authors (230, 231, 242-245). The vibrational assignments in the IR and Raman spectra have been reviewed by Gallinella, Fortunato, and Mirone (245). The in-phase C=C stretching mode gives rise to a very intense and strongly polarized Raman band at 1500 cm^{-1}. In hexachlorocyclopentadiene (246) this band is reported at 1572 cm^{-1}.

6.13 CYCLOHEXENES AND CYCLOHEXADIENES

The Raman spectra of cyclohexene has been extensively studied (31,136,233, 247-249). The vibrational assignment and valence force field of cyclohexene and cyclohexene-d_{10} have been given by Neto, DiLauro, Castellucci, and Califano (249). The C=C stretching frequency in the Raman spectrum of the liquid (Spectrum 38) occurs at 1656 cm^{-1}, which is comparable to that found for *cis* CHR=CHR noncyclic alkenes (1660-1654 cm^{-1}). An intense polarized band at 822 cm^{-1} is assigned to a symmetric ring stretch. The frequencies of the C=C stretch in various cyclohexene derivatives are listed in Table 6.14. In

Table 6.13 Characteristic Frequencies in the Raman Spectra of Substituted Cyclopentenes[a]

Type of Substitution	Frequency C=C Stretch (cm^{-1})	Symmetric Ring Stretch	References
1. Monosubstituted			
1-alkyl	1659–1650	898-880	39,221,236,238
1-$COOCH_3$	1631	903	39
1-C_6H_5	1627	890	39
1-CHO	1616	882	39
1-CN	1615	896	39
3-Methyl	1613	903	36,238
4-Methyl	1652	889	238
2. Disubstituted			
1,2-Dialkyl	1686–1671	916-909	236,239
2,3-Dimethyl	1655	915	239
2,4-Dimethyl	1666	907	37
cis-3,4-Dimethyl	1617	-	36
4,4-Dimethyl	1619	903	240
3. Trisubstituted			
1,2,3-Trimethyl	1693	888	36
1,2,4-Trimethyl	1693	891	37
2,3,3-Trimethyl	1665	921	36
2,3,4-Trimethyl	1668	892	36
4. Octachlorocyclopentene	1606	893	241

[a]Numbering of ring: 5, 4, 3, 2, 1 (cyclopentene ring; double bond between 1 and 2)

alkyl derivatives with no substituents on the double bond ν(C=C) is found near 1650 cm^{-1}; with one substituent on the double bond ν(C=C) falls in the range 1680–1660 cm^{-1}; while for disubstitution on the double bond ν(C=C) is close to 1680–1675 cm^{-1} (239).

Assignments in the vibrational spectra of 1,3-cyclohexadiene have been made by DiLauro, Neto, and Califano (254). In the Raman spectrum of the liquid (234,254–257) the symmetric (in-phase) C=C stretching vibration is assigned to a strong, polarized band at 1575 cm^{-1}. The antisymmetric (out-of-phase) C=C stretching mode occurs in the IR of the liquid at 1604 cm^{-1}. For alkyl-substituted cyclohexadienes (258,259) this frequency is found at 1577 cm^{-1} for 1-methyl-2,4-cyclohexadiene and in the range 1590–1586 cm^{-1} for 1-alkyl-2,6-cyclohexadienes.

Table 6.14 C=C Stretching Frequencies in the Raman Spectra of Cyclohexenes[a]

Type of Substitution	C=C Stretching Frequency (cm^{-1})	Reference
1. Cyclohexene	1656	249
2. Monosubstituted cyclohexenes		
1-Alkylcyclohexene	1675–1667	238,250
1-Chlorocyclohexene	1658	251
1-Bromocyclohexene	1650	251
1-Phenylcyclohexene	1632	238
1-Trimethylsilylcyclohexene	1620	252
4-Methylcyclohexene	1650	250
6-Methylcyclohexene	1648	239
3. Disubstituted cyclohexenes		
1,2-Dimethylcyclohexene	1675	222
1,2-Dichlorocyclohexene	1650	251
1,2-Dibromocyclohexene	1636	251
1,4-Dimethylcyclohexene	1680	250
1,6-Dimethylcyclohexene	1668	239
1-Methyl-6-hydroxycyclohexene	1670	39
1-Methyl-6-ethoxycyclohexene	1673	39
4,5-Dibromocyclohexene	1661	253

[a]Numbering of ring: (ring positions 4 3 / 5 2 / 6 1)

The vibrational spectra of 1,4-cyclohexadiene have been investigated by Gerding and Haak (260) and by Stidham (261). The symmetric C=C stretch occurs as a strong band in the Raman of the liquid at 1680 cm^{-1} and the antisymmetric C=C stretch as a strong band in the IR at 1639 cm^{-1}. The ring "breathing" vibration was assigned to a strong Raman band at 854 cm^{-1}.

6.14 CYCLOHEPTENES

In the Raman spectrum of liquid cycloheptene (243,247,262) a strong polarized band arising from the C=C stretching mode occurs at 1656 cm^{-1}. From an analysis of the IR and Raman spectra of liquid and solid cycloheptene together with a normal coordinate analysis Neto, DiLauro, and Califano (262) have concluded that cycloheptene exists in the chair conformation in the crystal. In the liquid, the predominant conformation is the chair, but it exists in equilibrium with a small amount of the boat conformation. In the Raman spectrum of 1-methylcycloheptene (250) the double bond stretching frequency shifts to 1672 cm^{-1}.

The vibrational spectra of 1,3-cycloheptadiene and cycloheptatriene have been investigated by Sobolev et al. (243). Intense polarized Raman bands were observed at 1613 cm^{-1} for the diene and at 1534 cm^{-1} for the triene.

6.15 CYCLOOCTENES AND HIGHER HOMOLOGUES

The data on the C=C stretching frequencies in the Raman spectra of unsaturated eight- and higher-membered rings are summarized below:

Compound	ν(C=C) (cm^{-1})	Reference
cis-Cyclooctene	1650	263
trans-Cyclooctene	1648	263
cis-1,5-Cyclooctadiene	1660	263,264
trans-1,5-Cyclooctadiene	1622	263
1,3-Cyclooctadiene	1626	263
1,3,5-Cyclooctatriene	1639,1612	265
Cyclooctatetraene	1655	266–268
1,4,7-Cyclononatriene	1640	269
1,5,9-*trans,trans,trans*-Cyclododecatriene	1677	270

6.16 C=C STRETCHING FREQUENCIES IN BICYCLIC SYSTEMS

Raman spectra have been reported for: bicyclo [2.2.1]-hept-2-ene (VII), (norbornylene) and its derivatives (271,272); bicyclo [2.2.2]-oct-2-ene (VIII), and its derivatives (273); and dicyclopentadiene (IX), [3a,4,7,7a-tetrahydro-4, 7-methanoindene] (271,274). The effect on ν(C=C) of fusion of a second ring onto a cycloalkene ring is illustrated in Table 6.15. When fusion of a second ring introduces additional strain, ν(C=C) decreases; for example, compare cyclopentene (1614 cm^{-1}) and bicyclo [2.2.1]-hept-2-ene (1568 cm^{-1}).

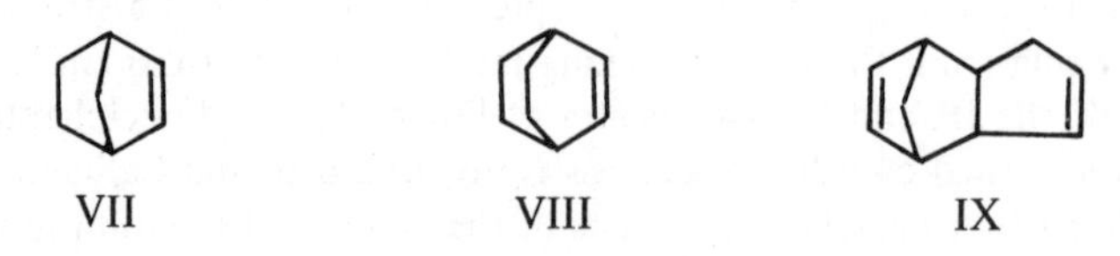

Table 6.15 C=C Stretching Frequencies in the Raman Spectra of Unsaturated Bicyclic Compounds

Name	Structure	ν(C=C) Frequency (cm^{-1})	Reference
1. Five-membered rings			
Cyclopentene		1614	236
Dihydrodicyclopentadiene		1611	271
Dicyclopentadiene		1611,1568	271,274
Bicyclo[2.2.1] hept-2-ene		1568	272
Bicyclo[2.2.1] hepta-2,5-diene		1575	272
2. Six-membered rings			
Cyclohexene		1656	249
Bicyclo [2.2.2] oct-2-ene		1614	273

REFERENCES

1. G. Herzberg, *Infrared and Raman Spectra,* D. Van Nostrand Co., Princeton, N.J., 1945, 325–328.
2. R. L. Arnett and B. L. Crawford, Jr., *J. Chem. Phys., 18,* 118 (1950).
3. B. L. Crawford, Jr., J. E. Lancaster, and R. G. Inskeep, *J. Chem. Phys., 21,* 678 (1953).
4. W. L. Smith and I. M. Mills, *J. Chem. Phys., 40,* 2095 (1964).
5. H. J. Becher and A. Adrian, *J. Mol. Struct., 6,* 479 (1970).
6. C. Brecher and R. S. Halford, *J. Chem. Phys., 35,* 1109 (1961).
7. D. A. Dows, *J. Chem. Phys., 36,* 2833 (1962).
8. M. Brith and A. Ron, *J. Chem. Phys., 50,* 3053 (1969).
9. T. Feldman, J. Romanko, and H. L. Welsh, *Can. J. Phys., 34,* 737 (1956).
10. M. de Hemptinne and J. Charette, *Bull. Cl. Sci., Acad. Roy. Belg., 39,* 622 (1953).
11. S. M. Blumenfeld, S. P. Reddy, and H. L. Welsh, *Can. J. Phys., 48,* 513 (1970).
12. J. E. Kilpatrick and K. S. Pitzer, *J. Res. NBS, 38,* 191 (1947).
13. D. C. McKean and J. L. Duncan, *Spectrochim. Acta, 27A,* 1879 (1971).
14. W. H. Fletcher and W. T. Thompson, *J. Mol. Spectrosc., 25,* 240 (1968).
15. P. Pulay and W. Meyer, *J. Mol. Spectrosc., 40,* 59 (1971).
16. R. Lespieau and M. Bourguel, *Compt. Rend., 190,* 1504 (1930).
17. M. Bourguel, *Compt. Rend., 193,* 934 (1931).

18. M. Bourguel, *Compt. Rend., 194,* 1736 (1932).
19. M. Bourguel, B. Gredy, and L. Piaux, *Compt. Rend., 195,* 129 (1933).
20. M. Bourguel, *Compt. Rend., 195,* 311 (1933).
21. M. Bourguel and L. Piaux, *Bull. Soc. Chim. Fr.,* (5), *2,* 1958 (1935).
22. H. Van Risseghem, B. Gredy, and L. Piaux, *Compt. Rend., 196,* 938 (1933).
23. B. Gredy, *Bull. Soc. Chim. Fr.,* (5), *2,* 1029 (1935).
24. B. Gredy, *Bull. Soc. Chim. Fr.,* (5), *2,* 1038 (1935).
25. B. Gredy, *Bull. Soc. Chim. Fr.,* (5), *2,* 1951 (1935).
26. B. Gredy, *Bull. Soc. Chim. Fr.,* (5), *4,* 415 (1937).
27. J. H. Hibben, *The Raman Effect and its Chemical Application,* Reinhold, New York, N. Y., 1939, 161–177.
28. K. W. F. Kohlrausch, *Ramanspektren,* Akad. Verlag. Becker and Erler Kom.-Ges., Leipzig, 1943, 302–319.
29. F. F. Cleveland, *J. Chem. Phys., 11,* 1 (1943).
30. F. F. Cleveland, *J. Chem. Phys., 11,* 227 (1943).
31. F. F. Cleveland, *J. Chem. Phys., 11,* 301 (1943).
32. H. Gerding and A. P. van der Vet, *Rec. Trav. Chim., 64,* 257 (1945).
33. J. Goubeau, *Angew. Chem., 59,* 87 (1947).
34. J. Goubeau, E. Köhler, E. Lell, E. Tschentscher, and M. Nordmann, *Beih. Angew. Chem.,* No. 56 (1948).
35. J. Goubeau and H. Seifert, *Monatsh. Chem., 79,* 469 (1948).
36. M. R. Fenske, W. G. Braun, R. V. Wiegand, D. Quiggle, R. H. McCormick, and D. H. Rank, *Anal. Chem., 19,* 700 (1947).
37. W. G. Braun, D. F. Spooner, and M. R. Fenske, *Anal. Chem., 22,* 1074 (1950).
38. G. S. Landsberg, P. A. Bazhulin, and M. M. Sushchinskii, *The Fundamental Parameters of the Raman Spectra of Hydrocarbons,* " *Akad. Nauk SSSR,* 1956.
39. L. Piaux, *Ann. Chim.,* (11), *4,* 147 (1935).
40. N. Sheppard and D. M. Simpson, *Quart. Rev., 6,* 1 (1952).
41. P. F. Gruzdev, *Zh. Fiz. Khim., 28,* 507 (1954).
42. L. M. Sverdlov, *Proc. Acad. Sci. USSR., Phys. Chem. Sect., 112,* 109 (1957).
43. D. G. Rea, *Anal. Chem., 32,* 1638 (1960).
44. H. Moser and U. Weber, *Proc. Int. Meet. Mol. Spectrosc., 4th, Bologna,* 1959, *3,* 1116 (1962).
45. L. M. Sverdlov, *Dokl. Akad. Nauk SSSR, 106,* 80 (1956).
46. N. Sheppard, *J. Chem. Phys., 17,* 74 (1949).
47. H. W. Schrötter and E. G. Hoffman, *Liebigs Ann. Chem., 672,* 44 (1964).

48. N. Sheppard, *J. Chem. Phys., 17,* 455 (1949).
49. L. M. Sverdlov and O. N. Vinogradova, *Dokl. Akad. Nauk SSSR, 100,* 45 (1955).
50. W. Lüttke and S. Braun, *Ber. Bunsenges. Phys. Chem., 71,* 34 (1967).
51. C. M. Pathak and W. H. Fletcher, *J. Mol. Spectrosc., 31,* 32 (1969).
52. W. C. Harris and I. W. Levin, *J. Mol. Spectrosc., 39,* 441 (1971).
53. C. M. Richards and J. R. Nielsen, *J. Opt. Soc. Am., 40,* 442 (1950).
54. L. M. Sverdlov, *Opt. Spektrosk., 1,* 753 (1956).
55. W. H. Snyder and H. S. Kimmel, *J. Mol. Struct., 4,* 473 (1969).
56. P. Torkington and H. W. Thompson, *Trans. Faraday Soc., 41,* 236 (1945).
57. B. Bak and D. Christensen, *Spectrochim. Acta, 12,* 355 (1958).
58. D. E. Mann and E. K. Plyler, *J. Chem. Phys., 23,* 1989 (1955).
59. J. R. Nielsen, C. W. Gullikson, and A. H. Woollett, *J. Chem. Phys., 23,* 1994 (1955).
60. J. R. Nielsen and J. C. Albright, *J. Chem. Phys., 26,* 1566 (1957).
61. C. J. Muelleman, K. Ramaswamy, F. F. Cleveland, and S. Sundaram, *J. Mol. Spectrosc., 11,* 262 (1963).
62. J. Yarwood and W. J. Orville-Thomas, *J. Chem. Soc., 1965,* 7481.
63. N. C. Craig and G. Y.-S. Lo, *J. Mol. Spectrosc., 23,* 307 (1967).
64. N. C. Craig, G. Y. -S. Lo, C. D. Needham, and J. Overend, *J. Am. Chem. Soc., 86,* 3232 (1964).
65. N. C. Craig, G. Y. -S. Lo, L. G. Piper, and J. C. Wheeler, *J. Phys. Chem., 74,* 1712 (1970).
66. G. A. Crowder and N. Smyrl, *J. Mol. Spectrosc., 40,* 117 (1971).
67. N. C. Craig and J. Overend, *J. Chem. Phys., 51,* 1127 (1969).
68. N. C. Craig and D. A. Evans, *J. Am. Chem. Soc., 87,* 4223 (1965).
69. D. E. Mann and E. K. Plyler, *J. Chem. Phys., 26,* 773 (1957).
70. W. F. Edgell and W. E. Byrd, *J. Chem. Phys., 18,* 892, 1310 (1950).
71. D. C. Smith, J. R. Nielsen, and H. H. Claassen, *J. Chem. Phys., 18,* 326 (1950).
72. J. B. Hatcher and D. M. Yost, *J. Chem. Phys., 5,* 992 (1937).
73. J. R. Nielsen, H. H. Claassen, and D. C. Smith, *J. Chem. Phys., 18,* 485 (1950).
74. J. R. Nielsen, C. Y. Liang, and D. C. Smith, *J. Chem. Phys., 20,* 1090 (1952).
75. R. Theimer and J. R. Nielsen, *J. Chem. Phys., 27,* 264 (1957).
76. R. Theimer and J. R. Nielsen, *J. Chem. Phys., 26,* 1374 (1957).
77. J. R. Nielsen and R. Theimer, *J. Chem. Phys., 30,* 103 (1959).
78. R. Theimer and J. R. Nielsen, *J. Chem. Phys., 30,* 98 (1959).

79. W. F. Edgell, *J. Am. Chem. Soc., 70,* 2816 (1948).
80. J. R. Nielsen, H. H. Claassen, and D. C. Smith, *J. Chem. Phys., 20,* 1916 (1952).
81. J. A. Rolfe and L. A. Woodward, *Trans. Faraday Soc., 50,* 1030 (1954).
82. D. E. Mann, N. Acquista, and E. K. Plyler, *J. Chem. Phys., 21,* 1949 (1953).
83. A. Monfils and J. Duchesne, *J. Chem. Phys., 18,* 1415 (1950).
84. J. R. Nielsen, H. H. Claassen, and D. C. Smith, *J. Chem. Phys., 18,* 812 (1950).
85. L. J. Bellamy, *Spectrochim. Acta, 13,* 60 (1958).
86. J. R. Scherer and J. Overend, *J. Chem. Phys., 32,* 1720 (1960).
87. E. C. Tuazon, W. G. Fateley, and F. F. Bentley, *Appl. Spectrosc., 25,* 374 (1971).
88. K. W. F. Kohlrausch and W. Stockmair, *Z. Phys. Chem., 29B,* 292 (1935).
89. M. de Hemptinne, *Trans. Faraday Soc., 42,* 5 (1946).
90. J. Evans and H. Bernstein, *Can. J. Chem., 33,* 1792 (1955).
91. C. W. Gullikson and J. R. Nielsen, *J. Mol. Spectrosc., 1,* 158 (1957).
92. L. M. Sverdlov, Yu. V. Klochkovskii, V. S. Kukina, and T. D. Mezhueva, *Opt. Spectrosc., 9,* 383 (1960).
93. G. Varsanyi, *Acta Chim. Acad. Sci. Hung., 35,* 61 (1963).
94. J. Charette and M. de Hemptinne, *Bull. Cl. Sci., Acad. Roy. Belg., 38,* 934 (1952).
95. L. Kahovec and K. W. F. Kohlrausch, *Z. Phys. Chem., 46B,* 165 (1940).
96. P. Torkington and H. W. Thompson, *J. Chem. Soc., 1944,* 303.
97. B. Trumpy, *Z. Phys., 90,* 133 (1934).
98. O. Paulsen, *Z. Phys. Chem., 28B,* 123 (1935).
99. H. J. Bernstein and D. A. Ramsay, *J. Chem. Phys., 17,* 556 (1949).
100. H. J. Bernstein and A. D. E. Pullin, *Can. J. Chem., 30,* 963 (1952).
101. A. Dadieu, K. W. F. Kohlrausch, and A. Pongratz, *Monatsh. Chem., 60,* 221 (1932).
102. J. C. Evans and H. J. Bernstein, *Can. J. Chem., 33,* 1171 (1955).
103. M. de Croes, M. Perlinghi, and R. van Riet, *Bull. Cl. Sci., Acad. Roy. Belg., 42,* 379 (1956).
104. J. M. Dowling, P. G. Puranik, A. G. Meister, and S. I. Miller, *J. Chem. Phys., 26,* 233 (1957).
105. S. I. Miller, A. Weber, and F. F. Cleveland, *J. Chem. Phys., 23,* 44 (1955).
106. R. H. Krupp, E. A. Piotrowski, F. F. Cleveland, and S. I. Miller, *Dev. Appl. Spectrosc., 2,* 52 (1962).
107. P. Joyner and G. Glockler, *J. Chem. Phys., 20,* 302 (1952).

108. H. W. Thompson and P. Torkington, *Proc. Roy. Soc. (London)*, *184A*, 21 (1945).
109. J. C. Evans, *J. Chem. Phys.*, *30*, 934 (1959).
110. M. de Hemptinne, C. Velghe, and R. van Riet, *Bull. Cl. Sci., Acad. Roy. Belg.*, *30*, 40 (1944).
111. A. Dadieu and K. W. F. Kohlrausch, *Monatsh. Chem.*, *55*, 58 (1930).
112. S. B. Sanyal, *Indian J. Phys.*, *24*, 151 (1950).
113. G. Allen and H. J. Bernstein, *Can. J. Chem.*, *32*, 1044 (1954).
114. J. R. Scherer, J. C. Evans, and J. Overend, *J. Chem. Phys.*, *33*, 314 (1960).
115. H. Wittek, *Z. Phys. Chem.*, *48B*, 1 (1940).
116. H. J. Bernstein, *J. Chem. Phys.*, *18*, 478 (1950).
117. D. E. Mann, J. H. Meal, and E. K. Plyler, *J. Chem. Phys.*, *24*, 1018 (1956).
118. F. E. Malherbe, G. Allen, and H. J. Bernstein, *Can. J. Chem.*, *31*, 1223 (1953).
119. R. Forneris and D. Bassi, *J. Mol. Spectrosc.*, *26*, 220 (1968).
120. E. J. Flourie and W. D. Jones, *Spectrochim. Acta*, *25A*, 653 (1969).
121. R. Forneris and M. Uehara, *J. Mol. Struct.*, *5*, 441 (1970).
122. A. Kirrmann, *Bull. Soc. Chim. Fr.*, *1948*, 163.
123. G. Kremer, *Bull. Soc. Chim. Fr.*, *1948*, 165.
124. A. Kirrmann and G. Kremer, *Bull. Soc. Chim. Fr.*, *1948*, 166.
125. A. Kirrmann and G. Oestermann, *Bull. Soc. Chim. Fr.*, *1948*, 168.
126. A. Kirrmann, *Bull. Soc. Chim. Fr.*, *1948*, 170.
127. A. Kirrmann, *Proc. Int. Congr. Pure Appl. Chem. (London)*, *11*, 495 (1947).
128. H. Gerding and G. W. A. Rijnders, *Rec. Trav. Chim.*, *65*, 143 (1946).
129. H. Gerding, H. J. Prins, and G. W. A. Rijnders, *Rec. Trav. Chim.*, *65*, 165 (1946).
130. A. D. Petrov and G. I. Nikishin, *Izv. Akad. Nauk SSSR, Otd. Khim. Nauk*, 1128 (1952); *Chem. Abstr.*, *48*, 1247c.
131. I. A. D'yakonov and T. V. Domareva, *J. Gen. Chem., USSR*, *25*, 899 (1955).
132. R. D. McLachlan and R. A. Nyquist, *Spectrochim. Acta*, *24A*, 103 (1968).
133. K. Radcliffe and J. L. Wood, *Trans. Faraday Soc.*, *62*, 2038 (1966).
134. W. B. Bacher and J. W. Wagner, *Z. Phys. Chem.*, *43B*, 191 (1939).
135. A. L. Verma and P. Venkateswarlu, *J. Mol. Spectrosc.*, *39*, 227 (1971).
136. R. Lespieau and M. Bourguel, *Bull. Soc. Chim. Fr.*, *47*, (4) 1365 (1930).
137. R. Vessiere, *Bull. Soc. Chim. Fr.*, *1960* , 369.

138. A. A. Petrov, *Zh. Obshch. Khim.*, *26,* 3319 (1956); *Chem. Abstr.*, *51,* 9471b.
139. K. K. Deb and D. K. Mukherjee, *Indian J. Phys.*, *37,* 339 (1963).
140. G. H. Griffith, L. A. Harrah, J. W. Clark, and J. R. Durig, *J. Mol. Struct.*, *4,* 255 (1969).
141. A. L. Verma, *J. Mol. Spectrosc.*, *39,* 247 (1971).
142. Y. P. Egorov, L. A. Leites, I. D. Kravtsova, and V. F. Mironov, *Proc. Acad. Sci. USSR, Div. Chem. Sci.*, 1114 (1963).
143. B. V. Thosar, *Z. Phys.*, *107,* 780 (1937).
144. W. G. Fateley, G. L. Carlson, and F. E. Dickson, *Appl. Spectrosc.*, *22,* 650 (1968).
145. Pao-shan Yu, V. N. Nikitin, and M. V. Vol'kenshtein, *Russ. J. Phys. Chem.*, *36,* 356 (1962).
146. G. L. Carlson, W. G. Fateley, and R. E. Witkowski, *J. Am. Chem. Soc.*, *89,* 6437 (1967).
147. K. W. F. Kohlrausch and R. Skrabal, *Monatsh. Chem.*, *70,* 377 (1937).
148. J. E. Katon and W. R. Feairheller, Jr., *J. Chem. Phys.*, *47,* 1248 (1967).
149. K. W. F. Kohlrausch, *Chem. Ber.*, *72,* 2054 (1939).
150. K. Noack and R. N. Jones, *Can. J. Chem.*, *39,* 2225 (1961).
151. W. R. Feairheller, Jr. and J. E. Katon, *J. Mol. Struct.*, *1,* 239 (1967).
152. A. J. Bowles, W. O. George, and D. B. Cunliffe-Jones, *J. Chem. Soc. (B)*, *1970,* 1070.
153. N. Jonathan, *J. Mol. Spectrosc.*, *6,* 205 (1961).
154. A. W. Reitz and R. Skrabal, *Monatsh. Chem.*, *70,* 398 (1937).
155. A. A. Petrov, V. A. Kolesova, and Y. I. Porfir'eva, *Zh. Obshch. Khim.*, *27,* 2076 (1957).
156. R. Golse and Le-Van-Thoi, *Compt. Rend.*, *230,* 210 (1950).
157. T. V. Yakovleva, A. A. Petrov, and V. S. Zavgorodnii, *Opt. Spectrosc.*, *12,* 106 (1962).
158. M. I. Batuev, E. N. Prilezhaeva, and M. F. Shostakovskii, *Bull. Acad. Sci. USSR, Cl. Sci. Chim.*, *1947,* 123.
159. A. Kirrmann and P. Chancel, *Bull. Soc. Chim. Fr.*, *1954,* 1338.
160. M. F. Shostakovskii, M. I. Batuev, I. A. Chekulaeva, and A. D. Matveeva, *Bull. Acad. Sci. USSR, Div. Chem. Sci.*, *1955,* 481.
161. P. P. Shorygin, T. N. Shkurina, M. F. Shostakovskii, and E. P. Gracheva, *Bull. Acad. Sci. USSR, Div. Chem. Sci.*, *1955,* 935.
162. E. M. Popov, N. S. Andreev, and G. I. Kagan, *Opt. Spectrosc.*, *12,* 17 (1962).
163. E. M. Popov and G. I. Kagan, *Opt. Spectrosc.*, *11,* 394 (1961).
164. J. Fabian, H. Kröber, and R. Mayer, *Spectrochim. Acta*, *24A,* 727 (1968).
165. S. G. Frankiss, *Spectrochim. Acta*, *22,* 295 (1966).

166. V. A. Shlyapochnikov and V. G. Osipov, *Bull. Acad. Sci. USSR, Ser. Chem., 1968,* 1773.
167. R. Gaufres and C. Roulph, *J. Mol. Struct., 9,* 107 (1971).
168. H. Hunziker and H. H. Günthard, *Spectrochim. Acta, 21,* 51 (1965).
169. A. Kirrmann, P. Federlin, and P. Bieber, *Bull. Soc. Chim. Fr., 1954,* 1466.
170. M. Yasumi, *J. Chem.,Soc. Jap., 63,* 983 (1942).
171. S. C. Sirkar and N. K. Roy, *J. Chem. Phys., 21,* 938 (1953).
172. N. K. Roy, *Indian J. Phys., 27,* 167 (1953).
173. A. W. Reitz and R. Sabathy, *Monatsh. Chem., 71,* 131 (1938).
174. J. Bragin, K. L. Kizer, and J. R. Durig, *J. Mol. Spectrosc., 38,* 289 (1971).
175. G. A. Crowder, *J. Mol. Spectrosc., 20,* 430 (1966).
176. S. B. Lie and P. Klaboe, *Spectrochim. Acta, 26A,* 1191 (1970).
177. V. F. Mironov, A. L. Kravchenko, and A. D. Petrov, *Dokl. Akad. Nauk SSSR, 155,* 843 (1964).
178. J. R. Durig, C. K. Tong, C. W. Hawley, and J. Bragin, *J. Phys. Chem., 75,* 44 (1971).
179. B. Gredy, *Bull. Soc. Chim. Fr., 3* (5), 1101 (1936).
180. J. W. McCutcheon, M. F. Crawford, and H. L. Welsh, *Oil and Soap, 18,* 9 (1941).
181. P. Devlin, J. Overend, and B. Crawford, Jr., *Spectrochim. Acta, 20,*23 (1964).
182. K. Noack and R. N. Jones, *Can. J. Chem., 39,* 2201 (1961).
183. H. Gerding and H. G. Haring, *Rec. Trav. Chim., 69,* 935 (1950).
184. D. A. Long and W. O. George, *Spectrochim. Acta, 20,* 1799 (1964).
185. S. B. Lie, P. Klaboe, D. H. Christensen, and G. Hagen, *Spectrochim. Acta, 26A,* 1861 (1970).
186. S. B. Lie, P. Klaboe, E. Kloster-Jensen, G. Hagen, and D. H. Christensen, *Spectrochim. Acta, 26A,* 2077 (1970).
187. A. I. Lebedeva and L. F. Almashi, *Dokl. Akad. Nauk SSSR, 86,* 75 (1952).
188. J. G. Aston, G. Szasz, H. W. Woolley, and F. G. Brickwedde, *J. Chem; Phys., 14,* 67 (1946).
189. C. M. Richards and J. R. Nielsen, *J. Opt. Soc. Am., 40,* 438 (1950).
190. D. J. Marais, N. Sheppard, and B. P. Stoicheff, *Tetrahedron, 17,* 163 (1962).
191. I. S. Borshagovskaya, Y. N. Panchenko, and Y. A. Pentin, *Opt. Spectrosc., 22,* 194 (1967).
192. Y. N. Panchenko, Tran Xuan Hoang, and Y. A. Pentin, *Vestn. Mosk. Univ., Ser II, 22,* 32 (1967).

193. Y. N. Panchenko, Y. A. Pentin, V. I. Tyulin, and V. M. Tatevskii, *Opt. Spectrosc., 16,* 536 (1964).
194. L. M. Sverdlov and N. V. Tarasova, *Opt. Spectrosc., 9,* 159 (1960).
195. N. V. Tarasova and L. M. Sverdlov, *Opt. Spectrosc., 18,* 336 (1964).
196. A. Dadieu and K. W. F. Kohlrausch, *Chem. Ber., 63,* 1657 (1930).
197, P. S. Srinivasan, *Proc. Indian Acad. Sci., 2A,* 105 (1935).
198. D. Craig, J. J. Shipman, and R. B. Fowler, *J. Am. Chem. Soc., 83,* 2885 (1961).
199. E. V. Sobolev and V. T. Aleksanyan, *J. Struct. Chem. USSR, 4,* 483 (1963).
200. S. Dzhessati, A. R. Kyazimova, V. I. Tyulin, and Y. A. Pentin, *Vestn. Mosk. Univ., Khim, 23,* 19 (1968).
201. N. V. Tarasova and L. M. Sverdlov, *Opt. Spectrosc., 21,* 176 (1966).
202. T. Kubota, *Bull. Chem. Soc. Jap., 13,* 678 (1938).
203. N. V. Tarasova and L. M. Sverdlov, *Russ. J. Phys. Chem., 42,* 840 (1968).
204. S. M. Makin, B. K. Kruptsov, V. M. Medvedeva, and L. N. Smirnova, *J. Gen. Chem. USSR, 8,* 2492 (1962).
205. M. I. Batuev, L. I. Shmonina, A. D. Matveeva, and M. F. Shostakovskii, *Bull. Acad. Sci. USSR, Div. Chem. Sci., 1961,* 474.
206. P. Sohar and G. Varsanyi, *J. Mol. Struct., 1,* 437 (1968).
207. R. M. Conrad and D. A. Dows, *Spectrochim. Acta, 21,* 1039 (1965).
208. J. C. Albright and J. R. Nielsen, *J. Chem. Phys., 26,* 370 (1957).
209. M. I. Batuev and A. D. Matveeva, *Izv. Akad. Nauk SSSR, Otd. Khim. Nauk, 1958,* 1393.
210. E. R. Lippincott, C. E. White, and J. P. Sibalia, *J. Am. Chem. Soc., 82,* 2537 (1960).
211. E. R. Lippincott and T. E. Kenney, *J. Am. Chem. Soc., 84,* 3641 (1962).
212. E. R. Lippincott, W. R. Feairheller, Jr., and C. E. White, *J. Am. Chem. Soc., 81,* 1316 (1959).
213. K. W. Hausser, *Z. Tech. Phys., 15,* 10 (1934).
214. J. Behringer and J. Brandmüller, *Ann. Phys., 4,* 234 (1959).
215. T. M. Ivanova, L. A. Yanovskaya, and P. P. Shorygin, *Opt. Spectrosc., 18,* 115 (1965).
216. P. P. Shorygin and T. M. Ivanova, *Opt. Spectrosc., 25,* 107 (1968).
217. C. Tric, *J. Chem. Phys., 51,* 4778 (1969).
218. R. W. Mitchell and J. A. Merritt, *Spectrochim. Acta, 27A,* 1609 (1971).
219. F. F. Cleveland, M. J. Murray, and W. S. Gallaway, *J. Chem. Phys., 15,* 742 (1947).
220. G. Chiurdoglu, J. Laune, and M. Poelmans, *Bull. Soc. Chim. Belge., 65,* 257 (1956).

221. V. T. Aleksanian (Aleksanyan), Kh. E. Sterin, A. A. Mel'nikov, and A. F. Plate, *Bull. Acad. Sci. USSR, Phys. Ser., 22,* 1062 (1958); also *Chem. Abstr. 53,* 861b (1959).
222. T. Hayashi, *Sci. Pap. Inst. Phys. Chem. Res. (Tokyo), 27,* 99 (1935).
223. R. W. Mitchell, E. A. Dorko, and J. A. Merritt, *J. Mol. Spectrosc., 26,* 197 (1968).
224. R. W. Mitchell and J. A. Merritt, *Spectrochim. Acta, 25A,* 1881 (1969).
225. M. Ito, *Spectrochim. Acta, 22,* 1581 (1966).
226. A. Krebs, B. Schrader, and F. Höfler, *Tetrahedron Lett.,* 5935 (1968).
227. F. Höfler, B. Schrader, and A. Krebs, *Z. Naturforsch., 24a,* 1617 (1969).
228. R. C. Lord and D. G. Rea, *J. Am. Chem. Soc., 79,* 2401 (1957).
229. J. R. Nielsen, M. Z. El-Sabban, and M. Alpert, *J. Chem. Phys., 23,* 324 (1955).
230. A. W. Reitz, *Z. Phys. Chem., 33B,* 179, 368 (1936).
231. A. W. Reitz, *Z. Phys. Chem., 38B,* 381 (1938).
232. H. Gerding and A. P. Van der Vet, *Rec. Trav. Chim., 64,* 257 (1945).
233. C. W. Beckett, N. K. Freeman, and K. S. Pitzer, *J. Am. Chem. Soc., 70,* 4227 (1948).
234. S. V. Markova, P. A. Bazhulin, and M. M. Sushchinskii, *Opt. Spektrosk., 1,* 41 (1956).
235. S. I. Subbotin, V. I. Tyulin, and V. M. Tatevskii, *Opt. Spektrosk., 17,* 381 (1964).
236. L. M. Sverdlov and E. P. Krainov, *Opt. Spectrosc., 6,* 214 (1959).
237. M. Petzuch, *Z. Naturforsch., 24a,* 637 (1969).
238. M. Mousseron, R. Richaud, and R. Granger, *Bull. Soc. Chim. Fr., 1946,* 222.
239. G. Chiurdoglu and A. Guillemonat, *Bull. Soc. Chim. Fr., 5,* (5) 1624, (1938).
240. J. L. Lauer, W. H. Jones, Jr., and H. C. Beachell, *J. Chem. Phys., 30,* 1489 (1959).
241. K. W. F. Kohlrausch and H. Wittek, *Chem. Ber., 75,* 227 (1942).
242. E. V. Sobolev, V. T. Aleksanyan, and V. A. Mironov, *Dokl. Akad. Nauk SSSR, 152,* 923 (1963).
243. E. V. Sobolev, V. T. Aleksanyan, E. M. Mil'vitskaya, and M. A. Pryanishnikova, *J. Gen. Chem. (USSR), 4,* 189 (1963).
244. H. P. Fritz and L. Schäfer, *Spectrochim. Acta, 21,* 211 (1965).
245. E. Gallinella, B. Fortunato, and P. Mirone, *J. Mol. Spectrosc., 24,* 345 (1967).
246. H. Gerding, H. J. Prins, and H. Van Brederode, *Rec. Trav. Chim., 65,* 168 (1946).
247. M. Godchot, E. Canals, and G. Cauquil, *Compt. Rend., 196,* 780 (1933).

248. K. Sakashita, *J. Chem. Soc. Jap., Pure Chem. Sec., 77,* 1094 (1956).
249. N. Neto, C. DiLauro, E. Castellucci, and S. Califano, *Spectrochim. Acta, 23A,* 1763 (1967).
250. M. Godchot, E. Canals, and G. Cauquil, *Compt. Rend., 197,* 1407 (1933).
251. G. Chiurdoglu, R. Ottinger, J. Reisse, and A. Toussaint, *Spectrochim. Acta, 18,* 215 (1962).
252. A. D. Petrov, V. F. Mironov, and V. G. Glukhovtser, *Zh. Obshch. Khim. 27,* 1535 (1957); *Chem. Abstr., 52,* 3668g.
253. H. Gerding and F. A. Haak, *Rec. Trav. Chim., 68,* 336 (1949).
254. C. DiLauro, N. Neto, and S. Califano, *J. Mol. Struct., 3,* 219 (1969).
255. K. W. F. Kohlrausch and R. Seka, *Chem. Ber., 68B,* 528 (1935).
256. J. W. Murray, *J. Chem. Phys., 3,* 59 (1935).
257. G. B. Bonino and R. Manzoni-Ansidei, *Nature, 135,* 873 (1935).
258. E. Canals, M. Mousseron, and F. Winternitz, *Compt. Rend., 219,* 210 (1944).
259. M. Mousseron and F. Winternitz, *Bull. Soc. Chim. Fr., 1946,* 232.
260. H. Gerding and F. A. Haak, *Rec. Trav. Chim., 68,* 293 (1949).
261. H. D. Stidham, *Spectrochim. Acta, 21,* 23 (1965).
262. N. Neto, C. DiLauro, and S. Califano, *Spectrochim. Acta, 26A,* 1489 (1970).
263. J. Goubeau, *Ann. Chim., 567,* 214 (1950).
264. P. J. Hendra and D. B. Powell, *Spectrochim. Acta, 17,* 913 (1961).
265. E. R. Lippincott and R. C. Lord, *J. Am. Chem. Soc., 79,* 567 (1957).
266. E. R. Lippincott and R. C. Lord, *J. Am. Chem. Soc., 68,* 1868 (1946).
267. M. St. C. Flett, W. T. Cave, E. E. Vago, and H. W. Thompson, *Nature, 159,* 739 (1947).
268. E. R. Lippincott, R. C. Lord, and R. S. McDonald, *J. Am. Chem. Soc., 73,* 3370 (1951).
269. S. J. Wilt and M. A. El-Sayed, *J. Am. Chem. Soc., 88,* 2911 (1966).
270. N. Neto, C. DiLauro, and S. Califano, *Spectrochim. Acta, 24A,* 385 (1968).
271. K. W. F. Kohlrausch and R. Seka, *Chem. Ber., 69,* 729 (1936).
272. V. T. Aleksanyan and Kh. E. Sterin, *Mater. Desyatogo Vses. Soveshch. Spektrosk., 1957, L'vov. Gos. Univ. im. I. Franko, Fiz. Sb. 1,* No. 3,59; *Chem. Abstr. 53,* 21158a (1959).
273. K. W. F. Kohlrausch, R. Seka, and O. Tramposch, *Chem. Ber., 75,* 1385 (1942.)
274. E. G. Treschova, V. M. Tatevskii, V. R. Skvarchenko, and R. Ya. Levina, *Opt. Spektrosk., 5,* 553, (1958).

CHAPTER SEVEN

ALIPHATIC ALDEHYDES AND KETONES

7.1 INTRODUCTION

Among some of the early Raman investigations of aliphatic aldehydes and ketones were those performed by Kohlrausch and Köppl (1) (aliphatic aldehydes and ketones), Milone (2) (ethyl ketones), and Biquard (3) (ketones). The mid-IR spectra of these compounds have been reviewed by Thompson and Torkington (4) and by Lecomte (5). The low frequency skeletal deformations in aliphatic ketones have been investigated by Lecomte, Josien, and Lascombe (6) (650–350 cm^{-1}), Katon and Bentley (7) (700–350 cm^{-1}), Shimanouchi, Abe, and Mikami (8) (700–100 cm^{-1}), and Doolittle (9) (400–200 cm^{-1}). Jones and Noack (10) have studied the conformational equilibria in diethyl ketone in both the IR and Raman. The assignment of the aldehyde C-H stretching vibration has been discussed by Pinchas (11) and by Saier, Cousins, and Basila (12,13). Worden (14) has assigned the observed bands in the IR and Raman of propionaldehyde and propionaldehyde-d_1.

Normal coordinate analyses of aldehydes and ketones have only been performed on the simpler members of each class. Normal vibration calculations for H_2CO,D_2CO, CH_3CHO, and CH_3CDO were reported by Cossee and Schachtschneider (15). The force field of acetaldehyde was also calculated by Epshtein (16). Acetone has been the subject of many investigations and the assignments of the skeletal bending modes are still a matter under debate.

Difficulties also arise in that many of the fundamentals, especially those due to the methyl groups, are almost degenerate. Mirone (17) has reviewed the work before 1964. Cossee and Schachtschneider (15) and Epshtein (16) derived approximate force fields for acetone and acetone-d_6. Dellepiane and Overend (18) measured and assigned the bands in the IR and Raman of $(CH_3)_2CO$, CH_3COCD_3, and $(CD_3)_2CO$. Allkins and Lippincott (19) conducted Raman polarization measurements on the three skeletal bending modes. Forel and Fouassier (20,21) calculated a valence force field for acetone and its d_3 and d_6 isotopic species and discussed most of the earlier assignments. For the skeletal bending modes in acetone, we shall adopt those assignments given in refs. 16–18 and 21 and also listed by Shimanouchi (22). For C_{2v} symmetry, the a_1 C-C-C bending vibration occurs at 390 cm^{-1}, the b_1 C=O in-plane bending at 530 cm^{-1}, and the b_2 C=O out-of-plane bending at 490 cm^{-1}.

7.2 ALIPHATIC ALDEHYDES

The characteristic frequencies in the laser Raman spectra of aliphatic aldehydes (e.g., Spectrum 39, Appendix 2) are given in Table 7.1. Generally, two characteristic bands appear in the region 2850–2700 cm^{-1}. Saier, Cousins, and Basila (13) have shown that this doublet arises from a Fermi resonance interaction between the aldehydic C-H stretching fundamental and the first overtone of the aldehydic C-H bending vibration. In the Raman spectra the intensity of the lower frequency band at 2720–2695 cm^{-1} is very strong. The higher frequency band at 2830–2810 cm^{-1} often appears as a shoulder on other CH stretching bands. The carbonyl stretching frequency in liquid aliphatic aldehydes occurs in the range 1740–1720 cm^{-1}. The relative intensity of ν(C=O) is generally weaker in the Raman spectrum than in the corresponding IR spectrum. The carbon-hydrogen in-plane wag gives rise to a Raman band at 1410–1380 cm^{-1}.

In aliphatic aldehydes with no branching at the α-carbon the C-C stretch of the C-C=O group occurs at 1120–1090 cm^{-1}. Those aldehydes containing a tertiary carbon atom adjacent to the carbonyl display an intense band in the range 800–770 cm^{-1} due to the symmetric skeletal stretch of the C_4 moiety. If a quaternary carbon atom is adjacent to the carbonyl, a band at 770–750 cm^{-1} appears, arising from the symmetric stretching of the C_5 skeleton. The C-C=O in-plane deformation is also sensitive to branching at the α-carbon. With no α-branching the frequency of this vibration is located in the region 530–510 cm^{-1}. Monosubstitution of an alkyl group at the α-carbon raises the frequency to 550–540 cm^{-1} and disubstitution to 600–580 cm^{-1}. The C=O out-of-plane wag (763 cm^{-1} for acetaldehyde) is very weak in the Raman and generally is not observed for higher aliphatic aldehydes.

Lucazeau and Novak (23–26) have investigated the IR and Raman spectra of a series of halogenated aldehydes. The Raman spectrum of fluoral, CF_3CHO, has been reported by Berney (27). Normal coordinate analyses for CX_3CHO (X = H, F, Cl, and Br) have been performed by Hagen (28). The introduction of a halogen into the α-position results in an increase in the car-

Table 7.1 Characteristic Frequencies in the Raman Spectra of Liquid Aliphatic Aldehydes

	Vibration Frequency (cm^{-1})					
Compound	Fermi Resonance Doublet		C=O Stretch	Aldehyde C–H Deformation	C–C Stretch Adjacent to C=O	In-plane C–C=O Deformation
A. No α-carbon branching						
Acetaldehyde	2838	2732	1715	1392	1111	512
Propionaldehyde	2823	2724	1722	1392	1089	512
n-Butyraldehyde	2817	2732	1718	1389	1112	511
n-Valeraldehyde	2820	2722	1721	1390	1116	522
Isovaleraldehyde	2825	2726	1721	1390	1125	515
n-Caproaldehyde	2821	2710	1720	1388	1112	516
n-Heptylaldehyde	2820	2712	1721	1388	1120	520
n-Octylaldehyde	2825	2708	1723	1388	1120	525
n-Nonylaldehyde	2810	2722	1721	1397	1111	523
n-Decylaldehyde	2810	2710	1722	1388	1118	522
B. With α-carbon branching						
Isobutyraldehyde	2811	2716	1721	1392	796[a]	543
sec-Valeraldehyde	2821	2719	1718	1392	771[a]	540
tert-Valeraldehyde	2816	2723	1720	1406	762[b]	594
tert-Caproaldehyde	–	–	1725	1389	749[b]	585
Relative Raman Intensity	Medium (shoulder)	Very strong	Very strong–strong	Strong–medium	Medium–very strong	Medium–weak

[a]Skeletal stretch of the C_4 group.
[b]Skeletal stretch of the C_5 group.

bonyl stretching frequency in the Raman, as shown in the following table:

Aldehyde	ν(C=O) (cm^{-1})	Reference
CF_3CHO	1770	27
CH_2ClCHO	1731	25
$(CH_3)_2CClCHO$	1754	26
$CHCl_2CHO$	1747	24
CCl_3CHO	1757	23,28
CBr_3CHO	1745	23

7.3 ALIPHATIC KETONES

Table 7.2 (pages 102,103) summarizes the Raman bands useful in identifying aliphatic ketones (e.g., Spectra 40,41). The carbonyl stretching vibration in aliphatic ketones occurs in the frequency range 1725–1700 cm^{-1}, which is 10–20 cm^{-1} lower than in aliphatic aldehydes. Normal coordinate analyses (15,21) have shown that this vibration is mainly 60–80% C=O stretch with the symmetric C-C(=O)-C stretch also involved. The antisymmetric C-C(=O)-C stretch occurs in the region 1170–1160 cm^{-1} for methyl ketones and in the region 1130–1100 cm^{-1} for ethyl ketones. The symmetric carbon-carbon stretches occur over a wider range, 800–700 cm^{-1}. Monoalkyl substitution at the α-carbon atom gives rise to a strong band at 750–710 cm^{-1}, which is due to the symmetric stretching of the four-carbon atom skeleton. Dialkyl substitution at the carbon adjacent to the carbonyl group is identified by a very strong band at 680–650 cm^{-1}, which is attributed to the symmetric stretch of the five-carbon skeleton.

Those ketones in which there are methylene groups adjacent to the carbonyl have a strong band arising from the methylene in-plane deformation in the range 1420–1410 cm^{-1}.

Jones and Noack (10) observed that the Raman spectra of diethyl ketone exhibited a pronounced temperature dependence which they attributed to the presence of at least two conformational isomers in the liquid state. Shimanouchi, Abe, and Mikami (8), in a study of methyl ethyl ketone, methyl *n*-propyl ketone, and diethyl ketone, concluded that only the trans conformation, with respect to the C-C-C-C axis, is stable in the solid state. They also noted that various rotational isomers exist in the liquid and gaseous states. In Table 7.2 the frequencies listed for the C-C=O and C-C(=O)-C in-plane deformations are those that have the strongest intensity in the Raman and can generally be attributed to the trans conformation which is usually present in the greatest concentration at room temperature. Methyl ketones exhibit a band at 600–580 cm^{-1} due to the C-C=O in-plane deformation of the trans form. A weaker Raman band at 540–520 cm^{-1} is attributed to this vibration from molecules in various gauche forms. For ethyl ketones, the strongest C-C=O in-plane deformation in the Raman occurs at 600–585 cm^{-1} whereas other conformations give rise to weaker bands at 640–620 cm^{-1} and 530–520 cm^{-1}. For monoalkyl substitution at the α-carbon, this deformation occurs at about 605 cm^{-1}; for dialkyl substitution at the α-carbon, it is located near 550 cm^{-1}. The frequency of the C-C(=O)-C in-plane deformation for most ketones is found in the range 430–390 cm^{-1}. The carbonyl out-of-plane deformation for aliphatic ketones occurs at 490–460 cm^{-1}, but in the Raman this band has a very weak intensity and is not a reliable characteristic frequency.

The effect of cyclization upon ν(C=O) is quite marked for the C_4 and C_5 rings, but for C_6 and higher homologues this frequency falls in the range for open chain aliphatic ketones. The Raman data for these compounds are summarized below:

Ketone	ν(C=O) (cm^{-1})	Reference
Cyclobutanone	1782	29,30
Cyclopentanone	1744	31
Cyclohexanone	1709	32
Cycloheptanone	1699	33
Cyclooctanone	1708	33

In the liquid Raman spectrum of 2,3-butanedione (biacetyl) (34) the symmetric C=O stretch appears as a very strong polarized band at 1720 cm^{-1}. The antisymmetric carbonyl stretch appears only in the IR at 1715 cm^{-1}. In acetylacetone (pentane-2,4-dione) (35) (Spectrum 43) the symmetric C=O stretch occurs as a polarized band of medium intensity at 1719 cm^{-1} in the Raman spectrum of the liquid and the antisymmetric C=O stretch as a weak depolarized band at 1697 cm^{-1}.

Table 7.2 Characteristic Frequencies in the Raman Spectra of Aliphatic Ketones in the Liquid State

	Vibration Frequency (cm^{-1})					
Compound	C=O Stretch	Antisymmetric C–C(=O)–C Stretch	Symmetric C–C(=O)–C Stretch	CH_2 Deformation of $-CH_2-C(=O)-$ Group	C–C=O In-plane Deformation	C–C(=O)–C In-plane Deformation
A. Methyl ketones						
2-Propanone	1706	1220	787	—	528	391
2-Butanone	1711	1165	763	1412	592	408
2-Pentanone	1710	1170	723	1416	591	404
2-Hexanone	1709	1166	724	1416	588	394
2-Heptanone	1709	1168	722	1416	592	399
2-Octanone	1710	1162	716	1412	592	406
2-Undecanone	1710	1160	717	1404	587	412
4-Methyl-2-pentanone	1709	1169	785	1416	593	422
5-Methyl-2-hexanone	1714	1166	748	1408	585	391
B. Ethyl ketones						
3-Pentanone	1710	1120	792	1418	589	406
3-Hexanone	1709	1111	742	1415	596	413
3-Heptanone	1710	1130	750	1417	592	390
3-Octanone	1709	1115	771	1414	595	410
3-Nonanone	1712	1119	784	1415	596	—
5-Methyl-3-hexanone	1711	1129	791	1414	591	420
6-Methyl-3-heptanone	1712	1110	785	1418	589	425

C. Other aliphatic ketones						
4-Heptanone	1707	1114	716	1410	529	424
2,6-Dimethyl-4-heptanone	1717	1133	779	1411	534	419
5-Nonanone	1717	1121	772	1413	540	422
D. Ketones with α-carbon branching						
3-Methyl-2-butanone	1709	1190	724[a]	1421	605	442
3-Methyl-2-pentanone	1708	1178	714[a]	—	603	429
2-Methyl-3-pentanone	1708	1115	736[a]	1416	604	—
2,4-Dimethyl-3-pentanone	1707	1114	741[a]	—	604	470
3,3-Dimethyl-2-butanone	1702	1132	671[b]	—	552	451
3,3-Dimethyl-2-pentanone	1701	1162	660[b]	—	544	—
Relative Raman intensity	Very strong–strong	Medium–weak	Strong–medium	Strong	Strong–medium	Medium–weak

[a]Symmetric stretch of the C_4 skeleton.
[b]Symmetric stretch of the C_5 skeleton.

REFERENCES

1. K. W. F. Kohlrausch and F. Köppl, *Z. Phys. Chem., 24B,* 370 (1934).
2. M. Milone, *Gazz. Chim. Ital., 64,* 876 (1934).
3. D. Biquard, *Bull. Soc. Chim., Fr., 7 (5),* 894 (1940).
4. H. W. Thompson and P. Torkington, *J. Chem. Soc., 1945,* 640.
5. J. Lecomte, *Bull. Soc. Chim., Fr., 1955,* 717.
6. J. Lecomte, M. L. Josien, and J. Lascombe, *Bull. Soc. Chim. Fr., 1956,* 1963.
7. J. E. Katon and F. F. Bentley, *Spectrochim. Acta, 19,* 639 (1963).
8. T. Shimanouchi, Y. Abe, and M. Mikami, *Spectrochim. Acta, 24A,* 1037 (1968).
9. R. E. Doolittle, *Appl. Spectrosc., 24,* 337 (1970).
10. R. N. Jones and K. Noack, *Can. J. Chem., 39,* 2214 (1961).
11. S. Pinchas, *Anal. Chem., 27,* 2 (1955).
12. E. L. Saier, L. R. Cousins, and M. R. Basila, *Anal. Chem., 34,* 824 (1962).
13. E. L. Saier, L. R. Cousins, and M. R. Basila, *J. Phys. Chem., 66,* 232 (1962).
14. E. F. Worden, Jr., *Spectrochim. Acta, 18,* 1121 (1962).
15. P. Cossee and J. H. Schachtschneider, *J. Chem. Phys., 44,* 97 (1966).
16. L. M. Epshstein, *J. Struct. Chem. (USSR), 8,* 234 (1967).
17. P. Mirone, *Spectrochim. Acta, 20,* 1646 (1964).
18. G. Dellepiane and J. Overend, *Spectrochim. Acta, 22,* 593 (1966).
19. J. R. Allkins and E. R. Lippincott, *Spectrochim. Acta, 25A,* 761 (1969).
20. M. T. Forel and M. Fouassier, *Spectrochim. Acta, 23A,* 1977 (1961).
21. M. T. Forel and M. Fouassier, *J. Chim. Phys., 67,* 1132 (1970).
22. T. Shimanouchi, *Tables of Molecular Vibrational Frequencies – Part 3,* NSRDS-NBS 17, 1968.
23. G. Lucazeau and A. Novak, *Spectrochim. Acta, 25A,* 1615 (1969).
24. G. Lucazeau and A. Novak, *J. Mol. Struct., 5,* 85 (1970).
25. G. Lucazeau and A. Novak, *J. Chim. Phys., 67,* 1614 (1970).
26. G. Lucazeau and A. Novak, *J. Chim. Phys., 68,* 252 (1971).
27. C. V. Berney, *Spectrochim. Acta, 25A,* 793 (1969).
28. G. Hagen, *Acta Chem. Scand., 25,* 813 (1971).
29. J. D. Roberts and W. Sauer, *J. Am. Chem. Soc., 71,* 3925 (1949).
30. K. Frei and H. Günthard, *J. Mol. Spectrosc., 5,* 218 (1960).
31. H. E. Howard-Lock and G. W. King, *J. Mol. Spectrosc., 35,* 393 (1970).
32. H. Fuhrer, V. B. Kartha, P. J. Kreuger, H. H. Mantsch, and R. N. Jones, *Chem. Rev., 72,* 439 (1972).
33. M. Godchot and G. Cauquil, *Compt. Rend., 208,* 1065 (1939).
34. J. R. Durig, S. E. Hannum, and S. C. Brown, *J. Phys. Chem., 75,* 1946 (1971).
35. E. E. Ernstbrunner, *J. Chem. Soc., (A), 1970,* 1558.

CHAPTER EIGHT

CARBOXYLIC ACIDS

AND DERIVATIVES

8.1 MONOCARBOXYLIC ACIDS

The Raman spectra of formic acid have been reported in the vapor state (1), the liquid state (2–4), and in aqueous solution (5–7). The low frequency Raman spectra of solid and liquid formic acid have been reviewed by Blumenfeld and Fast (8). In the vapor state above 160°C the monomer, HCOOH, predominates whereas at about 20°C the planar cyclic dimer, $(HCOOH)_2$, is the predominant form. In crystalline formic acid it has been shown (9) that two polymorphic forms exist, each having an infinite hydrogen-bonded chain configuration—α, formed when the liquid is frozen and β, resulting from freezing of vapor onto a cold window. In liquid formic acid, it has been proposed (4) that the assignments of the fundamental vibration frequencies can be explained by the existence of a helical-chain polymer similar in form to the β-chain polymorph. The carbonyl stretching, ν(C=O), and related vibrations for these species have been assigned as follows:

Formic Acid Species	ν(C=O) (cm^{-1})	ν(C-O) (cm^{-1})	δ(O=C-O) (cm^{-1})	Reference
Monomer (160–140°C)	1770 (IR)	1105	636	10
Dimer (26°C)	1754 (IR)	1218	697	11
Liquid (94°C)	1679 (R)	1194	675	4
Liquid (0°C)	1654 (R)	1208	679	4

Formic Acid Species	ν(C=O) (cm^{-1})	ν(C-O) (cm^{-1})	δ(O=C-O) (cm^{-1})	Reference
Aqueous solution (35–100% HCO_2H	1672 (R)	–	–	7
α-Crystalline (77°K)	1703,1609 (IR)	1255,1224	721	11
β-Crystalline	1620 (IR)	1248,1210	720	12

Acetic acid in the vapor state above 150°C exists nearly completely in the monomeric form. IR spectra and vibrational assignments have been reported for this species and its deuterated analogues (13–15). In the vapor state at 25°C, most of the IR bands could be assigned to the presence of the cyclic dimer (13,16). Numerous Raman investigations have been conducted for acetic acid in the liquid state (2,3,16–20) and in aqueous (17,21–24) and non-aqueous (17,24–27) solutions. Raman spectra of the deuterated acetic acids (CH_3COOD, CD_3COOH, and CD_3COOD) in the liquid state have also been reported (16,28–30). In liquid acetic acid, the predominant species is the cyclic dimer (I); however, the presence of open dimers (II) and higher chain polymers has been detected (20). In the crystalline form, acetic acid molecules are associated in the form of infinite hydrogen-bonded chains (31).

CH_3–C(=O--H–O)(–O–H--O=)C–CH_3

I

CH_3–C(=O)–O–H--O=C(CH_3)–HO

II

The effect of these changes in state upon ν(C=O) and other vibrations in acetic acid are:

Physical State	Species	ν(OH)	ν(C=O)	ν(C-O)	δ(O=C-O)	Reference
		Frequency (cm^{-1})				
Vapor (150°C)	Monomer	3583	1788 (IR)	1182	642	15
Vapor (25°C)	Cyclic dimer	3027	1730 (IR)	1292	621	16
Liquid (25°C)	Cyclic dimer	3032	1715 (IR), 1675 (R)	1283	624	16
Crystal (0°C)	Polymeric	2927	1659 (IR), 1645 (IR)	1268	629	31

In the condensed states, the strongest band in the carbonyl stretching range in the IR and Raman spectra of most carboxylic acids arises from the coupled carbonyl stretches of the cyclic centrosymmetric dimer, (III). Table 8.1 lists, for various classes of monocarboxylic acids, the frequency of the antisymmetric carbonyl stretch of the dimer that is active in the IR only and that of the symmetric carbonyl stretch that occurs only in the Raman. Other characteristic frequencies in the Raman spectra of dimeric carboxylic acids include bands due to the in-plane OH deformation at 1430–1410 cm^{-1} and to the C–O stretch at 1305–1270 cm^{-1} (see Spectrum 44, Appendix 2).

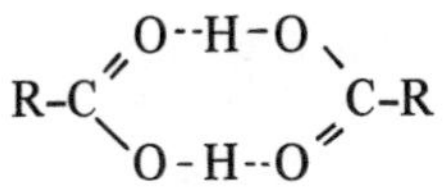

```
     O--H-O                 CH2 —— C=O
R-C//      \C-R              |      \
   \O-H--O//                 |       O
                             O - - H/
                           /
                       H3C

    III                        IV
```

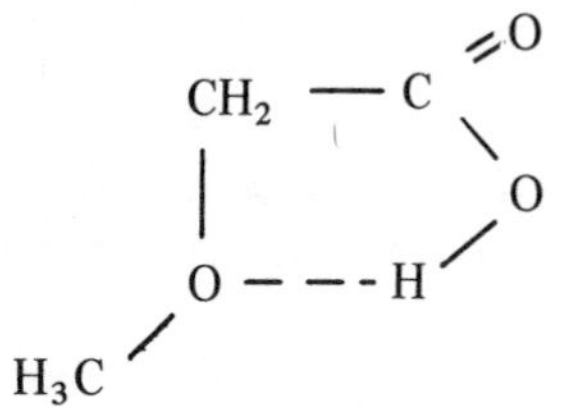

In carboxylic acids with proton-accepting groups in the α-position, such as α-keto, α-hydroxy, and α-alkoxy carboxylic acids, and in solutions of acids in solvents containing these proton-accepting groups, the possibility of types of intra- and intermolecular hydrogen bonding other than the cyclic dimer exists. Detailed IR investigations have been made on the various types of hydrogen bonding in α-keto and α-alkoxycarboxylic acids (54,55) and in α- and β-hydroxy acids (56). Colthup et al. (57) have stated than when the carbonyl group in the acid is not hydrogen bonded as in carboxyl-ether systems, for example, the *trans*-monomer form of methoxyacetic acid (IV), a band is found at 1760–1735 cm^{-1} in both the IR and Raman spectra. When the carbonyl is hydrogen-bonded but not dimerized, as in alcohol-carbonyl systems, a band active in both the IR and Raman appears at 1730–1705 cm^{-1}.

8.2 POLYCARBOXYLIC ACIDS

The carbonyl stretching frequencies in the IR and Raman spectra of some di- and tricarboxylic acids are listed in Table 8.2. Anhydrous crystalline oxalic acid can exist in two forms–in the α-form the -C=O and -OH groups of each molecule are associated with different oxalic acid molecules which form a planar sheet structure and in the β-form the molecules are linked together in long chains by cyclic hydrogen bonds similar to those found in monocar-

Table 8.1 Carbonyl Stretching Frequencies in the Vibrational Spectra of Carboxylic Acid Dimers

	IR Frequency (cm^{-1})			Raman Frequency (cm^{-1})		
Acid	ν_{asym}(C=O)	Physical State	Reference	ν_{sym}(C=O)	Physical State	Reference
CH_3-COOH	1715	Liquid	16	1675	Liquid	16
RCH_2-COOH	1729–1716	Liquid	32	1654–1649	Liquid	2
$(CH_3)_2CH$-COOH	1710	Liquid	33	1660	Liquid	33
$(CH_3)_3C$-COOH	1704	CCl_4 solution	34	1645	Liquid	2
Cyclopropane carboxylic acid	–	–	–	1645	Liquid	35
Cyclobutane carboxylic acid	1708	Liquid	36	1654	Liquid	36
Cyclopentane carboxylic acid	–	–	–	1645	Liquid	37
CF_3-COOH	1785	Liquid	38	1770	Liquid	38
$ClCH_2$-COOH	1742	Liquid	39	1725, 1669	Liquid	40
	–	–	–	1729	40% Aqueous solution	41
$CH_3CH(Cl)$-COOH	1730	Liquid	42	1664	Liquid	43
Cl_2CH-COOH	1799, 1739	Liquid	44	1678, 1741	Liquid	2
Cl_3C-COOH	1742	Solid	44	1687	Solid	40
$BrCH_2$-COOH	1726	CCl_4 solution	45	1715	Aqueous solution	46
ICH_2-COOH	1680	Mull	47	1669	Solid	47
$HOCH_2$-COOH	1730, 1710	Solid	42	1717	Aqueous solution	46
H_2C=CH-COOH	1705	Liquid	48	1660	Liquid	48
$H_2C=C(CH_3)$-COOH	1700	Liquid	42	1656	Liquid	49
OHC-COOH	1825	$CHCl_3$ solution	50	1746	Aqueous solution	50
HC≡C-COOH	1698	Solution ($CCl_4+C_6H_{14}$)	51	1676	Liquid	51
Benzoic acid	1697	Solid	52	1632	Solid	52
Alkyl-substituted benzoic acids	1710–1660	Liquid, solid	42	1686–1625	Liquid,solid	53

Table 8.2 Carbonyl Stretching Frequencies in the Vibrational Spectra of Polycarboxylic Acids

Acid	Formula	Physical State	Frequency, ν (C=O) (cm^{-1}) IR	Reference	Raman	Reference
1. Oxalic acid	$(COOH)_2$	α-Crystal	1756,1697	58	1782,1723	60
		β-Crystal	1732	58	1711	60
		Dihydrate	1688	58	1751	60
		Aqueous solution	1750	59	1750	59
2. Malonic acid	HOOC-CH_2-COOH	β-Crystal	1740,1696	61	1681,1645	61
		Aqueous solution	–	–	1738	62
3. Succinic acid	HOOC-$(CH_2)_2$-COOH	β-Crystal	1729,1690	63	1652	63
		Aqueous solution	–	–	1717,1654	64
4. Adipic acid	HOOC-$(CH_2)_4$-COOH	Crystal	1697	65	1648	65
5. Sebacic acid	HOOC-$(CH_2)_8$-COOH	Crystal	1711	66	1663	66
6. Maleic acid	*trans*- HOOC-CH=CH-COOH	Powder	1701	67	1723,1697	67
		Aqueous solution	–	–	1733–1709	67
7. Fumaric acid	*cis*-HOOC-CH=CH-COOH	Powder	1680	42	1681	67
		Aqueous solution	–	–	1727–1715	67
8. Tartaric acid	HOOC-CHOH-CHOH-COOH	α-Crystal	1740,1725	68	1745,1737	68
		meso-Aqueous solution	–	–	1732	62
9. Tricarballylic acid	HOOC-CH_2CH(COOH)CH_2-COOH	Aqueous solution	–	–	1724	62
10. Citric acid	HOOC-CH_2CH(OH)(COOH)CH_2-COOH	Aqueous solution	–	–	1731	62

boxylic acids (58). The α-form is characterized by strong bands at 1765 and 1697 cm^{-1} in the IR and at 1782 and 1723 cm^{-1} in the Raman. For the β-form a strong IR band occurs at 1732 cm^{-1} and a strong Raman band at 1711 cm^{-1}. The other saturated dicarboxylic acids generally crystallize in the linear hydrogen-bonded chain structure similar to β-oxalic acid. In aqueous solutions, polycarboxylic acids exhibit in their Raman spectra a very broad band located at 1750–1710 cm^{-1}.

8.3 CARBOXYLATE IONS

Detailed assignments in the IR and Raman spectra of the formate, acetate, and oxalate ions have been reviewed by Ito and Bernstein (69). In the carboxylate ion (RCO_2^-), the two carbon-oxygen stretching modes couple to form a symmetric (ν_{sym}) and an antisymmetric (ν_{asym}) stretching vibration. The antisymmetric vibration occurs in the range 1690–1550 cm^{-1} as a strong band in the IR and as a weak, depolarized band in the Raman. The symmetric vibration gives rise to a weak IR band and to a strong, polarized Raman band in the region 1440–1340 cm^{-1}. Spinner (70) has shown from a systematic study of the effect of substituents on the vibrational spectra of some substituted acetate ions that ν_{sym} (1583 cm^{-1} for solid $CH_3CO_2^-Na^+$) increases markedly with the electron-withdrawing effect of R and shows no mass effect and ν_{sym} (1416 cm^{-1} for aqueous solution of $CH_3CO_2^-Na^+$) is not noticeably affected by polar effects but decreases markedly as the mass of R increases. Below 1200 cm^{-1}, the spectra of the acetate ions resembled that of the corresponding acetic acid. In the vibrational spectra of alkali metal benzoates (71), ν_{asym} falls in range 1561–1552 cm^{-1} and ν_{sym} in the range 1427–1380 cm^{-1}.

The IR and Raman spectra of aqueous solutions of the sodium salts of malonic, succinic, maleic, and fumaric acids have been examined and assigned (72). For these dibasic anions, the intramolecular vibrational interactions among the four carbon-oxygen groups can lead to four different coupled modes which give rise to a complex band structure between 1670 and 1350 cm^{-1}. However, for the four anions studied, all exhibited a broad ill-defined Raman band of low intensity centered near 1570 cm^{-1}.

8.4 ESTERS

The complete vibrational assignments for methyl formate and methyl acetate have been given by Wilmshurst (73). Normal coordinate calculations have also been carried out for methyl formate (74–76) and for methyl acetate (75,76). In the IR spectrum of methyl formate in solution, the two most intense bands occur at 1734 cm^{-1} (C=O stretch) and at 1203 cm^{-1} (mainly C–O stretch). In the Raman spectrum of the liquid (Spectrum 45), ν(C=O) occurs as a strong polarized band at 1717 cm^{-1} and ν(C–O) is assigned to a very weak band at 1207 cm^{-1}. In general, for esters ν(C=O) is a reliable group frequency in both the IR and Raman spectra but the ν(C–O) band only occurs with sufficient intensity in the IR to be a dependable characteristic frequency. Generally, ν(C=O) occurs with a stronger relative intensity in the IR than in the Raman. In the Raman spectra of a series of seven *O*-alkyl formates, Kohlrausch and Pongratz (77) found ν(C=O) to be located in the narrow range 1720–1715 cm^{-1}. Other characteristic polarized Raman bands for these compounds occurred at 1383–1375 cm^{-1} (C–H in-plane deformation) and at 775–750 cm^{-1} (O–C=O in-plane deformation). For methyl formate and for those esters containing the -C(=O)-OCH_2- moiety, a strong polarized Raman band arising from an in-plane deformation occurred at 336–300 cm^{-1}. In the Raman spectra of chloroformates Cl–C(=O)–OR (77), ν(C=O) increases to 1780–1772 cm^{-1}, whereas for carbamates H_2N–C(=O)–OR (78) the carbonyl stretching frequency drops to the range 1694–1688 cm^{-1}.

The characteristic Raman frequencies for the esters of acetic acid (e.g., Spectrum 46) and other carboxylic acids are given in Table 8.3. For these *O*-alkyl saturated esters, the carbonyl stretching frequency falls in the range 1740–1730 cm^{-1}. The presence of alkyl branching on the carbon adjacent to the C=O group or that adjacent to the C–O bond lowers ν(C=O) by 5–15 cm^{-1}. The effect of α-halogen substitution on ν(C=O) has been studied by Cheng (79). The carbonyl stretching frequency increases with the number of chlorine atoms on the α-carbon. Substitution of a bromine atom has negligible effect since the field effect is offset by the mass effect of the atom.

In unsaturated esters the effect of conjugation is to lower the carbonyl stretching frequency. In the Raman spectra of a series of esters of propiolic acid (HC≡C–COOR), Sabathy (84) found that ν(C=O) was shifted to the range 1716–1708 cm^{-1}. The effect of an ethylenic bond adjacent to the carbonyl group is shown in Table 8.4 for a series of α, β-unsaturated esters (e.g., Spectrum 47). In aromatic esters, ν(C=O) is generally found near 1720 cm^{-1} in alkyl benzoates (89,90) and at 1738–1728 cm^{-1} in phthalate esters (91).

8.5 ACYCLIC ANHYDRIDES

The vibrational assignments in the IR and Raman spectra of acetic and perdeuteroacetic anhydride have been reviewed by Mirone, Fortunato, and Canziani (93). The coupling of the two C=O stretching vibrations gives rise

Table 8.3 Characteristic Frequencies in the Raman Spectra of *O*-Alkyl Saturated Esters

Type	Structure	ν(C=O) (cm^{-1})	Other Prominent Characteristic Raman Bands (cm^{-1})	Reference
1. Formates	HCOOR	1720–1715	1383–1375 (C–H in-plane deformation); 775–750 (O–C=O in-plane deformation); 336–300 (C–O–CH_2– in-plane deformation)	77, 80
2. Acetates	CH_3COOR	1741–1734	1459–1448 (CH_3 symmetric deformation); 847–825 (H_3C–C stretch; 644–634 (O–C=O) in-plane deformation)	79, 80
3. Propionates	CH_3CH_2COOR	1734–1727	855–845; 606–600; 458–434	46, 80, 81
4. *n*-Butyrates	$CH_3CH_2CH_2$-COOR	1734–1725	869–853; 603–598; 336–324	35, 80, 92
5. Higher straight-chain esters	R_1COOR (R_1=*n*-C_4H_9 to *n*-C_9H_{19})	1738–1731	—	80, 92
6. α-Alkyl branched	$R_1R_2CHCOOR$	1733–1728	844–822; 489–478	35, 80, 92
7. α,α-Dialkyl branched	$R_1R_2R_3CCOOR$	1729–1721	588–575	80
8. Cyclopropane-carboxylate	C_3H_5–COOR	1723–1716	1196–1191; 497–476	35
9. Cyclobutane-carboxylate	C_4H_7–COOR	1725–1716	918–912; 951–943	37
10. Cyclopentane-carboxylate	C_5H_9–COOR	1728–1720	1036–1010; 898–890	37
11. Chloroformates	ClCOOR	1780–1772	480–473	77

12.	Carbamates	H_2NCOOR	1694–1688	1124–1120; 1075–1070; 855–845; 660–658	78
13.	Chloroacetates	$ClCH_2COOR$	1748–1738	795–790; 702–695; 595–588	79
14.	Bromoacetates	$BrCH_2COOR$	1740–1732	718–711; 677–660; 552–551; 408–373	79
15.	Dichloroacetates	$Cl_2CHCOOR$	1756–1749	823–814; 783–769; 428–417; 241–226	79
16.	Trichloroacetates	Cl_3CCOOR	1769–1762	836–830; 758–740; 689–675; 442–433; 293–289; 234–199	79
17.	X–CH_2COOR	$X=NH_2, N(CH_3)_2$	1745–1730	—	46
		$X=OH, OCH_3$	1749–1734		
18.	CH_3–CHX–$COOR$	$X=NH_2, N(CH_3)_2$	1736–1726	—	43
		$X=OH, OCH_3$	1746–1733	—	
		$X=Cl$	1744–1736	—	
		$X=Br$	1738–1731	—	
19.	Oxalates	$(COOR)_2$ $R=CH_3$	1763	—	76, 82, 83
		$R=C_2H_5$	1761	—	
20.	Malonates	H_2C–$(COOR)_2$			
		$R=CH_3$	1741	—	82
		$R=C_2H_5$	1742	—	
21.	Succinates	$(CH_2COOR)_2$			
		$R=CH_3$	1734	—	82
		$R=C_2H_5$	1731	—	

Table 8.4 Characteristic Raman Bands for α, β-Unsaturated Esters

Type	Structure		ν(C=O) (cm^{-1})	ν(C=C) (cm^{-1})	Reference
1. Acrylates	H_2C=CHCOOR		1722–1714	1638–1632	35,85 86
2. Methacrylates	H_2C=C(CH_3)COOR		1723–1716	1639–1637	49
3. *trans*-Crotonates	(CH_3)(H)C=C(H)(COOR)	R=CH_3	1726	1662	85
		R=C_2H_5	1720	1663	
4. *cis*-Crotonates	(H)(CH_3)C=C(H)(COOR)	R=CH_3	1715	1644	87,88
		R=C_2H_5	1716	1644	
5. Fumarates	HC–COOR ‖ ROOC–CH	R=CH_3	1724	1650	87
		R=C_2H_5	1722	1657	
6. Maleates	HC–COOR ‖ HC–COOR	R=CH_3	1727	1650	87
		R=C_2H_5	1724	1653	

to two characteristic bands in the carbonyl stretching region. In the Raman spectrum of liquid acetic anhydride (Spectrum 48), the antisymmetric ν(C=O) vibration is assigned to a depolarized band at 1756 cm^{-1} and the symmetric ν(C=O) vibration to a polarized band at 1820 cm^{-1}. In other noncyclic saturated anhydrides, these bands occur at 1745–1738 cm^{-1} and at 1805–1799 cm^{-1}, respectively (92,94). In conjugated noncyclic anhydrides such as $(H_2C{=}CH{-}CO)_2O$, these vibrations occur near 1720 and 1775 cm^{-1} (95).

8.6 ACID HALIDES AND RELATED COMPOUNDS

The characteristic Raman frequencies for acid halides, carbonyl halides, and oxalyl halides are summarized in Table 8.5. A detailed vibrational assignment of CH_3COCl, CD_3COCl, and CH_2DCOCl has been given by Overend et al. (96). An analysis of the dependence of the carbonyl stretching frequency of acetyl, carbonyl, and benzoyl halides on mass and force constant effects has been presented by Overend and Scherer (97, 98).

Table 8.5 Characteristic Raman Frequencies for Acid Halides and Related Compounds

Type	Structure	ν(C=O) (cm^{-1})	Other Characteristic Raman Frequencies (cm^{-1})	Reference
Acid fluorides				
1. Acetyl fluoride	$H_3C-C(=O)-F$	1840,1799	—	99
2. Propionyl fluoride	$CH_3CH_2-C(=O)-F$	1837	—	99
3. Benzoyl fluoride	$C_6H_5-C(=O)-F$	1812	—	99,102
4. Toluoyl fluoride	$CH_3-C_6H_4-C(=O)-F$			
	ortho	1800	—	99
	meta	1803	—	
	para	1801	—	
Acid chlorides				
5. *n*-Alkyl	R-C(=O)-Cl	1810–1788	592–564, ν(C-Cl); 438–430, δ(O=C-Cl); 353–327, δ(C-C-Cl)	77,100
6. α-Alkyl branched	$R_1R_2CH-C(=O)-Cl$	1803–1788	433–430, δ(O=C-Cl)	77,100
7. α,α-Dialkyl branched	$R_1R_2R_3C-C(=O)-C$	1790–1774	418–413, δ(O=C-Cl)	77,100
8. Chloroacetyl chloride	$ClCH_2-C(=O)-Cl$	1806	452, δ(O=C-Cl)	101
9. Bromoacetyl chloride	$BrCH_2-C(=O)-Cl$	1794	442, δ(O=C-Cl)	100
10. Dichloroacetyl chloride	$Cl_2CH-C(=O)-Cl$	1803,1777	460, δ(O=C-Cl)	101
11. Trichloroacetyl chloride	$Cl_3C-C(=O)-Cl$	1806	—	101
12. Cyclopropane-carboxylic acid chloride	$C_3H_5C(=O)-Cl$	1770	433, δ(O=C-Cl)	35
13. Cyclobutane-carboxylic acid chloride	$C_4H_9-C(=O)-Cl$	1789	425, δ(O=C-Cl)	37
14. Cyclopentane-carboxylic acid chloride	$C_5H_7-C(=O)-Cl$	1791	427, δ(O=C-Cl)	37
15. Benzoyl chloride	$C_6H_5-C(=O)-Cl$	1774,1731	—	102

(continued)

Table 8.5 (continued)

Type	Structure	ν (C=O) (cm^{-1})	Other Characteristic Raman Frequencies (cm^{-1})	Reference
Acid Bromides				
16. *n*-Alkyl	R–C(=O)–Br	1812–1788	570–530, ν (C–Br); 356–318, δ (O=C–Br); 304–258, δ (C–C–Br)	99,100
17. Benzoyl bromide	C_6H_5–C(=O)=Br	1772	—	102
18. Toluoyl bromide	$CH_3C_6H_4$–C(=O)–Br			
	ortho	1775	—	99
	meta	1754,1728	—	
	para	1770	—	
Acid iodides				
19. Propionyl iodide	CH_3CH_2–C(=O)–I	1806	—	99
20. Benzoyl iodide	C_6H_5–C(=O)–I	1752	—	102
Carbonyl halides				
21. Carbonyl fluoride	F–C(=O)–F	1944,1909	—	103
22. Carbonyl chlorofluoride	F–C(=O)–Cl	1858,1832	—	103
23. Phosgene	Cl–C(=O)–Cl	1807	—	103
Oxalyl halides				
24. Oxalyl fluoride	F–C(=O)–C(=O)–F	1872	—	104
25. Oxalyl chloride	Cl–C(=O)–C(=O)–Cl	1778	—	105
26. Oxalyl bromide	Br–C(=O)–C(=O)–Br	1763	—	106

REFERENCES

1. L. G. Bonner and J. S. Kirby-Smith, *Phys. Rev., 57,* 1078 (1940).
2. K. W. F. Kohlrausch, F. Köppl, and A. Pongratz, *Z. Phys. Chem., 21B,* 242 (1933).
3. B. D. Saksena, *Proc. Indian Acad. Sci., 12A,* 312 (1940).
4. G. E. Tomlinson, B. Curnutte, and C. E. Hathaway, *J. Mol. Spectrosc., 36,* 26 (1970).
5. J. Gupta, *Indian J. Phys., 10,* 117 (1936).
6. P. Koteswaram, *Z. Physik., 112,* 395 (1939).
7. N. G. Zarakhani and M. I. Vinnik, *Zh. Fiz. Khim., 37,* 2550 (1963).
8. S. M. Blumenfeld and H. Fast, *Spectrochim. Acta, 24A,* 1449 (1968).
9. Y. Mikawa, R. J. Jakobsen, and J. W. Brasch, *J. Chem. Phys., 45,* 4750 (1966).
10. R. C. Millikan and K. S. Pitzer, *J. Chem. Phys., 25,* 1305 (1956).
11. R. C. Millikan and K. S. Pitzer, *J. Am. Chem. Soc., 80,* 3515 (1958).
12. Y. Mikawa, J. W. Brasch, and R. J. Jakobsen, *J. Mol. Spectrosc., 24,* 314 (1967).
13. W. Weltner, Jr., *J. Am. Chem. Soc., 77,* 3941 (1955).
14. J. K. Wilmshurst, *J. Chem. Phys., 25,* 1171 (1956).
15. M. Haurie and A. Novak, *J. Chim. Phys., 62,* 137 (1965).
16. M. Haurie and A. Novak, *J. Chim. Phys., 62,* 146 (1965).
17. P. Koteswaram, *Z. Physik., 110,* 118 (1938).
18. I. Cucurezeanu, *Rev. Phys. Acad. Rep. Pop. Roum. 2,* 243 (1957); *Chem. Abstr., 53,* 2787c (1959).
19. H. Dunken and P. Fink, *Z. Chem., 2,* 117 (1962).
20. M. Haurie and A. Novak, *Compt. Rend., 264B,* 694 (1967).
21. P. Traynard, *Bull. Soc. Chim. Fr., 1947,* 316.
22. S. Feneant, *Compt. Rend., 235,* 1292 (1952).
23. A. A. Glagoleva and A. A. Ferkhmin, *Zh. Obshch. Khim, 28,* 289 (1958); *Chem. Abstr., 52,* 10718i (1958).
24. P. Waldstein and L. A. Blatz, *J. Phys. Chem., 71,* 2271 (1967).
25. P. Koteswaram, *Curr. Sci., 8,* 70 (1939).
26. P. Koteswaram, *Indian J. Phys., 14,* 333 (1941).
27. M. Haurie and A. Novak, *J. Chim. Phys., 64,* 679 (1967).
28. W. R. Angus, A. H. Leckie, and C. L. Wilson, *Nature, 135,* 913 (1935).
29. W. R. Angus, A. H. Leckie, and C. L. Wilson, *Proc. Roy. Soc., 155A,* 183 (1936).
30. W. Engler, *Z. Phys. Chem., 32B,* 471 (1936).

31. M. Haurie and A. Novak, *Spectrochim. Acta, 21,* 1217 (1965).
32. P. J. Cornish and D. Chapman, *J. Chem. Soc., 1957,* 1746.
33. A. Bellocq, R. Martegoutes, M. Dubien, J. Belloc, P. Dizabo, and C. Garrigou-Lagrange, *J. Chim. Phys., 66,* 449 (1969).
34. R. H. Gillette, *J. Am. Chem. Soc., 58,* 1143 (1936).
35. K. W. F. Kohlrausch and R. Skrabal, *Monatsh. Chem., 70,* 377 (1937).
36. L. Bardet, J. Maillols, R. Granger, and E. Fabregue, *J. Mol. Struct., 10,* 343 (1971).
37. K. W. F. Kohlrausch and R. Skrabal, *Monatsh. Chem., 70,* 44 (1937).
38. N. Fuson, M. Josien, E. A. Jones, and J. R. Lawson, *J. Chem. Phys., 20,* 1627 (1952).
39. R. E. Kagarise, *J. Phys. Chem., 61,* 499 (1957).
40. I. D. Poliakova and S. S. Raskin, *Opt. Spectrosc., 3,* 220 (1959).
41. J. T. Edsall, *J. Chem. Phys., 4,* 1 (1936).
42. M. St. C. Flett, *J. Chem. Soc., 1951,* 962.
43. O. Burkard and L. Kahovec, *Monatsh. Chem., 71,* 333 (1938).
44. S. Bratoz, D. Hadzi, and N. Sheppard, *Spectrochim. Acta, 8,* 249 (1956).
45. J. Bellanto and J. R. Barcelo, *Spectrochim. Acta, 16,* 1333 (1960).
46. L. Kahovec and K. W. F. Kohlrausch, *Monatsh. Chem., 68,* 359 (1936).
47. J. E. Katon and T. P. Carll, *J. Mol. Struct., 7,* 391 (1971).
48. W. R. Feairheller, Jr. and J. E. Katon, *Spectrochim. Acta, 23A,* 2225 (1967).
49. A. Kirrmann, P. Federlin, and P. Bieber, *Bull. Soc. Chim. Fr., 1954,* 1466.
50. G. Fleury and V. Tabacik, *J. Mol. Struct., 10,* 359 (1971).
51. J. E. Katon and N. T. McDevitt, *Spectrochim. Acta, 21,* 1717 (1965).
52. A. V. Sechkarev and A. K. Petrov, *Opt. Spectrosc., 19,* 503 (1965).
53. E. Herz and H. Wittek, *Monatsh. Chem., 74,* 271 (1943).
54. M. Oki and M. Hirota, *Bull. Chem. Soc. Jap., 34,* 374 (1961).
55. M. Josien, M. Joussot-Dubien, and J. Vizet, *Bull. Soc. Chim. Fr., 1957,* 1148.
56. J. Bolard, *J. Chim. Phys., 62,* 887 (1965).
57. N. B. Colthup, L. H. Daly, and S. E. Wiberley, *Introduction to Infrared and Raman Spectroscopy,* Academic Press, New York, 1964, 258–259.
58. L. J. Bellamy and R. J. Pace, *Spectrochim. Acta, 19,* 435 (1963).
59. L. Bardet, G. Fleury, and V. Tabacik, *Compt. Rend., 270B,* 1277 (1970).
60. L. Kahovec, K. W. F. Kohlrausch, and J. Wagner, *Z. Phys. Chem., 49B,* 145 (1941).
61. V. Ananthanarayanan, *Proc. Indian Acad. Sci., 51,* 328 (1960).
62. J. T. Edsall, *J. Chem. Phys., 5,* 508 (1937).
63. M. Suzuki and T. Shimanouchi, *J. Mol. Spectrosc., 28,* 394 (1968).

64. L. Bardet, J. Maillols, and H. Maillols, *Compt. Rend., 270B,* 158 (1970).
65. M. Suzuki and T. Shimanouchi, *J. Mol. Spectrosc., 29,* 415 (1969).
66. V. Ananthanarayanan, *Spectrochim. Acta, 20,* 197 (1964).
67. M. Harrand and C. Fegly, *J. Chim. Phys., 64,* 991 (1967).
68. J. Bolard, *J. Chim. Phys., 62,* 894 (1965).
69. K. Ito and H. J. Bernstein, *Can. J. Chem., 34,* 170 (1956).
70. E. Spinner, *J. Chem. Soc., 1964,* 4217.
71. J. H. S. Green, W. Kynaston, and A. S. Lindsey, *Spectrochim. Acta, 17,* 486 (1961).
72. C. B. Baddiel, C. D. Cavendish, and W. O. George, *J. Mol. Struct., 5,* 263 (1970).
73. J. K. Wilmshurst, *J. Mol. Spectrosc., 1,* 201 (1957).
74. H. Susi and J. R. Scherer, *Spectrochim. Acta, 25A,* 1243 (1969).
75. E. M. Popov, G. A. Kogan, M. I. Struchkova, and V. N. Zheltova, *J. Struct. Chem. USSR, 12,* 61 (1971).
76. P. Matzke, O. Chacon, and C. Andrade, *J. Mol. Struct., 9,* 255 (1971).
77. K. W. F. Kohlrausch and A. Pongratz, *Z. Phys. Chem., 22B,* 373 (1933).
78. K. W. F. Kohlrausch and A. Pongratz, *Z. Phys. Chem., 27B,* 176 (1934).
79. H. Cheng, *Z. Phys. Chem., 24B,* 293 (1934).
80. K. W. F. Kohlrausch, F. Köppl, and A. Pongratz, *Z. Phys. Chem., 22B,* 359 (1933).
81 M. E. High, *Phys. Rev., 38,* 1837 (1931).
82. K. W. F. Kohlrausch and A. Pongratz, *Chem. Ber., 66,* 1355 (1933).
83. J. K. Wilmshurst and J. F. Horwood, *J. Mol. Spectrosc., 21,* 48 (1966).
84. R. Sabathy, *Z. Phys. Chem. 41B,* 183 (1938).
85. A. J. Bowles, W. O. George, and D. B. Cunliffe-Jones, *J. Chem. Soc. (B), 1970,* 1070.
86. W. R. Feairheller, Jr. and J. E. Katon, *J. Mol. Struct., 1,* 239 (1968).
87. A. Dadieu, A. Pongratz, and K. W. F. Kohlrausch, *Monatsh. Chem., 60,* 250 (1932).
88. R. Vessiere, *Bull. Soc. Chim. Fr., 1960,* 369.
89. K. Matsuno and K. Han, *Bull. Chem. Soc. Jap., 8,* 333 (1933).
90. L. Kahovec and J. Wagner, *Monatsh. Chem., 74,* 279 (1943).
91. R. A. Nyquist, *Appl. Spectrosc., 26,* 81 (1972).
92. O. Ballaus, *Monatsh. Chem., 74,* 85 (1942).
93. P. Mirone, B. Fortunato, and P. Canziani, *J. Mol. Struct., 5,* 283 (1970).
94. K. W. F. Kohlrausch, A. Pongratz, and R. Seka, *Chem. Ber., 66,* 1 (1933).
95. L. Kahovec and K. W. F. Kohlrausch, *Z. Elektrochem., 43,* 285 (1937).
96. J. Overend, R. A. Nyquist, J. C. Evans, and W. J. Potts, *Spectrochim. Acta, 17,* 1205 (1961).

97. J. Overend and J. R. Scherer, *Spectrochim. Acta, 16,* 773 (1960).
98. J. Overend and J. R. Scherer, *J. Chem. Phys., 32,* 1296 (1960).
99. H. Seewann-Albert and L. Kahovec, *Acta Phys. Austriaca, 1,* 352 (1948).
100. H. Seewann-Albert, *Acta Phys. Austriaca, 1,* 359 (1948).
101. H. Cheng, *Z. Phys. Chem., 26B,* 288 (1934).
102. E. Herz, L. Kahovec, and K. W. F. Kohlrausch, *Monatsh. Chem., 74,* 253 (1943).
103. A. H. Nielsen, T. G. Burke, P. J. H. Woltz, and E. A. Jones, *J. Chem. Phys., 20,* 596 (1952).
104. J. L. Hencher and G. W. King, *J. Mol. Spectrosc., 16,* 168 (1965).
105. J. L. Hencher and G. W. King, *J. Mol. Spectrosc., 16,* 158 (1965).
106. K. G. Kidd and G. W. King, *J. Mol. Spectrosc., 28,* 411 (1968).

CHAPTER NINE

AMIDES, AMINO ACIDS, PEPTIDES, AND PROTEINS

9.1 INTRODUCTION

The Raman spectra of formamide and acetamide have been studied by Kohlrausch and Pongratz (1), Kahovec et al. (2,3), and Rao (4). A series of the normal primary amides (formamide to *n*-valeramide) were examined by Reitz and Wagner (5). The Raman spectra of *N*-alkyl- and *N, N*-dialkyl acetamides were reported by Kohlrausch and Seka (6) and by Sannie and Poremski (7).

The work on the characteristic frequencies in the IR spectra of amides has been adequately summarized by Bellamy (8,9). Spectra-structure correlations for 76 amides in the IR range 700–250 cm^{-1} were carried out by Katon, Feairheller, and Pustinger (10). More recently, Hallam et al. (11,12) have investigated the solvent shifts in the vibrational frequencies of a series of primary, secondary, and tertiary amides. A review of the spectroscopic evidence for conformational isomerism of the amide group has been given by Hallam and Jones (13).

Normal coordinate calculations for formamide have been performed by Suzuki (14) and by Puranik and Sirdeshmukh (15). Suzuki (16) has also reported a normal coordinate treatment of acetamide. The following normal coordinate analyses have been done for *N*-alkyl amides: *N*-methylformamide (17), the trans form of *N*-methylacetamide (18,19), and the cis isomer of *N*-methylacetamide (20). The normal vibrations of the tertiary amides, *N, N*-dimethylformamide and *N, N*-dimethylacetamide have been investigated by Chalapathi and Ramiah (21).

9.2 PRIMARY AMIDES

The characteristic frequencies in the laser Raman spectra of primary alkyl amides (see Spectra 49, 50; Appendix 2) are listed in Table 9.1. The nomenclature for the amide bands is that proposed by Miyazawa, Shimanouchi, and Mizushima (22) in their work on monosubstituted amides. For the primary alkyl amines, there are two nitrogen-hydrogen stretches whose positions depend on the physical state of the sample. In concentrated solutions and in the liquid and solid states, the molecules are associated by hydrogen bonding; the antisymmetric N-H stretch occurs near 3350 cm^{-1} and the symmetric N-H stretch occurs near 3180 cm^{-1}. In dilute solutions of nonpolar solvents, the free nonhydrogen-bonded N-H stretches occur near 3500 and 3400 cm^{-1}, respectively.

Two other bands characteristic of primary amides are labelled Amide I and Amide II in Table 9.1. The Amide I band is often referred to as the carbonyl absorption. However Suzuki, in his potential energy distributions (14,16), has shown that in formamide this band arises from a mixed vibration involving 64% carbonyl stretch and 33% C-N stretch and in acetamide this vibration consists of 59% C=O stretch and 28% C-N stretch. In the associated state, this vibration occurs near 1650 cm^{-1}, whereas in the nonbonded state (dilute solution) the frequency shifts to the range 1715–1675 cm^{-1}. The Amide II arises in primary amides mainly from the symmetric NH_2 bending (scissors) vibration. The assignment of Suzuki gives 85% NH_2 bending and 17% C=O stretch for the same band in acetamide. In the associated state, the Amide I band occurs at 1650–1620 cm^{-1} and in the unassociated state at 1620–1585 cm^{-1}. In concentrated solutions, all four bands arising from the Amide I and Amide II vibrations can often be observed because of the presence of free and associated molecules in equilibrium.

Other bands that may be useful in identifying primary amides from their laser Raman spectra include those assigned to the C-N stretch, NH_2 rocking, and the skeletal deformations of the N-C=O and C-C=O moieties. The C-N stretch for formamide is found at 1312 cm^{-1}; however, in acetamide and higher primary alkyl amides it is located at 1430–1390 cm^{-1}. The NH_2 rocking vibration occurs in the range 1150–1100 cm^{-1}. Bands due primarily to the in-plane N-C=O deformation are at 600–550 cm^{-1} and those for the C-C=O deformation are found in the region 500–450 cm^{-1} of the Raman spectrum.

Table 9.1 Characteristic Frequencies in the Raman Spectra of Primary Alkyl Amides

Compound	Physical State	Frequency Vibration (cm^{-1})							
		N-H Antisymmetric Stretch	N-H Symmetric Stretch	Amide I	Amide II	C-N Stretch	NH_2 Rocking	O=C–N Deformation	CCO Deformation
Formamide	Liquid	3330	3190	1673	1600	1312	1096	607	—
Acetamide	Liquid	3355	3165	1660	1615	1389	1120	568	446
	Solid	3350	3152	1672	1589	1402	1148	585	450
Propionamide	Liquid	—	3168	1658	1602	1412	1125	555	453
	Solid	3346	3170	1671	1591	1419	1146	581	476
n-Butyramide	Solid	3325	3167	1673	1582	1430	1149	570	460
Isobutyramide	Solid	3337	3147	1636	1576	1433	1148	—	500
n-Valeramide	Solid	3324	3172	1686	1587	1420	1150	—	470
Relative Raman intensity		Medium–weak	Medium	Strong–medium	Medium	Medium–weak	Medium	Medium	Medium

9.3 SECONDARY AMIDES

The characteristic vibrations in the laser Raman spectra of secondary amides (e.g., Spectra 51,52) are summarized in Table 9.2. LaPlanche and Rogers (23) examined the nuclear magnetic resonance spectra of a series of *N*-monosubstituted amides and found that *N*-methyl-, *N*-ethyl-, and *N*-isopropyl-formamides existed about 10% in the cis configuration in the liquid state at room temperature and that the corresponding acetamides, propionamides, and isobutyramides were entirely in the trans form. The IR data reviewed by Hallam and Jones (13) also show the trans form is always predominant in the *N*-alkyl amides. For this reason, the characteristic frequencies listed in Table 9.2 are those of the trans forms of the amides and were obtained from the Raman spectra of the neat liquid at room temperature.

The nitrogen-hydrogen stretch of the secondary amides is located near 3300 cm^{-1} in the associated state; for the nonbonded state the vibration is at 3491-3421 cm^{-1} for the trans form and at 3440-3404 cm^{-1} for the cis form. The Amide I band for the trans form of *N*-monosubstituted amides is at 1680–1630 cm^{-1} in the solid and liquid states and in the range 1700-1650 cm^{-1} in dilute solution. The cis form also has its Amide I band in this region. For these secondary amides, the Amide I band from the potential energy distribution is attributed mainly to the C=O stretching mode (70–80%); however, the C–N stretching (30–10%) and N–H bending (20–10%) modes also contribute. The Amide II bands are located in the range 1570–1510 cm^{-1} for the trans-associated form and at 1550–1500 cm^{-1} in the free form. No Amide II band in this region is found for the cis configuration. Normal coordinate calculations for the trans form of *N*-methylformamide (17) show that the NH bending mode contributes 60% of the potential energy and the C–N stretch 32% for this vibration. In the trans configuration of *N*-methylacetamide (18), these values are 63% and 44%, respectively. In the cis CONH group there is no coupling of the C–N stretching and N–H bending modes. For this configuration of *N*-methylacetamide (20) the potential energy of the 1450 cm^{-1} vibration is associated predominantly with N–H bending (78%) and the band at 1350 cm^{-1} with the C–N stretching mode (71%). In the trans form, the interaction of the two vibrations raises the Amide II band to frequencies greater than the CH_3 and CH_2 deformation frequencies.

This interaction of the N–H bending and C–N stretching also produces another characteristic frequency called the Amide III band which is lower in frequency than the pure C–N stretch as found in the cis configuration. In the liquid and solid (associated) states, this band is found at 1310–1250 cm^{-1} and in the free state it is located at 1250–1200 cm^{-1}. Potential energy distributions given for this vibration in the trans form of *N*-methylformamide are C–N stretch, 33%, and N–H bending, 36%. For trans *N*-methylacetamide, these values are 35 and 29%, respectively.

Table 9.2 Characteristic Frequencies in the Raman Spectra of Secondary Alkyl Amides[a]

	Vibration Frequency (cm^{-1})				
Compound	N-H Stretch	Amide I	Amide II	Amide III	Amide IV
N-Methylformamide	3300	1658	1550	1250	771
N-Ethylformamide	3300	1659	1542	1248	775
N-Methylacetamide	3308	1655	1550	1300	629
N-Ethylacetamide	3290	1652	1560	1295	616
N-Propylacetamide	3299	1655	—	1296	618
N-Butylacetamide	3300	1652	1565	1295	639
N-Methylpropionamide	3300	1648	—	1290	600
Relative Raman intensity	Strong-medium	Strong	Weak	Very strong-strong	Strong-medium

[a]These frequencies are for the trans configuration in the liquid state.

Another characteristic band in Raman spectra of monosubstituted amides is observed near 630 cm^{-1} in amides of the type R-CH_2-CO-NHR and near 780 cm^{-1} for amides of the type H-CO-NHR. This is called the Amide IV band and arises principally from the O=C-N bending mode. The Amide V (N-H out-of-plane bending) and the Amide VI (C=O out-of-plane bending) generally appear as very weak lines in the Raman and do not lead to useful correlations. The former vibration appears in the IR in the range 750-700 cm^{-1} and the latter vibration near 600 cm^{-1} for amides of the form R-CH_2-CO-NHR.

9.4 TERTIARY AMIDES

The spectra-structure correlations for tertiary alkyl amides (e.g., Spectra 53, 54) are contained in Table 9.3. The Amide I band for *N, N*-dialkylsubstituted amides occurs in the region 1670-1630 cm^{-1}. Unlike the similar carbonyl stretch vibration in primary and secondary amides, the Amide I band in tertiary aliphatic amides is not very sensitive to changes in phase or concentration of solution. The symmetric C-N stretch of the C-N-C moiety occurs at a frequency of 870-820 cm^{-1} in tertiary amides of type H-CO-NR_2 and in the range 750-700 cm^{-1} for amides of type R′-CO-NR_2. The only other characteristic frequency in the Raman spectra of tertiary amides was found near 650 cm^{-1} for *N, N*-substituted formamides and at 620-590 cm^{-1} in the higher tertiary amides. Those bands arise from the O=C-N in-plane deformation.

Table 9.3 Characteristic Frequencies in the Raman Spectra of Tertiary Alkyl Amides

	Vibration Frequency (cm^{-1})			
Compound	Amide I	C–N Symmetric Stretch	O=C–N Deformation	C–H Stretch of O=C–H
N,N-Dimethylformamide	1665	868	662	2860
N,N-Diethylformamide	1665	825	645	2878
N,N-Dibutylformamide	1667	822	642	2880
N,N-Dimethylacetamide	1640	748	590	–
N,N-Diethylacetamide	1638	710	589	–
N,N-Dipropylacetamide	1640	692	598	–
N,N-Dibutylacetamide	1645	692	605	–
N,N-Diisopropylacetamide	1637	678	601	–
N,N-Dimethylpropionamide	1640	760	598	–
N,N-Diethylpropionamide	1642	738	605	–
N,N-Di-*n*-propylpropionamide	1642	740	604	–
N,N-Di-*n*-butylpropionamide	1641	768	600	–
N,N-Dimethylbutyramide	1640	728	615	–
N,N-Diethyl-*n*-butyramide	1638	713	612	–
N,N-Diethyldodecanamide	1644	705	610	–
Relative Raman Intensity	Strong	Strong–medium	Medium–weak	Strong

A summary of the characteristic Raman frequencies for aliphatic amides is contained in Table 9.4.

9.5 UREA

The Raman spectrum of urea, $H_2N–CO–NH_2$, has been measured in the crystalline state (5) and in aqueous solution (24). Assignments in the vibrational spectra of urea have been discussed by Stewart (25). Normal coordinate analyses for the planar (26,27) and nonplanar vibrations (27) have been performed. The most intense Raman band at 1008 cm^{-1} in aqueous solution is assigned to the a_1 CN stretching vibration. Bands at 1680 and 1604 cm^{-1} were assigned to vibrations involving both the C=O and the NH_2 deformation. Three bands in the region 3496–3235 cm^{-1} are attributed to various N–H stretchings. In the crystalline state the ν(NH) occurs at 3426–3232 cm^{-1}.

Table 9.4 Summary of Characteristic Frequencies in the Raman Spectra of Alkyl Amides

Type	Vibration		Frequency (cm^{-1}) Associated Form	Nonbonded Form
I. Primary	NH_2 antisymmetric stretch		ca. 3350	ca. 3500
	NH_2 symmetric stretch		ca. 3180	ca. 3400
	Amide I		ca. 1650	1715–1675
	Amide II		1650–1620	1620–1585
	C-N stretch		1430–1390	1430–1390
	NH_2 rocking		1150–1100	1150–1100
	In-plane N-C=O deformation		600–550	600–550
	In-plane C-C=O deformation		500–450	500–450
II. Secondary	N-H stretch	trans	ca. 3300	3491-3421
		cis	ca. 3200	3440–3404
	Amide I	trans	1680–1630	1700–1650
		cis	1680–1630	1700–1650
	Amide II	trans	1570–1510	1550–1500
		cis	1490–1440	ca. 1450
	Amide III	trans	1310–1250	1250–1200
		cis	1350–1310	ca. 1350
	Amide IV H-CO-NHR		ca. 780	ca. 780
	$R'-CH_2-CO-NHR$		ca. 630	ca. 630
III. Tertiary	C-H stretch for $H-CO-NR_2$		2880–2860	2880–2860
	Amide I		1670–1630	1670–1630
	C-N-C symmetric stretch			
	$H-CO-NR_2$		870–820	870–820
	$R'-CO-NR_2$		750–700	750–700
	O=C-N in-plane deformation			
	$H-CO-NR_2$		ca. 650	ca. 650
	$R'-CO-NR_2$		620–590	620–590

9.6 AMINO ACIDS, PEPTIDES, AND PROTEINS

Edsall et al. (28–40) first used Raman spectroscopy in a classic extensive investigation of the structure of amino acids and related compounds in aqueous solution. Each amino acid can exist in three forms–as the anion ($H_2NCHRCOO^-$), the dipolar ion ($^+H_3NCHRCOO^-$), and the cation

Table 9.5 Raman Spectra of Glycine in Aqueous Solution[a]

Assignment	Anion Form $H_2N\text{-}CH_2\text{-}COO^-$ Frequency (cm^{-1})	Anion Form Relative Intensity	Dipolar Form $^+H_3N\text{-}CH_2\text{-}COO^-$ Frequency (cm^{-1})	Dipolar Form Relative Intensity	Cation Form $^+H_3N\text{-}CH_2\text{-}COOH$ Frequency (cm^{-1})	Cation Form Relative Intensity
CO_2^- rock; NH_3^+ torsion	507	8	502	8	494	5
CO_2^- wag	575	2	577	2	—	—
CO_2^- scissor	—	—	665	1	644	2
C-C stretch	899	8	896	11	869	15
—	968	2	—	—	—	—
C-N stretch	—	—	1027	4	1042	4
NH_2 twist; NH_3^+ rock	1100	4	1118	4	1123	3
CH_2 twist	1169	1	—	—	—	—
—	1315	5	—	—	—	—
CH_2 wag	1343	10	1327	14	1308	3
CO_2^- symmetric stretch	1407	28	1410	20	—	—
CH_2 scissor	1440	—	1440	8	1433	7
CO_2^- antisymmetric stretch[b]	1611	21	1615	19	1644	34
C=O stretch	—	—	—	—	1744	19
CH_2 stretch	—	—	2879	1	—	—
CH_2 stretch	2935	30	2968	19	2975	35
CH_2 stretch	—	—	3011	5	3011	13
NH_2 stretch	3320	8	—	—	—	—
NH_2 stretch	3385	4	—	—	—	—

[a]Frequencies reported in ref. 40, vibrational assignments in refs. 41 and 42.
[b]The H_2O deformation and NH_3^+ degenerate deformation also lie in this region

($^+H_3NCHRCOOH$). The vibrational assignments in the Raman spectra for the three forms of glycine in H_2O are given in Table 9.5. Suzuki et al. (41) have carried out a normal coordinate treatment of glycine and its deuterated analogues.

In general, the N-H stretching vibrations of the uncharged amino group of the amino acid anions give rise to two Raman bands—one of moderate intensity near 3370 cm^{-1} (antisymmetric NH_2 stretch) and another of high intensity near 3305 cm^{-1} (symmetric NH_2 stretch). Upon deuterium substitution these frequencies are displaced to 2500 cm^{-1} (antisymmetric ND_2 stretch) and 2435 cm^{-1} (symmetric ND_2 stretch). The N-H stretching frequency for the $-NH_3^+$ group (near 2970 cm^{-1}) is obscured in the Raman by the more intense C-H stretching vibrations. The N-D stretching frequency of $-ND_3^+$ occurs at 2200–2175 cm^{-1} (weak, broad band). The characteristic C=O stretching frequency for the unionized carboxyl group in the cation form of α-amino acids occurs in the range 1743–1729 cm^{-1} (H_2O solution). The ionized carboxyl group gives rise to an antisymmetric CO_2^- stretch at 1600–1570 cm^{-1} and a symmetric CO_2^- stretch at 1415–1400 cm^{-1}. In the Raman spectra of aqueous solutions the antisymmetric stretch is often masked by an H_2O deformation band near 1630 cm^{-1} but in D_2O solution it occurs as a strong band at 1566–1560 cm^{-1} for anionic amino acids and at 1623–1570 cm^{-1} for dipolar ions. The symmetric CO_2^- stretch occurs as an intense Raman band and is affected very slightly by ionization or deuteration of the amino group.

Amino acids with aromatic side groups give rise to strong Raman bands due to ring vibrations that are insensitive to changes in environment and state of aggregation. In tryptophan (I) bands due to the indole ring occur at 1582, 1553, 1363, 1338, 1014, 879, 761, 577, and 544 cm^{-1} (43). In phenylalanine (II) the intense Raman band characteristic of monosubstituted benzenes occurs at 1006 cm^{-1}. In tyrosine two bands characteristic of the *p*-hydroxyphenyl ring occur at 858 and 836 cm^{-1}.

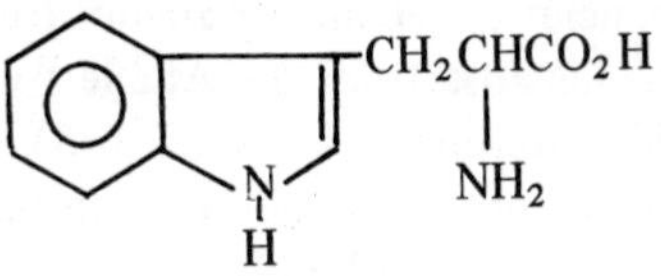

CH_2CHCO_2H (indole ring, N–H) with NH_2 — I

CH_2CHCO_2H (benzene ring) with NH_2 — II

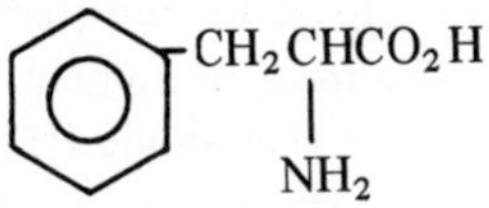

Sulfur-containing amino acids exhibit characteristic Raman bands due to S-H, C-S, and S–S linkages which are quite prominent in the spectra. Garfinkel and Edsall (38) have followed the ionization of the -SH group in aqueous solutions of cysteine, $HSCH_2CH(NH_2)CO_2H$, as a function of pH by observing the intensity of the S-H stretching frequency near 2575 cm^{-1} and the S-H deformation near 875 cm^{-1}. In acidic solutions (cation form) and at pH 5.7-7.4 (isoelectric form) these bands are strong, but at pH 9.36 the intensity drops to about one half its former value and at pH 11.93 the bands were undetectable. The strong Raman band due to the C–S stretch at 690–680 cm^{-1} was almost unaffected by the change in pH. In the Raman spectrum of the cationic form of cystine $[HO_2CCH(NH_3^+)CH_2S-]_2$, strong bands found at 504 and 667 cm^{-1} are attributed to the S–S stretch and C–S stretch, respectively (34). For the methionine cation, $CH_3S(CH_2)_2CH(NH_3^+)CO_2H$, in aqueous solution three intense Raman bands due to the C–S stretch of several conformations occur at 724, 701, and 655 cm^{-1} (43) analogous to those found in 2-thiabutane.

Amino acids that have been examined by Raman spectroscopy in single crystal form include α-glycine (44), γ-glycine (44,45), and β-alanine (46), while L-proline and L-hydroxyproline (47) were examined as powdered solids. All these amino acid crystals and solids were found to exist in the dipolar ion form.

Garfinkel and Edsall (37) first reported the Raman spectra of glycylglycine and four other dipeptides in aqueous solution. More recently, Raman spectra have been reported for di-, tri-, tetra-, and pentaglycine (48), for di-, tri-, and hexa-L-alanine (49) and for tri-, tetra-, and penta-L-proline (50) in both aqueous solution and in the solid state. Raman investigations of synthetic polypeptides since the introduction of laser sources include: polyglycine (48); poly-L-alanine (51,52); poly-L-proline (50); poly-L-lysine, poly-L-lysine hydrochloride, and poly-ϵ-carbobenzoxy-L-lysine (53); poly-γ-benzyl-L-glutamate, poly-L-leucine, poly-L-valine, and poly-L-serine (54); poly-L-glutamic acid and poly-L-ornithine (55); block copolymers of L-alanine and D,L-lysine (56), and poly(hydroxybutylglutamine-co-glycine) (57). The vibrational assignments in the spectra of these oligo- and polypeptides, as well as the application of Raman spectroscopy in the determination of polypeptide conformation, are discussed in detail in a review by Koenig (58). The peptide CONH group exhibits the characteristic Raman frequencies of secondary amides (see Tables 9.2 and 9.4). The most prominent bands arise from the Amide I vibration near 1660 cm^{-1} and the Amide III vibration near 1260 cm^{-1} (43). These bands also occur in the Raman spectra of proteins together with bands characteristic of the substituent amino acids. Lord and Yu (43) found a close correspondence between the Raman spectrum of completely hydrolyzed lysozyme in acidic solution and a spectrum formed by the superposition of the constituent amino acids. Other proteins examined by Raman spectroscopy include ribonuclease (59), pepsin (60), α-chymotrypsin (59,60), hemoglobin (61,62), and cytochrome *c* (62,63). For the latter two heme proteins good quality resonance Raman spectra could be obtained at protein concentrations ranging from 10^{-3} to 10^{-5} *M* and distinctive changes in the Raman spectra could be observed as a result of changes in oxidation state or in the state of ligation.

REFERENCES

1. K. W. F. Kohlrausch and A. Pongratz, *Z. Phys. Chem., 27B,* 176 (1934).
2. L. Kahovec and H. Wassmuth, *Z. Phys. Chem., 48B,* 1 (1940).

3. L. Kahovec and K. Knollmueller, *Z. Phys. Chem., 51B,* 49 (1941).
4. A. L. S. Rao, *J. Indian Chem. Soc., 18,* 337 (1941).
5. A. W. Reitz and J. Wagner, *Z. Phys. Chem., 43B,* 339 (1939).
6. K. W. F. Kohlrausch and R. Seka, *Z. Phys. Chem., 43B,* 355 (1939).
7. C. Sannie and V. Poremski, *Bull. Soc. Chim. Fr., 6* (5) 1629 (1939).
8. L. J. Bellamy, *Infrared Spectra of Complex Molecules,* Methuen, London 1954.
9. L. J. Bellamy, *Advances in Infrared Group Frequencies,* Methuen, London 1968.
10. J. E. Katon, W. R. Feairheller, Jr., and J. V. Pustinger, Jr., *Anal. Chem., 36,* 2126 (1964).
11. E. A. Cutmore and H. E. Hallam, *Spectrochim. Acta, 25A,* 1767 (1969).
12. H. E. Hallam, *Spectrochim. Acta, 25A,* 1785 (1969).
13. H. E. Hallam and C. M. Jones, *J. Mol. Struct., 5,* 1 (1970).
14. I. Suzuki, *Bull. Chem. Soc. Jap., 33,* 1359 (1960).
15. P. G. Puranik and L. Sirdeshmukh, *Proc. Indian Acad. Sci., 56,* 115 (1962).
16. I. Suzuki, *Bull. Chem. Soc. Jap., 35,* 1279 (1962).
17. I. Suzuki, *Bull. Chem. Soc. Jap., 35,* 540 (1962).
18. T. Miyazawa, T. Shimanouchi, and S. Mizushima, *J. Chem. Phys. 29,* 611 (1958).
19. V. V. Chalapathi and K. V. Ramiah, *Proc. Indian Acad. Sci., 64,* 148 (1966).
20. T. Miyazawa, *J. Mol. Spectrosc., 4,* 155 (1960).
21. V. V. Chalapathi and K. V. Ramiah, *Proc. Indian Acad. Sci., 68,* 109 (1968).
22. T. Miyazawa, T. Shimanouchi, and S. Mizushima, *J. Chem. Phys., 24,* 408 (1956).
23. L. A. LaPlanche and M. T. Rogers, *J. Am. Chem. Soc., 86,* 337 (1964).
24. J. W. Otvos and J. T. Edsall, *J. Chem. Phys., 7,* 632 (1939).
25. J. E. Stewart, *J. Chem. Phys., 26,* 248 (1957).
26. J. L. Duncan, *Spectrochim. Acta, 27A,* 1197 (1971).
27. Y. Saito, K. Machida, and T. Uno, *Spectrochim. Acta, 27A,* 991 (1971).
28. J. T. Edsall, *J. Chem. Phys., 4,* 1 (1936).
29. J. T. Edsall, *J. Phys. Chem., 41,* 133 (1937).
30. J. T. Edsall, *J. Chem. Phys., 5,* 225 (1937).
31. J. T. Edsall, *J. Chem. Phys., 5,* 508 (1937).
32. J. T. Edsall and H. Scheinberg, *J. Chem. Phys., 8,* 520 (1940).

33. J. T. Edsall, *J. Am. Chem. Soc., 65,* 1767 (1943).
34. J. T. Edsall, J. W. Otvos, and A. Rich, *J. Am. Chem. Soc., 72,* 474 (1950).
35. D. Garfinkel and J. T. Edsall, *J. Am. Chem. Soc., 80,* 3807 (1958).
36. M. Takeda, R. E. S. Iavazzo, D. Garfinkel, I. H. Scheinberg, and J. T. Edsall, *J. Am. Chem. Soc., 80,* 3813 (1958).
37. D. Garfinkel and J. T. Edsall, *J. Am. Chem. Soc., 80,* 3818 (1958).
38. D. Garfinkel and J. T. Edsall, *J. Am. Chem. Soc., 80,* 3823 (1958).
39. D. Garfinkel, *J. Am. Chem. Soc., 80,* 3827 (1958).
40. S. A. S. Ghazanfar, D. V. Myers, and J. T. Edsall, *J. Am. Chem. Soc., 86,* 3439 (1964).
41. S. Suzuki, T. Shimanouchi, and M. Tsuboi, *Spectrochim. Acta, 19,* 1195 (1963).
42. K. Krishnan and R. A. Plane, *Inorg. Chem., 6,* 55 (1967).
43. R. C. Lord and N-T. Yu, *J. Mol. Biol., 50,* 509 (1970).
44. R. S. Krishnan and P. S. Narayanan, in *Crystallography and Crystal Perfection,* G. N. Ramachandran, Ed., Academic Press, New York, 1963.
45. K. Balasubramanian, R. S. Krishnan, and Y. Iitaka, *Bull. Chem. Soc. Jap., 35,* 1303 (1962).
46. R. S. Krishnan and R. S. Katiyar, *Bull. Chem. Soc. Jap., 42,* 2098 (1969).
47. M. J. Deveney, A. G. Walton, and J. L. Koenig, *Biopolymers, 10,* 615 (1971).
48. M. Smith, A. G. Walton, and J. L. Koenig, *Biopolymers, 8,* 29 (1969).
49. P. L. Sutton and J. L. Koenig, *Biopolymers, 9,* 615 (1970).
50. W. B. Rippon, J. L. Koenig, and A. G. Walton, *J. Am. Chem. Soc., 92,* 7455 (1970).
51. J. L. Koenig and P. L. Sutton, *Biopolymers, 8,* 167 (1969).
52. E. W. Small, B. Fanconi, and W. L. Peticolas, *J. Chem. Phys., 52,* 4369 (1970).
53. J. L. Koenig and P. Sutton, *Biopolymers, 9,* 1229 (1970).
54. J. L. Koenig and P. Sutton, *Biopolymers, 10,* 89 (1971).
55. J. L. Koenig and B. Frushour, *Biopolymers, 11,* 1871 (1972).
56. A. Lewis and H. A. Scheraga, *Macromolecules, 4,* 539 (1971).
57. A. Lewis and H. A. Scheraga, *Macromolecules, 5,* 450 (1972).
58. J. L. Koenig, *Raman Spectroscopy of Biological Molecules, J. Poly. Sci.,* Part D, *6,* 59 (1972).
59. R. C. Lord and N-T. Yu, *J. Mol. Biol., 51,* 203 (1970).
60. M. C. Tobin, *Science, 161,* 68 (1968).

61. T. G. Spiro and T. C. Strekas, *Biochim. Biophys. Acta, 263,* 830 (1972).
62. T. G. Spiro and T. C. Strekas, *Proc. Nat. Acad. Sci. USA, 69,* 2622 (1972).
63. T. G. Spiro and T. C. Strekas, *Biochim. Biophys. Acta, 278,* 188 (1972).

CHAPTER TEN

C=N AND N=N COMPOUNDS

10.1 IMINES AND IMIDO-ETHERS

The C=N stretching frequencies in both the IR and Raman spectra of compounds containing the C=N group have been reviewed by Fabian, Legrand, and Poirier (1). The Raman data for ν(C=N) in aldimines (azomethines) and imido-ethers are collected in Table 10.1. Aldimines containing only alkyl substituents generally possess a Raman band near 1670 cm^{-1} for the C=N stretching frequency (2,3). The presence of an aryl group on the imino carbon or the nitrogen atom lowers ν(C=N) by 10–20 cm^{-1} (4). The presence of a methyl group attached to the nitrogen atom is characterized by a strong Raman band at 1405-1401 cm^{-1} arising from the symmetric methyl deformation (2). The vibrational assignments of diphenylketimine, $(C_6H_5)_2C$=NH, and its deuterated derivative, $(C_6H_5)_2C$=ND, in the IR and Raman have been discussed by Perrier-Datin and Lebas (5). In the Raman spectrum of the liquid a band at 3265 cm^{-1} is assigned to the N-H stretch and a band at 1598 cm^{-1} to ν(C=N). Cantarel (6) found that ketimines containing the $(C_6H_5)_2C$=N- group were characterized by Raman bands near 1620 [ν(C=N)], 1490, and 1274 cm^{-1}. In the Raman spectra of imido-ethers, the ν(N-H) occurs as a band in the range 3360–3327 cm^{-1} (7).

10.2 AMIDINES

Shigorin and Syrkin (8,9) have studied the Raman spectra of *N*-monosubstituted and *N*-disubstituted amidines. *N*-Monosubstituted midines, such as I

(R=H, alkyl), exhibit two Raman bands in the range 1658–1632 cm^{-1} ($CHCl_3$ solution) that are attributed to the C=N stretching vibrations of the syn and anti geometric isomers. *N*-Disubstituted amidines, such as II, exhibit only one Raman band in the range 1633–1619 cm^{-1} (in $CHCl_3$ solution) that is attributed to ν(C=N) of the syn form. The effects of intermolecular association upon ν(C=N) were studied in the case of *N,N'*-diethyl acetamidine, (III). In dioxane solution, only one Raman band was found at 1675 cm^{-1} whereas in hexane solution a band was observed at 1592 cm^{-1}. The former Raman band is assigned to the monomer and the latter to a cyclic dimer of the amidine. In the Raman spectrum of the liquid, these two bands are observed but a stronger band is located at 1635 cm^{-1} which is assigned to an acyclic dimer.

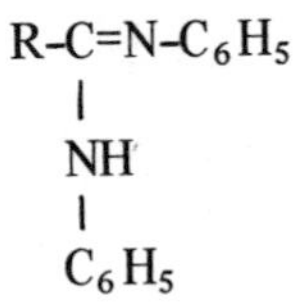

I	II	III
$R{-}C{=}N{-}C_6H_5$	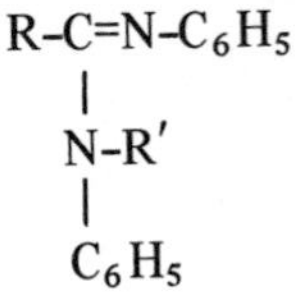 $R{-}C{=}N{-}C_6H_5$	$CH_3{\diagdown}C{=}N{-}C_2H_5$
|	|	|
NH	N-R′	NH
|	|	|
C_6H_5	C_6H_5	C_2H_5

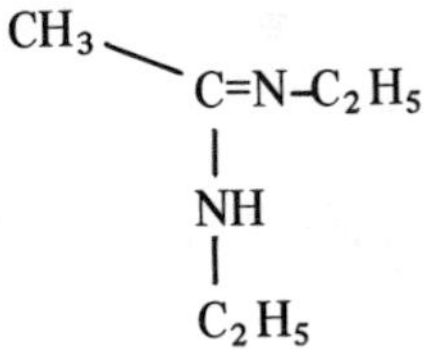

10.3 OXIMES AND GLYOXIMES

The frequencies of the C=N stretching mode in aldoximes (e.g., Spectrum 55) and ketoximes (e.g., Spectrum 56) are given in Table 10.2. For aliphatic aldoximes these stretching frequencies are 10–25 cm^{-1} lower than those of the corresponding aldimines, $R{-}CH{=}N{-}CH_3$. This decrease has been attributed (10) to the presence of intermolecular association, of the type shown in IV, in the aldoxime. These compounds also exhibit a strong Raman band in the range 1335–1330 cm^{-1} that has been attributed to the O-H in-plane deformation (11). In benzaldoxime, the difference in ν(C=N) for the two crystalline polymorphs is ascribed (10) to different types of intermolecular association. In the α-form, the association is V, and in the β-form it is VI. In aromatic aldoximes, such as salicylaldoximes, where intramolecular association can occur, ν(C=N) decreases to 1617 cm^{-1}.

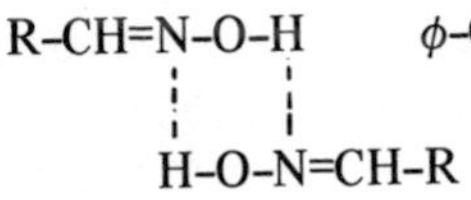

IV	V	VI
R–CH=N–O–H	ϕ–CH=N–OH - - - N=CH–ϕ	ϕ–CH=N–OH - - - O–N=CH–ϕ
⋮ ⋮	|	|
H–O–N=CH–R	OH	H

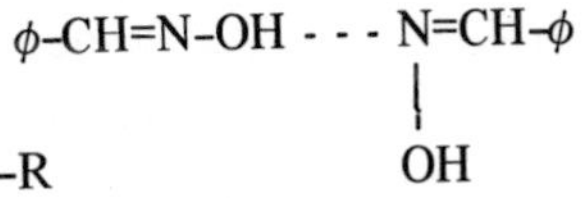

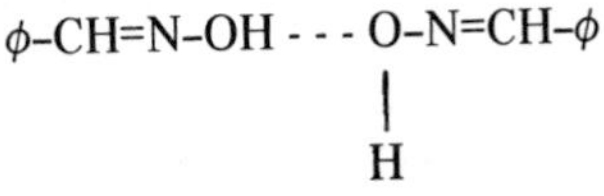

Table 10.1 ν (C=N) in the Raman Spectra of Imines and Related Compounds

Type	Structure	ν(C=N) (cm^{-1})	Reference
1. Aldimines			
C-Alkyl *N*-alkyl	H(R)C=N–R′	1673–1666	2,3
C-Alkyl *N*-cyclohexyl	H(R)C=N–C_6H_{11}	1667–1665	4
C-Phenyl *N*-methyl	H(C_6H_5)C=N–CH_3	1654	4
C-Phenyl *N*-phenyl	H(C_6H_5)C=N–C_6H_5	1639	4
2. Imido-ethers			
C-Dialkoxy	(H_5C_2O)(H_5C_2O)C=NH	1658	7
C-Alkyl *C*-alkoxy	RO(R′)C=NH	1655–1652	7
C-Phenyl *C*-alkoxy	RO(C_6H_5)C=NH	1653–1648	7

Detailed vibrational assignments in the IR and Raman spectra of acetone oxime, $(CH_3)_2C{=}NOH$, and its deuterated analogue, $(CH_3)_2C{=}NOD$, have been made by Harris and Bush (15). In the Raman spectrum of the solid, ν(C=N) occurs at 1675 cm^{-1} and in the liquid at 1669 cm^{-1}. A weak broad Raman band at 3118 cm^{-1} in the solid was assigned to the O-H stretching vibration. This band was absent in the liquid Raman spectrum. The ν(C=N) of other aliphatic ketoximes occurs in the range 1670–1650 cm^{-1}. Conjugation of the C=N bond with a phenyl group decreases the ν(C=N) about 20 cm^{-1}.

The Raman spectra of glyoxime (glyoxal dioxime), $(HO{-}N{=}CH)_2$, have been studied in the solid state and in aqueous solution by Bardet and Boudet (16). The symmetric stretching of the coupled C=N groups occurs as a strong band

Table 10.2 ν(C=N) in the Raman Spectra of Oximes

Type	Structure	ν (C=N) (cm^{-1})	Reference
1. Aldoximes			
Aliphatic	(H)(R)C=N–OH	1660–1649	10,12,13
Chloroaldoxime	(H)(CCl_3)C=N–OH	1619	10
Benzaldoxime	(H)(C_6H_5)C=N–OH	1635 (α-crystalline) 1645 (β-crystalline) 1629 (molten)	10,14
Salicylaldoxime	o-HO–C_6H_4–C(H)=N–OH	1617	10,13
2. Ketoximes			
Dialkyl (R=CH_3, C_2H_5, etc.)	(CH_3)(R)C=N–OH	1666–1655	10,13,14
Dialkyl (R,R′ ≠CH_3)	(R)(R′)C=N–OH	1658–1652	12
Acetophenone oxime	(CH_3)(C_6H_5)C=N–OH	1623	14

at 1627 cm^{-1} in the solid and at 1636 cm^{-1} in aqueous solution. The band corresponding to the antisymmetric stretching of the C=N groups is attributed to a weak band at 1610 cm^{-1} in the IR spectrum of the solid. In the Raman spectrum of methylglyoxime, HO–N=C(CH_3)–CH=N–OH, Kahovec and Kohlrausch (10) assigned strong bands at 1630 and 1516 cm^{-1} to the C=N stretching frequencies. In dimethyl glyoxime, these bands were shifted to 1645 and 1506 cm^{-1}

10.4 HYDRAZONES AND SEMICARBAZONES

The C=N stretching frequencies in the Raman spectra of a series of hydrazones are listed in Table 10.3. An extensive review of the spectra (IR, Raman, and UV) and structure of compounds containing the hydrazone group has been presented by Kitaev, Buzykin, and Troepol'skaya (17). In the IR spectra of hydrazones ν(C=N) occurs as a medium to weak band in the region 1680-1570 cm^{-1} and in aryl hydrazones this absorption is often masked by bands arising from vibrations of the phenyl rings in the same region. However, in the Raman spectra ν(C=N) occurs as a very strong to strong band (1660-1610 cm^{-1}) and in the aryl compounds the band due to ν(C=N) can readily be distinguished from the bands arising from the aryl ring system.

A discussion of the structure and Raman spectra of semicarbazones and thiosemicarbazones has been provided by Raevskii et al. (21,22). The Raman

Table 10.3 Frequency of ν(C=N) in the Raman Spectra of Hydrazones (Solid State)

Type	Structure	ν(C=N) (cm^{-1})	Reference
1. Hydrazones of			
a. Acetone	$(CH_3)(CH_3)C=N-NH_2$	1651	18,19
b. Methyl ethyl ketone	$(CH_3)(C_2H_5)C=N-NH_2$	1649	18
2. Acetaldehyde dimethylhydrazone	$(H)(CH_3)C=N-N(CH_3)(CH_3)$	1612, 1631	20
3. Phenylhydrazones of			
a. Aliphatic aldehydes	$(H)(R)C=N-N(H)(C_6H_5)$	1625-1618	19
b. Aliphatic ketones	$(H_3C)(R)C=N-N(H)(C_6H_5)$	1638-1629	19
4. Methylphenylhydrazones of aliphatic ketones	$(CH_3)(R)C=N-N(CH_3)(C_6H_5)$	1644-1635	19
5. Methylethylketone-benzylhydrazone	$(CH_3)(C_2H_5)C=N-N(H)(CH_2-C_6H_5)$	1637	19

data on the position of ν(C=N) are summarized in Table 10.4. The C=N stretching mode in the Raman spectra of crystalline aliphatic semicarbazones appears as a very strong band located in the narrow range 1665–1650 cm^{-1}. For aliphatic thiosemicarbazones, ν(C=N) occurs in the range 1652–1642 cm^{-1}. For aromatic derivatives of semi- and thiosemicarbazones, ν(C=N) decreases in frequency to 1620–1610 cm^{-1}.

10.5 ACYCLIC AZINES

The Raman spectrum of methanal azine (acetaldazine), $CH_3CH=N-N=CHCH_3$, has been the subject of several investigations (18,23–25). A study by Ogilvie

Table 10.4 ν(C=N) in the Raman Spectra of Semicarbazones and Thiosemicarbazones[a]

Parent Compound $R_1R_2C=O$		Semicarbazone $R_1R_2C=N-NH-C(=O)-NH_2$	Thiosemicarbazone $R_1R_2C=N-NH-C(=S)-NH_2$
R_1	R_2	ν(C=N) (cm^{-1})	ν(C=N) (cm^{-1})
H	CH_3	1645	—
H	C_6H_5	1612	1612
CH_3	CH_3	1650	1643
CH_3	*n*-Alkyl	1664–1660	1651–1642
CH_3	Cyclopropyl	1633	1626
CH_3	C_6H_5	1612	1621
CH_3	*p*-ClC_6H_4	1624	1616
CH_3	$H_5C_2O-C(=O)-$	1628	1627
CH_3	$H_5C_2O-C(=O)-CH_2-$	1654	1645
CH_3	$H_5C_2O-C(=O)-CH_2CH_2-$	1672	1660
Cyclopentanone		1650	1664
Cyclohexanone		1643	1644

[a]Values of ν(C=N) for solid state taken from refs. 21 and 22.

and Cole (25) of the IR and Raman spectra of methanal azine and several of its deuterated analogues in the vapor and solid phases has confirmed that the molecule exists in the *s*-transoid configuration. In the Raman spectrum of the vapor (300°K), the symmetric stretching of the C=N groups occurs as a band at 1615 cm^{-1} and the antisymmetric C=N stretching vibration occurs in the IR spectrum at 1637 cm^{-1}. In a series of aldazines and ketazines, Kitaev et al. (24) found that the antisymmetric C=N stretching vibrations occurred as intense bands in the IR spectra at 1663-1636 cm^{-1} but were completely inactive in the Raman. On the other hand, the symmetric C=N stretching vibrations that led to intense Raman bands at 1625-1608 cm^{-1} were IR-inactive. The vibrational spectra of the fluorinated azine $F_2C{=}N{-}N{=}CF_2$ were measured and assigned by King et al. (26). The structure was found to be trans-planar and ν_{sym}(C=N) appeared as a very strong polarized band in the Raman spectrum of the liquid (-40°C) at 1758 cm^{-1}, whereas $\nu_{antisym}$(C=N) was assigned to an intense band in the IR spectrum of the gas (25°C) at 1747 cm^{-1}

10.6 AZO COMPOUNDS

Electron diffraction studies of azomethane, $CH_3N{=}NCH_3$ (27) have shown that the molecule possesses the trans (C_{2h}) structure that has a center of symmetry. Therefore, the N=N stretching vibration is inactive in the IR and active in the Raman. This vibration is assigned to a band at 1576 cm^{-1} in the Raman spectrum of the liquid (-60°C) (23) and to a band at 1573 cm^{-1} in the spectrum of a CCl_4 solution (28). The Raman spectra of several esters of azodicarboxylic acid, ROOC-N=N-COOR, have been reported (29-31). A band near 1555 cm^{-1} was assigned to the N=N stretching vibration in these compounds. No such band was observed in the IR spectra (31).

The vibrational assignments in the spectra of azobenzene, $C_6H_5{-}N{=}N{-}C_6H_5$, have been reviewed by Kübler, Lüttke, and Weckherlin (32). In the Raman spectrum of azobenzene, a very strong band occurs at 1442 cm^{-1} in both the solid (33) and in CCl_4 and methanol solutions (34) which has been assigned to the N=N stretching vibration of the trans form of azobenzene. Hacker (33-35) has examined the resonance Raman spectra of fourteen derivatives of azobenzene and azonaphthalene and assigned the N=N stretching vibration to a strong polarized band in the range 1440-1380 cm^{-1}

REFERENCES

1. J. Fabian, M. Legrand, and P. Poirier, *Bull. Soc. Chim. Fr., 1956,* 1499.
2. L. Kahovec, *Z. Phys. Chem., 43B,* 364 (1939).

3. L. Kahovec, *Acta Phys. Austriaca, 1,* 307 (1948).
4. A. Kirrman and P. Laurent, *Bull. Soc. Chim. Fr., 6, (5),* 1657 (1939).
5. A. Perrier-Datin and J-M. Lebas, *Spectrochim. Acta, 25A,* 169 (1969).
6. R. Cantarel, *Compt. Rend., 210,* 480 (1940).
7. K. W. F. Kohlrausch and R. Seka, *Z. Phys. Chem., 38B,* 72 (1937).
8. D. N. Shigorin and Ya. K. Syrkin, *Izv. Akad. Nauk SSSR, Ser. Fiz.,9,* 225 (1945); *Chem. Abstr., 40,* 1831 (1946).
9. D. N. Shigorin and Ya. K. Syrkin, *Zh. Fiz. Khim., 23,* 241 (1949); *Chem. Abstr., 43,* 6081h (1949).
10. L. Kahovec and K. W. F. Kohlrausch, *Monatsh. Chem., 83,* 614 (1952).
11. D. Hadzi and L. Premru, *Spectrochim. Acta, 23A,* 35 (1967).
12. M. Milone, *Gazz. Chim. Ital., 67,* 527 (1937).
13. L. Kahovec and K. W. F. Kohlrausch, *Chem. Ber., 75,* 1541 (1942).
14. G. B. Bonino and R. Manzoni-Ansidei, *Z. Phys. Chem., 22B,* 169 (1933).
15. W. C. Harris and S. F. Bush, *J. Chem. Phys., 56,* 6147 (1972).
16. L. Bardet and A. Boudet, *Compt. Rend., 271B,* 710 (1970).
17. Yu. P. Kitaev, B. I. Buzykin, and T. V. Troepol'skaya, *Russ. Chem. Rev., 39,* 441 (1970).
18. A. Kirrmann, *Compt. Rend., 217,* 148 (1943).
19. A. E. Arbuzov, Yu. P. Kitaev, R. R. Shagidullin, and L. E. Petrova, *Bull. Acad. Sci. USSR, Chem. Ser., 1967,* 1822.
20. A. Kirrmann, *Bull. Soc. Chim. Fr., 1956,* 1751.
21. O. A. Raevskii, R. R. Shagidullin, and Yu. P. Kitaev, *Bull. Acad. Sci., USSR, Chem. Ser., 1966,* 200.
22. O. A. Raevskii, R. R. Shagidullin, and Yu. P. Kitaev, *Dokl. Acad. Sci., USSR, 170,* 853 (1966).
23. W. West and R. B. Killingsworth, *J. Chem., Phys., 6,* 1 (1938).
24. Yu. P. Kitaev, L. E. Nivorozhkin, S. A. Plegontov, O. A. Raevskii, and S. Z. Titova, *Dokl. Acad. Sci., USSR, 178,* 162 (1968).
25. J. F. Ogilvie and K. C. Cole, *Spectrochim. Acta, 27A,* 877 (1971).
26. S. T. King, J. Overend, R. A. Mitsch, and P. H. Ogden, *Spectrochim. Acta, 26A,* 2253 (1970).
27. A. Almenningen, I. M. Anfinsen, and A. Haaland, *Acta Chem. Scand., 24,* 1230 (1970).
28. L. Kahovec, K. W. F. Kohlrausch, A. W. Reitz, and J. Wagner, *Z. Phys. Chem., 39B,* 431 (1938).
29. A. Simon and H. Wagner, *Naturwissenschaften, 47,* 540 (1960).
30. S. S. Dubov and V. A. Ginsburg, *Zh. Vses. Khim. Obshchest. im. D. I. Mendeleeva, 7,* 583 (1962); *Chem. Abstr., 58,* 4041b (1963).
31. S. S. Dubov and O. G. Strukov, *Tr. Soveshch. po Fiz. Metodam Issled. Organ. Soedin. i Khim. Protsessov, Akad. Nauk. Kirg. SSR, Inst. Org.*

Khim., Frunze 1962, 94 (1964) *Chem. Abstr., 62,* 4785e (1965).

32. R. Kübler, W. Lüttke, and S. Weckherlin, *Z. Elektrochem.,* 64, 650 (1960).
33. J. Brandmüller, H. Hacker, and H. W. Schrötter, *Chem. Ber., 99,* 765 (1966).
34. H. Hacker, *Spectrochim. Acta, 21,* 1989 (1965).
35. J. Brandmüller and H. Hacker, *Z. Physik, 184,* 14 (1965).

CHAPTER ELEVEN

CUMULENES

AND HETEROCUMULENES

11.1 ALLENES ($R_2C{=}C{=}CR_2$)

The Raman spectrum of allene (propadiene) has been reported by several investigators (1-4). A complete analysis of the vibrational spectra has been made by Lord and Venkateswarlu (5) and by Sverdlov (6). The strongest bands in the Raman of the liquid occur at 3067 cm^{-1} (C-H stretch), 2996 cm^{-1} (C-H stretch), 1440 cm^{-1} (CH_2 bend), 1076 cm^{-1} (C=C=C symmetric stretch), 848 cm^{-1} (CH_2 wag), and 356 cm^{-1} (C=C=C bend). The antisymmetric C=C=C stretch is forbidden by symmetry in the Raman but appears as an intense band at 1957 cm^{-1} in the IR.

In the Raman spectrum of methylallene (1,2-butadiene) (1,7-9) and other monosubstituted *n*-alkyl derivatives of allene (1,2,10), two strong bands appear near 1130 and 1100 cm^{-1}; in phenylallene (propadienyl benzene) these bands are located at 1080 and 1065 cm^{-1}. A very weak Raman band at 1964-1958 cm^{-1} arising from the antisymmetric C=C=C stretch was also observed in these compounds. In the Raman spectra of halopropadienes ($H_2C{=}C{=}CH{-}X$, X = Cl, Br, I) (11,12) the symmetric C=C=C stretching frequency occurs as a very strong band in the range 1095-1076 cm^{-1}. Disubstituted allenes whose Raman spectra have been observed include 1,1-dimethylallene (1,9), 1,3-dimethylallene (2), and 1-hydroxymethyl-11-*n*-butylallene (13).

11.2 BUTATRIENE ($H_2C{=}C{=}C{=}CH_2$)

The complete IR and Raman spectra of butatriene have been reported and vibrational assignments made by Miller and Matsubara (14). A force field calculation for butatriene is given by Cyvin and Hagen (15). In the Raman, two intense polarized bands are observed at 2079 and 878 cm^{-1} and are assigned to the symmetric C=C stretches having the approximate form ←C–C→ ←C–C→ and ←C←C–C→C→, respectively.

11.3 CARBON SUBOXIDE AND SUBSULFIDE, (X=C=C=C=X, X = O,S)

The Raman spectra of carbon suboxide (C_3O_2) have been obtained in the liquid state (16–19), in the solid state (19), and in the gas phase (20). The strongest Raman bands (polarized) in the liquid arise from the C=O symmetric stretch located at 2200 cm^{-1} and the C=C symmetric stretch at 830 cm^{-1}. In the Raman spectrum of carbon subsulfide (S=C=C=C=S) (21,22) the analogous vibrational bands occur at 1663 and 485 cm^{-1}, respectively.

11.4 KETENES (R_2C=C=O)

Kopper (23) has reported the Raman spectrum of ketene (H_2C=C=O) in the liquid state and the vibrational assignments have been reviewed by Moore and Pimentel (24) and by Fletcher and Thompson (25). The pseudo-antisymmetric C=C=O stretch is assigned to a weak Raman band at 2049 cm^{-1} and the pseudo-symmetric C=C=O stretch to a very strong band at 1130 cm^{-1}. Some characteristic frequencies in the spectra of substituted ketenes are:

Compound	ν_{asym}(C=C=O) (cm^{-1}) (weak Raman, strong IR)	ν_{sym}(C=C=O) (cm^{-1}) (strong Raman, weak IR)	Reference
R_3SiCH=C=O	2112–2085	1295–1269	26
$(CH_3)_2C$=C=O	2112	1374	27
$(CF_3)_2C$=C=O	2197	1420	28
$(C_6H_5)_2C$=C=O	2105	1120	29

11.5 ISOCYANATES (R–N=C=O)

Vibrational assignments in the IR and Raman spectra of isocyanic acid (H–N=C=O) have been discussed by Herzberg and Reid (30). The pseudo-antisymmetric NCO stretch, ν_{asym}(N=C=O), occurs in the IR of the vapor at 2274 cm^{-1} and the pseudo-symmetric NCO stretch, ν_{sym}(N=C=O), occurs in the Raman of the liquid at 1318 cm^{-1}. Assignments of the fundamental modes of vibration of methyl, ethyl, and isopropyl isocyanate have been made by Hirschmann, Kniseley, and Fassel (31), those of *t*-butyl isocyanate by Koster (32), and those of phenyl isocyanate by Stephenson, Coburn, and Wilcox (33). Characteristic frequencies for these and other isocyanates (e.g., Spectrum 57) are listed in Table 11.1. In general, ν_{asym}(N=C=O) occurs as a very strong IR band near 2275 cm^{-1} and ν_{sym}(N=C=O) as a strong polarized Raman band near 1430 cm^{-1}.

11.6 ISOTHIOCYANATES (R–N=C=S)

Assignments in the vibrational spectra of isothiocyanic acid (H–N=C=S) have been reviewed by Orville-Thomas (40). In the gas phase in the IR, the pseudo-antisymmetric stretch, ν_{asym}(N=C=S), occurs at a frequency of 1963 cm^{-1} and the pseudo-symmetric stretch, ν_{sym}(N=C=S), at 860 cm^{-1}. In the Raman spectrum of an ether solution of HNCS, Goubeau and Gott (41) reported strong bands at 2017 and 801 cm^{-1}. In the Raman spectrum of polycrystalline HNCS, Durig et al. (42) reported ν_{asym} (N=C=S) at 1934 cm^{-1} and ν_{sym}(N=C=S) at 851 cm^{-1}.

The Raman spectrum of methyl isothiocyanate has been widely studied (41,43–47). The ν_{sym}(N=C=S) frequency is assigned (46,47) to a very strong band at 652 cm^{-1} whereas in the ν_{asym}(N=C=S) region there occur two bands at 2112 (strong) and 2199 cm^{-1} (medium) that have been attributed to Fermi resonance between the pseudo-antisymmetric stretching band and an overtone of a frequency near 1080 cm^{-1}. Other Raman data for alkyl (34,43,44,47) and aromatic (33,34,47) isothiocyanates have been reported. Detailed vibrational assignments have also been made for allyl (47), benzyl (47), and phenyl (33,47) isothiocyanates.

According to Ham and Willis (47), alkyl isothiocyanates are characterized by two bands, both in the IR and Raman, near 2100 (strong) and 2200 cm^{-1} (medium). The ν_{sym}(N=C=S) in the Raman is located at 690-650 cm^{-1} (very strong). Aromatic isothiocyanates exhibit three or more bands in the IR (2060, 2100, and 2180 cm^{-1}) of which only the first two appear in the Raman. Other characteristic bands of the aromatic NCS group include a band at 1243 cm^{-1} that is strong and polarized in the Raman but occurs only weakly in the IR, and a band at 927 cm^{-1} that is strong in the IR but weak in the Raman.

Raman spectra have also been reported for the following isothiocyanates: $Si(NCS)_4$ (36,48), SiH_3NCS (49), Cl_3SiNCS (32), phenoxy- and 2-chlorophenoxyisothiocyanatesilanes (50), GeH_3NCS (51), $PO(NCS)_3$ and $P(NCS)_3$ (52), and $POCl_2(NCS)$ (53).

Table 11.1 Characteristic Frequencies in the Raman Spectra of Isocyanates (R-N=C=O)

Compound	Frequency (cm^{-1}) ν_{asym}(N=C=O)	ν_{sym}(N=C=O)	Ref.
H-NCO	2274	1318	30
R-NCO (R = alkyl group)	2300–2250	1450–1400	31, 32
C_6H_5-NCO	2274	1443	33,34
trans-OCN-CH=CH-NCO	2250	1436	35
$(CH_3)_{4\text{-}n}Si(NCO)_n$ (n = 1–4)	2282–2260	1474–1435	36.38
$(C_6H_5O)_{4\text{-}n}Si(NCO)_n$ (n = 1–4)	–	1478–1465	37
$Cl_{(4\text{-}n)}Si(NCO)_n$ (n = 1 or 3)	2337–2296	1475	32
$Ge(NCO)_4$	2304	1432	38
$ClSO_2NCO$	2258	1357	39
Relative Raman Intensity	Weak or absent- (strong in IR)	Very strong- strong	

11.7 CARBODIIMIDES (R-N=C=N-R)

Vibrational frequencies of carbodiimides in the IR and Raman spectra have been reviewed by Mogul (54). The Raman spectrum of dipropylcarbodiimide has been reported by Kahovec and Kohlrausch (43). In the IR spectra of alkyl derivatives (55) a single strong band is found between 2140 and 2125 cm^{-1} which is due to the antisymmetric stretching of the -N=C=N- group. Compounds that contain an aromatic group directly attached to the diimide group exhibit two IR bands between 2150 and 2100 cm^{-1}. In the Raman spectra of alkyl carbodiimides, the NCN symmetric stretch appears near 1460 cm^{-1} which is also the region in which the antisymmetric methyl deformation mode appears.

Raman spectra of various silyl substituted carbodiimides have been reported (56,57). The ν_{asym}(NCN) band appears in the IR at 2260–2180 cm^{-1} and the ν_{sym} (NCN) is assigned to a band in Raman near 1500 cm^{-1}. In $H_3GeNCNGeH_3$, the Raman active NCN stretch occurs at 1419 cm^{-1} (58).

11.8 AZIDES (R-N=N=N)

In hydrazoic acid, the antisymmetric N=N=N vibration occurs in the IR spectrum of the gas as a strong band at 2140 cm^{-1}; the symmetric N=N=N stretch is assigned to a strong band in the IR at 1260 cm^{-1} (gas) and to a Raman band at 1300 cm^{-1} in the liquid (16,59). In the Raman spectrum of liquid CH_3N_3 (60), ν_{asym} (N_3) occurs as a medium-strong band at 2104 cm^{-1} and ν_{sym} (N_3) as a very strong band at 1276 cm^{-1} (16). Sheinker and Syrkin (61) examined the vibrational spectra (IR and Raman) of several organic azides and reported that the two characteristic frequencies of the N_3 group are for ν_{asym} (N_3), 2169-2080 cm^{-1} and for ν_{sym} (N_3), 1343-1177 cm^{-1}. In an investigation of the IR spectra of several organic acid azides, carbamyl azides, vinyl azides, and alpha-azido ethers, thioethers, and amines, Lieber et al. (62) reported that the antisymmetric and symmetric stretching frequencies of the azido group were found in the ranges 2162-2095 cm^{-1} and 1258-1206 cm^{-1}, respectively. Some of these derivatives exhibited anomalous splitting of the N_3 stretching bands that was attributed to Fermi resonance interaction with combination tones.

Raman spectra have also been reported for the following azido containing compounds: $(CH_3)_3SiN_3$ (63); trimeric diethylaluminum azide (64); germyl azide, GeH_3N_3 (65); and cyanogen azide, NCN_3 (66).

11.9 *N*-SULFINYLAMINES (THIONYL AMINES) (R-N=S=O)

The vibrational assignments for HNSO and DNSO have been made by Richert (67). In the gas phase IR, the pseudo-antisymmetric stretch, ν_{asym}(OSN), is assigned a strong band at 1261 cm^{-1} and the pseudo-symmetric stretch, ν_{sym} (OSN), to a weak band at 1090 cm^{-1}. The IR and Raman spectra of XNSO where X = F, Cl, Br, and I were reported by Eysel (68). For these halogen substituted sulfinylamines, in the liquid Raman spectra, a weak band due to ν_{asym} (NSO) occurred in the region 1247-1214 cm^{-1} and a strong band from ν_{sym} (NSO) at 1028-989 cm^{-1}. In the Raman spectrum of thionylaniline, C_6H_5-NSO, (33) the NSO pseudo-antisymmetric stretch was assigned a frequency of 1306 cm^{-1} (strong) and the NSO pseudo-symmetric stretch one at 1155 cm^{-1} (very strong). In the IR spectra of a series of sulfinyl-hydrazines, Klamann, Kramer, and Weyerstahl (69) found ν_{asym} (NSO) at 1289-1275 cm^{-1} and ν_{sym} (NSO) at 1125-1095 cm^{-1}.

REFERENCES

1. M. Bourguel and L. Piaux, *Bull. Soc. Chim. Fr., 51 (4),* 1041 (1932).
2. H. Kopper and A. Pongratz, *Monatsh. Chem., 62,* 78 (1933).
3. J. W. Linnett and W. H. Avery, *J. Chem. Phys., 6,* 686 (1938).

4. R. C. Lord and J. Ocampo, *J. Chem. Phys., 19,* 260 (1951).
5. R. C. Lord and P. Venkateswarlu, *J. Chem. Phys., 20,* 1237 (1952).
6. L. M. Sverdlov, *Opt. Spektrosk., 2,* 540 (1957).
7. G. J. Szasz, J. S. McCartney, and D. H. Rank, *J. Am. Chem. Soc., 69,* 3150 (1947).
8. L. M. Sverdlov and M. G. Borisov, *Opt. Spectrosc., 9,* 227 (1960).
9. M. G. Borisov and L. M. Sverdlov, *Opt. Spectrosc., 24,* 37 (1968).
10. L. Piaux, M. Gaudemar, and L. Henry, *Bull. Soc. Chim., Fr., 1956,* 794.
11. R. A. Nyquist, G. Y. S. Lo, and J. C. Evans, *Spectrochim. Acta, 20,* 619 (1964).
12. R. A. Nyquist, T. L. Reder, F. F. Stec, and G. J. Kallos, *Spectrochim. Acta, 27A,* 897 (1971).
13. D. Cossart, G. Taieb, C. Troyanoevsky, R. Cagnard, and J. Otto, *Compt. Rend., 260,* 1127 (1965).
14. F. A. Miller and I. Matsubara, *Spectrochim. Acta,, 22,* 173 (1966).
15. S. J. Cyvin and G. Hagen, *Acta Chem. Scand., 23,* 2037 (1969).
16. W. Engler and K. W. F. Kohlrausch, *Z. Phys. Chem., 34B,* 214 (1936).
17. H. D. Rix, *J. Chem. Phys., 22,* 429 (1954).
18. D. A. Long, F. S. Murfin, and R. L. Williams, *Proc. Roy. Soc., 223A,* 251 (1954).
19. W. H. Smith and G. E. Leroi, *J. Chem. Phys., 45,* 1767 (1966).
20. W. H. Smith and J. J. Barrett, *J. Chem. Phys., 51,* 1575 (1969).
21. W. H. Smith and G. E. Leroi, *J. Chem. Phys., 45,* 1778 (1966).
22. J. B. Bates and W. H. Smith, *Spectrochim. Acta, 27A,* 409 (1971).
23. H. Kopper, *Z. Phys. Chem., 34B,* 396 (1936).
24. C. B. Moore and G. Pimentel, *J. Chem. Phys., 38,* 2816 (1963).
25. W. H. Fletcher and W. T. Thompson, *J. Mol. Spectrosc., 25,* 240 (1968).
26. A. N. Lazarev, T. F. Tenisheva, and L. L. Shchukovskaya, *Russ. J. Phys. Chem., 43,* 949 (1969).
27. W. H. Fletcher and W. B. Barish, *Spectrochim. Acta, 21,* 1647 (1965).
28. F. A. Miller and F. E. Kiviat, *Spectrochim. Acta, 25A,* 1577 (1969).
29. S. Nadzhimutdinov, N. A. Slovokhotova, and V. A. Kargin, *Russ. J. Phys. Chem., 40,* 479 (1966).
30. G. Herzberg and C. Reid, *Discuss. Faraday Soc., 9,* 92 (1950).
31. R. P. Hirschmann, R. N. Kniseley, and V. A. Fassel, *Spectrochim. Acta, 21,* 2125 (1965).
32. D. F. Koster, *Spectrochim. Acta, 24A,* 395 (1968).
33. C. V. Stephenson, W. G. Coburn, Jr., and W. S. Wilcox, *Spectrochim. Acta, 17,* 933 (1961).

34. A. Dadieu, *Monatsh. Chem.*, *57,* 29 (1931).
35. G. L. Carlson, *Spectrochim. Acta*, *20,* 1781 (1964).
36. J. Goubeau, E. Heubach, D. Paulin, and I. Widmaier, *Z. Anorg. Allg. Chem.*, *300,* 194 (1959).
37. J. Prejzer, *Rocz. Chem.*, *41,* 647 (1967).
38. F. A. Miller and G. L. Carlson, *Spectrochim. Acta*, *17,* 977 (1961).
39. I. Kanesaka and K. Kawai, *Bull. Chem. Soc. Jap.*, *43,* 3298 (1970).
40. W. J. Orville-Thomas, *J. Chem. Soc.*, *1952,* 2383.
41. J. Goubeau and O. Gott, *Ber. Deut. Chem. Ges.*, *73B,* 127 (1940).
42. J. R. Durig, C. M. Player, Jr., J. Bragin, and W. C. Harris, *Mol. Cryst. Liq. Cryst.*, *13,* 97 (1971).
43. L. Kahovec and K. W. F. Kohlrausch, *Z. Phys. Chem.*, *37B,* 421 (1937).
44. R. Vogel-Högler, *Acta Phys, Austriaca*, *1,* 311 (1948).
45. F. A. Miller and W. B. White, *Z. Elektrochem.*, *64,* 701 (1960).
46. A. J. Costoulas and R. L. Werner, *Aust. J. Chem.*, *12,* 601 (1959).
47. N. S. Ham and J. B. Willis, *Spectrochim. Acta*, *16,* 279 (1960).
48. G. L. Carlson, *Spectrochim. Acta*, *18,* 1529 (1962).
49. E. A. V. Ebsworth, R. Mould, R. Taylor, G. R. Wilkinson, and L. A. Woodward, *Trans. Faraday Soc.*, *58,* 35 (1962).
50. W. Rodziewicz and Z. Michalowski, *Rocz. Chem.*, *43,* 465 (1969).
51. G. Davidson, L. A. Woodward, K. M. Mackay, and P. Robinson, *Spectrochim. Acta*, *23A,* 2383 (1967).
52. K. Oba, F. Watari, and K. Aida, *Spectrochim. Acta*, *23A,* 1515 (1967).
53. R. Schmitt and K. Dehnicke, *Z. Anorg. Allg. Chem.*, *298,* 152 (1959).
54. P. H. Mogul, *Nucl. Sci. Abstr.*, *21,* 47014 (1967).
55. G. D. Meakins and R. J. Moss, *J. Chem. Soc.*, *1957,* 993.
56. J. Pump, E. G. Rochow, and U. Wannagat, *Monatsh. Chem.*, *94,* 588 (1963).
57. E. A. V. Ebsworth and M. J. Mays, *Spectrochim. Acta*, *19,* 1127 (1963).
58. S. Cradock and E. A. V. Ebsworth, *J. Chem. Soc. (A)*, *1968,* 1423.
59. E. H. Eyster and R. H. Gillette, *J. Chem. Phys.*, *8,* 369 (1940).
60. L. Kahovec, K. W. F. Kohlrausch, A. W. Reitz, and J. Wagner, *Z. Phys. Chem.*, *39B,* 431 (1938).
61. Yu. N. Sheinker and Ya. K. Syrkin, *Izv. Akad. Nauk. SSSR, Ser. Fiz.*, *14,* 478 (1959); *Chem. Abstr. 45,* 3246g (1951).
62. E. Lieber, C. N. R. Rao, A. E. Thomas, E. Oftedahl, R. Minnis, and C. V. N. Nambury, *Spectrochim. Acta*, *19,* 1135 (1963).
63. H. Bürger, *Monatsh. Chem.*, *96,* 1710 (1965).
64. J. Müller and K. Dehnicke, *Z. Anorg. Allg. Chem.*, *348,* 261 (1966).

65. S. Cradock and E. A. V. Ebsworth, *J. Chem. Soc. (A), 1968,* 1420.
66. B. Bak, O. Bang, F. Nicolaisen, and O. Rump, *Spectrochim. Acta, 27A,* 1865 (1971).
67. H. Richert, *Z. Anorg. Allg. Chem., 309,* 171 (1961).
68. H. H. Eysel, *J. Mol. Struct., 5,* 275 (1970).
69. D. Klamann, U. Kramer, and P. Weyerstahl, *Chem. Ber., 95,* 2694 (1962).

CHAPTER TWELVE

C≡C AND C≡N COMPOUNDS

12.1 ALKYNES

The vibrational spectra of acetylene (ethyne) and its deuterated analogues have been reviewed by Herzberg (1) and more recently by Scott and Rao (2), Lafferty, Plyler, and Tidwell (3), and Tidwell and Plyler (4). The Raman spectrum of liquid acetylene has been reported by Glockler and Renfew (5) and the laser Raman spectra of the two modifications of the acetylene crystal by Ito, Yokoyama, and Suzuki (6). The frequencies of the fundamental vibrations of acetylene in these different phases are listed in Table 12.1.

The Raman spectra of a large number of acetylene derivatives have been investigated by Bourguel and Daure (7), Gredy (8) and Murray, Cleveland, and co-workers (9–12). The interpretation of the Raman spectra of alkyl substituted acetylenes has been reviewed by Sheppard and Simpson (13). Characteristic frequencies in the Raman spectra of various monosubstituted acetylenes are presented in Table 12.2. The most prominent Raman band in the spectra of monosubstituted acetylenes (see Spectrum 59, Appendix 2) arises from the C≡C stretching vibration and occurs in the range 2160–2100 cm^{-1}. For propyne (methyl acetylene) it occurs at a frequency of 2142 cm^{-1} but for higher straight chain alkyl substituents it is located at 2118 cm^{-1}. The presence of the isopropyl group adjacent to the triple bond, as in 3-methyl-1-butyne (isopropylacetylene) (Spectrum 60), has little effect on this frequency but there is a lowering of the vibrational frequency to 2105 cm^{-1} with the presence of the *tert*-butyl group (3,3-dimethyl-1-butyne). Conjugation of the triple bond with a double bond (e.g., 1-buten-3-yne) or with a phenyl group (e.g., ethynylben-

Table 12.1 Fundamental Vibrations of Acetylene

Symmetry Species	Vibration	Gas IR	Gas Raman	Liquid Raman	Crystal I Raman[a]	Crystal II Raman[b]
Σ_g^+	C-H stretch	*ia*	3374	3341	3332, 3324	3324, 3315
Σ_g^+	C≡C stretch	*ia*	1974	1961	1960, 1956	1962, 1951
Σ_u^+	C-H stretch	3289	*ia*	–	–	–
Π_g	C≡C-H bend	*ia*	612	625	626	660, 639 628
Π_u	C≡C-H bend	730	*ia*	–	–	–

[a] At 173°K.
[b] At 77°K.
[c] *ia* - Inactive vibration.

zene) also lowers the C≡C stretching frequency. In most cases conjugation with the carbonyl group has little effect. The propargyl halides (X-CH_2-C≡C-H) have higher C≡C stretching frequencies than the monoalkyl acetylenes (17).

The acetylenic carbon-hydrogen stretching frequency of monosubstituted alkynes occurs in the Raman in the region 3335–3300 cm^{-1}. Another characteristic frequency occurs at 675–620 cm^{-1} and is attributed to the C≡C-H deformation. Derivatives that possess axial symmetry about the triple bond axis exhibit one band since the two C≡C-H deformations are degenerate (e.g., propyne), whereas a splitting into two bands ($\Delta\nu$ about 10–40 cm^{-1}) occurs in those molecules that are not symmetric about this axis (e.g., propargyl halides). Alkyl acetylenes also have a strong Raman band at 355–335 cm^{-1} that is assigned to a skeletal deformation involving mainly the C-C≡C- group (7).

In symmetrically disubstituted acetylenes the C≡C stretching band is forbidden under the selection rules in the IR and is allowed only in the Raman.

Table 12.2 Characteristic Frequencies in the Raman Spectra of Monosubstituted Acetylenes (Liquid State)

Type of Compound	Characteristic Frequency (cm^{-1})			Other Identifying Substituent Vibrations	Reference
	≡C-H Stretch	C≡C Stretch	C≡C-H Deformation		
CH_3-C≡C-H	3305	2142	643	≡C-C stretch at 931 cm^{-1} (s) C≡C-C bend at 336 cm^{-1} (vs)	14,15
CH_3-$(CH_2)_n$-C≡C-H (n = 1-5)	3309-3302	2118	642-626	≡C-C stretch 840-810 cm^{-1} C≡C-C bend 348-336 cm^{-1}	9,10
$(CH_3)_2CH$-C≡C-H	3304	2119	630	—	16
$(CH_3)_3C$-C≡C-H	3307	2105	633	C_5 skeletal breathing vibration of *tert*-butyl group at 690 cm^{-1} (s)	10
C_5H_9-C≡C-H	3303	2116	628	—	8
C_6H_{11}-C≡C-H	3308	2118	633	—	8
R_1O-CR_2R_3-C≡C-H (R_1, R_2, R_3 = H or alkyl)	3317-3302	2115-2102	660-650; 639-620	—	7,8,10
X-CH_2-C≡C-H (X = F, Cl, Br, I)	3335-3305	2135-2125	675-640; 640-635	—	17
C_6H_5-C≡C-H	3288	2113	621	Ring "breathing" vibration at 994 cm^{-1} (vs)	9,18
CH_2=CH-C≡C-H	3305	2099	629	C=C stretch at 1593 cm^{-1} (s)	13
CH_2=C(CH_3)-C≡C-H	3315	2099	633	C=C stretch at 1615 cm^{-1} (s)	19
CH_3-CH=CH-C≡C-H	3300	2101	640	C=C stretch at 1617 cm^{-1} (s)	19
CF_3-C≡C-H	3316	2156	696	—	20
Relative Raman intensity	Medium	Very strong	Weak		

In asymmetrically disubstituted compounds the band is permitted in the IR but, if the triple bond is three or more positions from the end of the chain, the band is scarcely detectable (21); in the Raman spectra it is always prominent. In the Raman of disubstituted acetylenes (e.g., Spectrum 61) there are, in most cases, two or more bands near 2200 cm^{-1} rather than the expected single band corresponding to the C≡C stretch found in the Raman spectra of monosubstituted acetylenes. These additional bands have been attributed to overtones or combination bands whose intensity has been enhanced by resonance with a fundamental (10,21). The most prominent bands occurring in the C≡C stretching region (2260–2190 cm^{-1}) for various disubstituted acetylenes are listed in Table 12.3. Dialkyl acetylenes are characterized by strong Raman bands near 2235 and 2300 cm^{-1}. Those derivatives containing a methyl group attached to the triple bond also have a strong Raman band near 375 cm^{-1} (8). In some cases within a homologous series of derivatives, the primary, secondary, and tertiary compounds can be distinguished using the triple bond frequencies. This is illustrated in Table 12.3 for some acetylenic alcohols and is discussed for other series in more detail by Gredy (8).

The Raman spectrum of diacetylene (butadiyne) has been investigated by Timm and Mecke (27) and by Jones (28). The totally symmetric stretching frequency of the -C≡C-C≡C- group gives rise to a very strong polarized Raman band at 2172 cm^{-1} in the liquid. A series of five symmetrically disubstituted diacetylenes (R-C≡C-C≡C-R, where R = CH_3-, C_2H_5-, C_3H_7-, C_4H_9-, and C_5H_{11}-) has been studied by Meister and Cleveland (29). The symmetric C≡C stretching frequency appears in the Raman at 2264 cm^{-1} for 2,4-hexadiyne (dimethyldiacetylene) and in the range 2257-2251 cm^{-1} for the other four derivatives. A strong band also appears at 484–475 cm^{-1} in the Raman spectra of these compounds and is attributed to a skeletal deformation mode.

12.2 NITRILES

The Raman spectra of a number of nitriles and their characteristic frequencies have been reported by Reitz et al. (30–32). The corresponding frequencies in the IR have been discussed by Kitson and Griffith (33) and by Hidalgo (34). The Raman intensities of the ν(C≡N) band in aliphatic and aromatic nitriles have been studied by Jesson and Thompson (35). A theoretical study of the factors causing small frequency shifts in the characteristic ν(C≡N) band of nitriles has been made by Besnainou, Thomas, and Bratoz (36). Analysis of the skeletal deformation vibrations in the vibrational spectra of nitriles has been performed by Fujiyama (37). Detailed assignments for all the fundamental frequencies in the vibrational spectra have been proposed for methylcyanide (acetonitrile, ethanenitrile) (38,39), ethylcyanide (propionitrile, pro-

Table 12.3 Characteristic Frequencies in the Raman Spectra of Disubstituted Acetylenes in the Liquid State

Type of Compound	Frequencies (cm^{-1})	Relative Intensity	Frequencies (cm^{-1})	Relative Intensity	Reference
CH_3-C≡C-CH_3	2236	Strong	2313	Medium	22
R-C≡C-CH_3 (R = *n*-alkyl)	2239–2233	Strong	2316–2300	Medium	8,13
R-C≡C-R′	2235–2231	Strong	2301–2290	Medium	7,8,9
C_6H_5-C≡C-CH_3	2214	Very strong	2254	Very strong	9
C_6H_5-C≡C-R	2222–2202	Very strong	2263–2233	Very strong	8,9
CH_2=CH-C≡C-CH_3	2236	Very strong	—	—	19
CH_3-C≡C-Cl	2235	Strong	2263	Very strong	9,10
R-C≡C-Br	2219–2201	Very strong	—	—	9,10
R-C≡C-I	2212–2190	Very strong	—	—	9,10
R-C≡C-C(=O)-CH_3	2212–2205	Very strong	—	—	9
R-C≡C-CH_2OH	2228–2226	Strong	2290–2289	Medium	8
R-C≡C-CH(CH_3)-OH	2250	Strong	2314–2310	Very weak	8
R-C≡C-C(CH_3)$_2$-OH	2239–2237	Very strong	—	—	8
CF_3-C≡C-CH_3	2271	Very very strong	2325	Very strong	23
CF_3-C≡C-CF_3	2305	Very strong	—	—	24
N≡C-C≡C-X (X = Cl, Br, I)	2194–2131	Very strong	—	—	25
N≡C-C≡C-C≡N	2119	Very strong	—	—	25
X-C≡C-Y (X,Y = halogens)	2234–2118	Very strong			26

panenitrile) (40,41) 2-chloro- and 2-bromopropionitrile (41), isopropyl cyanide (2-cyanopropane) (42), allyl cyanide (3-butenenitrile) (43,44), *cis*- and *trans*-crotononitrile (45), and various members of the cyanoethylene series (46–49).

The triple bond stretching frequencies found in various types of nitriles are listed in Table 12.4. In the Raman spectra of saturated nitriles (e.g., Spectra 63,64), the C≡N stretching band occurs near 2245 cm^{-1}. Alkyl substitution on the carbon atom adjacent to the triple bond lowers this frequency (CH_3CN– 2249 cm^{-1}; R-CH_2-CN – 2242 cm^{-1}; R_2CH-CN – 2238 cm^{-1}; R_3C-CN – 2236 cm^{-1}). Successive chlorine substitution first raises the C≡N stretching frequency and then lowers it (CH_3CN – 2249 cm^{-1}; Cl-CH_2-CN – 2258 cm^{-1}; Cl_2CH-CN – 2253 cm^{-1}; Cl_3C-CN – 2250 cm^{-1}). It has been shown (36) that the nitrile band is not greatly affected by vibrational perturbations (i.e., mechanical coupling) but is affected by electronic perturbations. In the Raman it occurs as an intense band but in the IR its intensity is quite variable. The inductive effect determines the behavior of the nitrile band in the methyl- and halogen-substituted acetonitriles. The presence of an unsaturated nonaromatic group adjacent to the C≡N bond shifts the nitrile stretching band to lower frequencies and enhances the Raman intensity (e.g., for acrylonitrile (propenenitrile) ν(C≡N) = 222 cm^{-1}; the Raman intensity increases almost threefold compared to that in the saturated nitriles). Derivatives in which the nitrile group is attached to an aromatic ring (e.g., Spectrum 65) have C≡N stretching bands in the range 2240–2221 cm^{-1}. The frequency shifts depend upon both the nature and position of the ring substituents. Substitution of benzonitrile with a nitro group shifts the frequency towards that of saturated nitriles whereas substitution by an amino group lowers the frequency. The Raman intensity of the ν(C≡N) band in aromatic nitriles is more than tenfold that found in the simple alkyl nitriles.

Other bands in the Raman that are characteristic of nitriles include: (1) a band of medium to strong intensity in aromatic and aliphatic nitriles between 385–350 cm^{-1} that has been attributed to an in-plane deformation of the C-C≡N group (34), and (2) for aliphatic nitriles a very strong Raman band between 200–160 cm^{-1} that arises from a skeletal deformation mode (37).

12.3 ISONITRILES

Dadieu (57) has reported that the frequency of the $-\overset{+}{N}\equiv\overset{-}{C}$ stretching mode occurs at 2161 cm^{-1} for methyl isonitrile (methyl isocyanide) and at 2146 cm^{-1} for ethyl isonitrile. Ugi and Meyr (61) have studied the IR spectra of eighteen isonitriles and found that in the IR aliphatic isonitriles exhibit the isonitrile stretching in the range 2146–2134 cm^{-1}, whereas for the aromatic

Table 12.4 Characteristic Frequencies in the Liquid Spectra of Nitriles

Unconjugated Nitriles			Conjugated Nitriles		
Structure	Frequency ν (C≡N) (cm^{-1})	Reference	Structure	Frequency ν (C≡N) (cm^{-1})	Reference
HCN	2094	57	CH_2=CH-CN	2222	32
ClCN	2206	55,56			
BrCN	2187	55,56	CH_2=C(CH_3)-CN	2230	32,67
ICN	2158	55,56			
CH_3CN	2249	38	CH_2=CCl-CN	2234	47
Cl-CH_2CN	2258	54	*cis*-**CH_3CH=CH-CN**	2223	45
Cl_2CH-CN	2253	54			
Cl_3C-CN	2250	54	*trans*-CH_3CH=CH-CN	2231	45
R-CH_2-CN	2251-2240	30,31,40,41			
R_1R_2CH-CN	2247-2232	30,31,42	(CH_3)$_2$C=CH-CN	2216	32
$R_1R_2R_3$C-CN	2236	30,31	Br_2C=CBr-CN	2222 (solid)	49
CH_3-CHCl-CN	2255	41	**I_2C=CI-CN**	2207 (solid)	49
CH_3-CHBr-CN	2252	41	*trans* NC-CX=CX-CN (X = Cl,Br)	2230 (solid)	48
$(CH_2)_nC$-CN (n = 2,3,4)	2232	30	*trans*-**NC-CI=CI-CN**	2208 (solid)	48
CH_2=CH-CH_2-CN	2251	43,44			
NC-(CH_2)$_n$-CN (n = 1-8)	2278-2249	52,53	*trans* NC-CH=CH-CN	2230	46
ClC(CN)$_3$	2262	50			
C(CN)$_4$	2288	51			
			C_6H_5-CN	2230	**30,57,58**
			R-C_6H_4-CN (R = alkyl or halogen)	2240-2221	30,57
			NC-CN	2322	32
			X-C≡C-CN **(X = Cl,Br,I)**	2297-2270	59
			NC-C≡C-C≡C-CN	2235	60

derivatives the range is 2125–2109 cm^{-1}. The calculation of the force constants and normal coordinates of methyl isocyanide has been reviewed by Duncan (39).

12.4 CYANAMIDES

The Raman spectrum of cyanamide ($H_2N-C{\equiv}N$) has been investigated by Kahovec and Kohlrausch (62,63) and by Davies and Jones (64). A complete assignment of the fundamental frequencies in both the IR and Raman spectra of $H_2N-C{\equiv}N$ and $D_2N-C{\equiv}N$ has been made by Fletcher and Brown (65). In the Raman the polarized C≡N stretching band occurs at 2259 cm^{-1} with a shoulder centered at 2235 cm^{-1} that has been attributed to an overtone band. In dimethylcyanamide ν(C≡N) occurs at 2220 cm^{-1} in the Raman spectrum of the liquid (66). In a series of dialkyl cyanamides the C≡N stretching frequency occurred in the region 2220–2200 cm^{-1} (62).

12.5 THIOCYANATES

Raman spectra have been reported for methyl thiocyanate (62,68) and for ethyl thiocyanate (57,68). Detailed vibrational assignments have been made for methyl (69–71), ethyl (71), and isopropyl (71) thiocyanates. In alkyl thiocyanates the C≡N stretching frequency occurs in the region 2156–2140 cm^{-1} (71,72) and in aromatic thiocyanates at 2175–2160 cm^{-1} (73). Hirschmann, Kniseley, and Fassel (71) have shown that the complex band structure in the carbon-sulfur stretching region of alkyl thiocyanates (725–550 cm^{-1}) can be explained on the basis of rotational isomerism about the C–S bond between the alkyl group and the thiocyanate group. For *n*-alkyl derivatives bands near 680 and 650 cm^{-1} are attributed to C–S stretching that occurs completely within the –S – C≡N group of the isomers and bands near 700 and 621 cm^{-1} to C – S stretching that involves the alkyl carbon adjacent to the sulfur of the thiocyanate group.

REFERENCES

1. G. Herzberg, *Infrared and Raman Spectra of Polyatomic Molecules,* Van Nostrand, New York, 1945, pp. 288–293.
2. J. F. Scott and K. N. Rao, *J. Mol. Spectrosc., 20,* 438 (1966).
3. W. J. Lafferty, E. K. Plyler, and E. D. Tidwell, *J. Chem. Phys., 37,* 1981 (1962).

4. E. D. Tidwell and E. K. Plyler, *J. Opt. Soc. Am., 52,* 656 (1962).
5. G. Glockler and M. M. Renfrew, *J. Chem. Phys., 6,* 340 (1938).
6. M. Ito, T. Yokoyama, and M. Suzuki, *Spectrochim. Acta, 26A,* 695 (1970).
7. M. Bourguel and P. Daure, *Bull. Soc. Chim., Fr., 47 (4),* 1349 (1930).
8. B. Gredy, *Ann. Chim., XI Ser., 4,* 5 (1935).
9. M. J. Murray and F. F. Cleveland, *J. Am. Chem. Soc., 60,* 2664 (1938); *61,* 3546 (1939); *62*, 3185 (1940); *63,* 1718 (1941).
10. F. F. Cleveland and M. J. Murray, *J. Chem. Phys., 9,* 390 (1941); *11,* 450 (1943); *12*, 156 (1944).
11. F. F. Cleveland, M. J. Murray, and H. J. Taufen, *J. Chem. Phys., 10,* 172 (1942).
12. F. F. Cleveland, *J. Chem. Phys., 11,* 1 (1943).
13. N. Sheppard and D. M. Simpson, *Quart. Rev., 6,* 1 (1952).
14. P. N. Daykin, S. Sundaram, and F. F. Cleveland, *J. Chem. Phys., 37,* 1087 (1962).
15. J. L. Duncan, *Spectrochim. Acta, 20,* 1197 (1964).
16. B. Gredy, *Bull. Soc. Chim., Fr., 2 (5),* 1951 (1935).
17. R. A. Nyquist and J. C. Evans, *Spectrochim. Acta, 19,* 1153 (1963).
18. G. W. King and S. P. So, *J. Mol. Spectrosc., 36,* 468 (1970).
19. T. V. Yakovleva and A. A. Petrov, *Opt. Spectrosc., 11,* 320 (1961).
20. C. V. Berney, L. R. Cousins, and F. A. Miller, *Spectrochim. Acta, 19,* 2019 (1963).
21. J. H. Wotiz and F. A. Miller, *J. Am. Chem. Soc., 71,* 3441 (1949).
22. S. Sportouch and R. Gaufres, *J. Chim. Phys., 67,* 394 (1970).
23. E. C. Tuazon and W. G. Fateley, *J. Chem. Phys., 53,* 3178 (1970).
24. F. A. Miller and R. P. Bauman, *J. Chem. Phys., 22,* 1544 (1954).
25. P. Klaboe and E. Kloster-Jensen , *Spectrochim. Acta, 23A,* 1981 (1967).
26. P. Klaboe and E. Kloster-Jensen, *Spectrochim. Acta, 26A,* 1567 (1970).
27. B. Timm and R. Mecke, *Z. Physik, 94,* 1 (1935).
28. A. V. Jones, *Proc. Roy. Soc., 211A,* 285 (1952).
29. A. G. Meister and F. F. Cleveland, *J. Chem. Phys., 12,* 393 (1944).
30. A. W. Reitz and R. Skrabal, *Monatsh. Chem., 70,* 398 (1937).
31. A. W. Reitz and R. Sabathy, *Monatsh. Chem., 71,* 100 (1938).
32. A. W. Reitz and R. Sabathy, *Monatsh. Chem., 71,* 131 (1938).
33. R. E. Kitson and N. E. Griffith, *Anal. Chem., 24,* 334 (1952).
34. A. Hidalgo, *Compt. Rend., 249,* 395 (1959).
35. J. P. Jesson and H. W. Thompson, *Proc. Roy. Soc., 268A,* 68 (1962).
36. S. Besnainou, B. Thomas, and S. Bratoz, *J. Mol. Spectrosc., 21,* 113 (1966).

37. T. Fujiyama, *Bull. Chem. Soc. Jap., 44,* 1194 (1971).
38. H. W. Thompson and R. L. Williams, *Trans. Faraday Soc., 48,* 502 (1952).
39. J. L. Duncan, *Spectrochim. Acta, 20,* 1197 (1964).
40. N. E. Duncan and G. J. Janz, *J. Chem. Phys., 23,* 434 (1955).
41. P. Klaboe and J. Grundnes, *Spectrochim. Acta, 24A,* 1905 (1968).
42. P. Klaboe, *Spectrochim. Acta, 26A,* 87 (1970).
43. G. H. Griffith, L. A. Harrah, J. W. Clark, and J. R. Durig, *J. Mol. Struct., 4,* 255 (1969).
44. A. L. Verma, *J. Mol. Spectrosc., 39,* 247 (1971).
45. J. R. Durig, C. K. Tong, C. W. Hawley, and J. Bragin, *J. Phys. Chem., 75,* 44 (1971).
46. P. Devlin, J. Overend, and B. Crawford, Jr., *Spectrochim. Acta, 20,* 23 (1964).
47. S. B. Lie and P. Klaboe, *Spectrochim. Acta, 26A,* 1191 (1970).
48. P. Klaboe, S. B. Lie, D. H. Christensen, and G. Hagen, *Spectrochim. Acta, 26A,* 1861 (1970).
49. S. B. Lie, P. Klaboe, E. Kloster-Jensen, G. Hagen, and D. H. Christensen, *Spectrochim. Acta, 26A,* 2077 (1970).
50. F. Miller and W. K. Baer, *Spectrochim. Acta, 19,* 73 (1963).
51. R. E. Hester, K. M. Lee, and E. Mayer, *J. Phys. Chem., 74,* 3373 (1970).
52. K. W. F. Kohlrausch and G. P. Ypsilanti, *Z. Phys. Chem., 29B,* 274 (1934).
53. T. Fujiyama and T. Shimanouchi, *Spectrochim. Acta, 20,* 829 (1964).
54. H. C. Cheng, *Z. Phys. Chem., 26B,* 288 (1934).
55. W. West and M. Farnsworth, *J. Chem. Phys., 1,* 402 (1933).
56. J. Wagner, *Z. Phys. Chem., 48B,* 309 (1941).
57. A. Dadieu, *Monatsh. Chem., 57,* 437 (1931).
58. K. W. F. Kohlrausch and A. Pongratz, *Monatsh. Chem., 63,* 427 (1933).
59. P. Klaboe and E. Kloster-Jensen, *Spectrochim. Acta, 23A,* 1981 (1967).
60. F. Miller and D. H. Lemmon, *Spectrochim. Acta, 23A,* 1415 (1967).
61. I. Ugi and R. Meyr, *Chem. Ber., 93,* 239 (1960).
62. L. Kahovec and K. W. F. Kohlrausch, *Z. Phys. Chem., 37B,* 421 (1937).
63. L. Kahovec and K. W. F. Kohlrausch, *Z. Phys. Chem., 193,* 188 (1944).
64. M. Davies and W. J. Jones, *Trans. Faraday Soc., 54,* 1454 (1958).
65. W. H. Fletcher and F. B. Brown, *J. Chem. Phys., 39,* 2478 (1963).
66. F. B. Brown and W. H. Fletcher, *Spectrochim. Acta, 19,* 915 (1963).
67. J. Bragin, K. L. Kizer, and J. R. Durig, *J. Mol. Spectrosc., 38,* 289 (1971).
68. R. Vogel-Högler, *Acta Phys. Austriaca, 1,* 311, (1948).

69. N. S. Ham and J. B. Willis, *Spectrochim. Acta, 16,* 279 (1960).
70. F. A. Miller and W. B. White, *Z. Elektrochem., 64,* 701 (1960).
71. R. P. Hirschmann, R. N. Kniseley, and V. A. Fassel, *Spectrochim. Acta, 20,* 809 (1964).
72. E. Lieber, C. N. R. Rao, and J. Ramachandran, *Spectrochim. Acta, 13,* 296 (1959).
73. G. L. Caldow and H. W. Thompson, *Spectrochim. Acta, 13,* 212 (1958).

CHAPTER THIRTEEN

BENZENE

AND ITS DERIVATIVES

13.1 BENZENE

A comprehensive study of the IR and Raman spectra of benzene and five deuterated benzenes has been made by Ingold et al. (1–21). The IR spectrum of crystalline benzene was reported by Mair and Hornig (22) whose assignments for the fundamental vibrational frequencies of benzene confirmed those of Ingold et al. (21), except for the two b_{2u} species. The Raman spectra of benzene and its deuterated derivatives have also been thoroughly discussed by Langseth and Lord (23), Kohlrausch (24), and Varsanyi (25). The numerous normal coordinate analyses for benzene and its derivatives have been reviewed by Szoke in Chapter 2 of ref. 25.

The symmetry of the benzene molecule is D_{6h} and the distribution of the fundamental vibrations and their activity (R = Raman active, IR = infrared active, ia = inactive) are: Γ (planar vibrations) = $2a_{1g}$ (R) + $2b_{1u}$ (ia) + $2b_{2u}$ (ia) + $4e_{2g}$ (R) + $3e_{1u}$ (IR) + a_{2g} (ia) + a_{2u} (IR) and Γ (nonplanar vibrations) = $2b_{2g}$ (ia) + e_{1g} (R) + $2e_{2u}$ (ia). The vibrational assignments in the IR and Raman spectra of liquid benzene (Spectrum 66, Appendix 2) are given in Table 13.1. Three notations have frequently been used to number the vibrational modes of benzene and its derivatives. The notation proposed by Wilson (26) will be used in this review and the correspondence between this system and those used by Herzberg (27) and by Randle and Whiffen (28) is shown in Table 13.1. The Cartesian displacements of the atoms during the fundamental vibrations of benzene are illustrated in Figure 13.1 for the planar vibrations (29) and in Figure 13.2 for the nonplanar vibrations (30).

Table 13.1 Assignments in the Vibrational Spectra of Liquid Benzene

Symmetry	Herzberg Number	Wilson Number	Randle and Whiffen Number	Approximate Description of Vibration	IR Frequency (cm^{-1})	IR Relative Intensity	Raman Frequency (cm^{-1})	Raman Relative Intensity
Class								
a_{1g}	1	2	z_2	CH stretch	Inactive	—	3062	vs
	2	1	r	Ring "breathing"	Inactive	—	992	s
a_{2g}	3	3	e	CH deformation	Inactive	—	1326	vw
a_{2u}	4	11	f	CH deformation	671	s	Inactive	—
b_{1u}	5	13	q	CH stretch	3068[a]	vw	Inactive	—
	6	12	p	Ring deformation	1010[a]	w	Inactive	—
b_{2g}	7	5	j	CH deformation	Inactive	—	Inactive	—
	8	4	v	Ring deformation	Inactive	—	Inactive	—
b_{2u}	9	14	o	Ring stretch	1310	w	Inactive	—
	10	15	c	CH deformation	1150	w	Inactive	—
e_{1g}	11,11′	10a,10b	g,i	CH deformation	Inactive	—	849	m
e_{1u}	12,12′	20b,20a	z_4,z_1	CH stretch	3080, 3030[b]	s	Inactive	—
	13,13′	19a,19b	m,n	Ring stretch + deformation	1478	s	Inactive	—
	14,14′	18a,18b	b,d	CH deformation	1038	s	Inactive	—
e_{2g}	15,15′	7b,7a	z_5,z_3	CH stretch	Inactive	—	3047	s
	16,16′	8b,8a	l,k	Ring stretch	Inactive	—	1606, 1585[b]	s
	17,17′	9b,9a	u,a	CH deformation	Inactive	—	1178	s
	18,18′	6b,6a	s,t	Ring deformation	Inactive	—	606	s
e_{2u}	19,19′	17a,17b	h,y	CH deformation	975	w	Inactive	—
	20,20′	16a,16b	w,x	Ring deformation	404	w	Inactive	—

[a] Observed only in solid.
[b] Fermi resonance.

	a_{1g}	a_{1g}
Wilson Vibration Number:	2	1
Frequency in cm^{-1}:	3062	992

	b_{1u}	b_{1u}
Wilson Vibration Number:	13	12
Frequency in cm^{-1}:	3068	1010

	b_{2u}	b_{2u}
Wilson Vibration Number:	14	15
Frequency in cm^{-1}:	1310	1150

	b_{2g}
Wilson Vibration Number:	3
Frequency in cm^{-1}:	1326

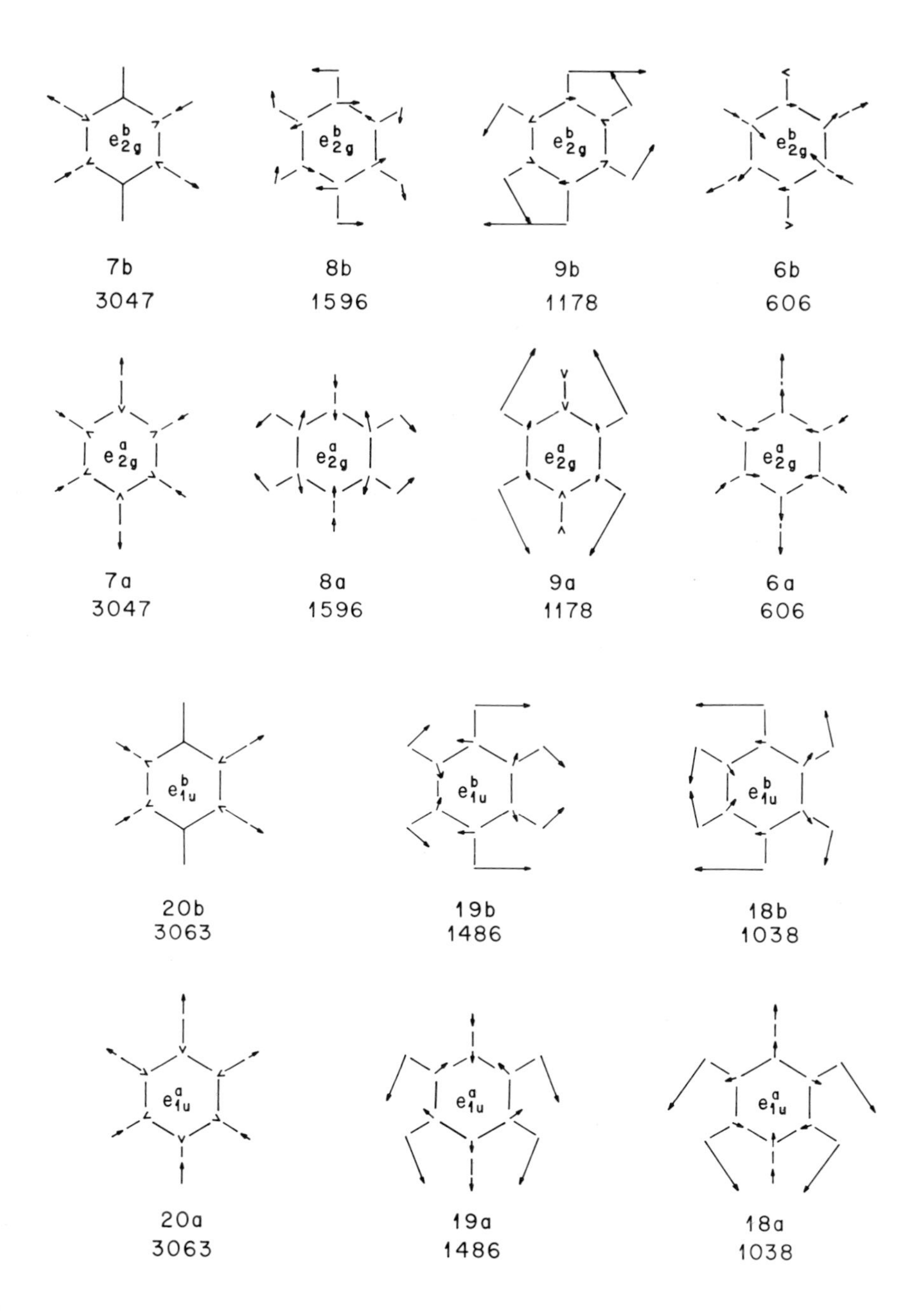

Figure 13.1 Planar vibrations of benzene. Cartesian displacements adapted from ref. 29.

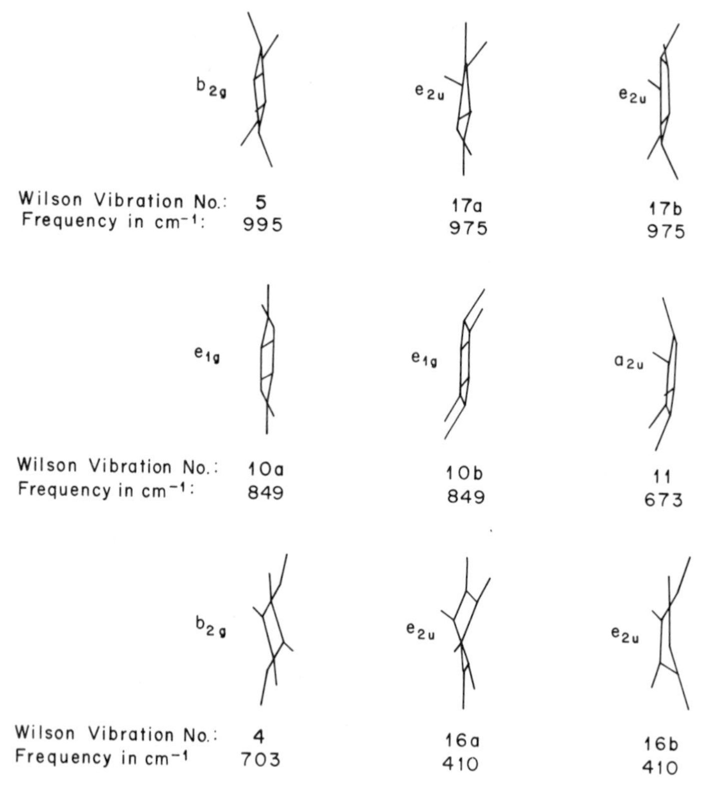

Figure 13.2 Nonplanar vibrations of benzene. Cartesian displacements adapted from ref. 30.

13.2 SUBSTITUTED BENZENES – INTRODUCTION

The literature concerning the characteristic frequencies in the IR and Raman spectra of substituted benzenes is voluminous. The early work on the Raman spectra of these derivatives has been discussed by Kohlrausch (24). A review of the correlations in both the IR and Raman has been made by Jones and Sandorfy (31) and by Colthup, Daly, and Wiberley (32). A systematic and detailed analysis of the vibrational spectra of substituted benzenes has more recently been performed by Varsanyi (25). Bogomolov has carried out a classification of the characteristic vibrations of mono- (33), ortho- (34), para- (35), meta- (36), and 1,2,3-trisubstituted benzenes (37). Scherer et al. (29, 30, 38–44) have investigated the normal vibrations of the complete series of chlorinated benzenes and their fully deuterated analogues. Green et al. have recently reported a series of fifteen papers on the vibrational spectra of various benzene derivatives. These include: nitrobenzene, the benzoate ion, alkali metal benzoates, and salicylates (45); benzonitrile (46); anisole, ethylbenzene, phenetole, methyl phenyl sulfide, and ethyl phenyl sulfide (47); methylphenyl sulfide, diphenyl sulfide, diphenyl disulfide, and diphenyl sulfoxide (48); phenylphosphine, dichlorophenylphosphine, and some related compounds (49); *p*-disubstituted benzenes (50); 4-fluoro- and 4-bromobenzenethiol, 4-chloro- and 4-bromophenylmethyl sulfide (51); *m*-disubstituted benzenes (52); *o*-disubstituted benzenes (53); monosubstituted nitrobenzenes (54); 1,3,5 and 1, 2, 3-trisubstituted benzenes (55); 1,2,4-trisubstituted benzenes (56); dinitrobenzenes (57); monosubstituted phenols (58), and the dimethylphenols, 2,6-, 2,5-, and 3,4-dichlorophenol (59).

Upon substitution of the benzene molecule the symmetry is in most cases lowered and the degenerate vibrations are split into their components. The symmetry of each of the various types of monoatomic substituted benzenes is given in Figure 13.3. The choice of axes for each symmetry group is in accord with the recommendations of the Joint Commission for Spectroscopy (60). Table 13.2 summarizes the symmetry species and their activity for each of the fundamental benzene vibrations and their decomposition under D_6h symmetry.

Characteristic vibrations in substituted benzenes are those in which the carbon atom(s) linking the benzene ring with the substituent(s) and the atom(s) of the substituent(s) undergo insignificant displacements during the vibration. The vibration itself is localized on the atoms of the benzene ring that are removed from the point(s) of substitution. Nonspecific vibrations are those that are essentially localized on the bond joining the ring with the substituent(s) and on the angles nearest to it, all of which undergo considerable displacements.

The effect of substitution on the splitting of each doubly degenerate vibration of benzene can occur in three ways. First, if this vibration is not subject to perturbation by the substituent(s), both vibrations will be specific in frequency, form, and displacements to the benzene vibration and the values of their frequencies will be close to one another and to the benzene frequency. Second, if the perturbation is small only one of the degenerate vibrations will be specific in frequency, form, and atomic displacements, its frequency being close to the corresponding benzene frequency. The second degenerate benzene vibration will be specific only in frequency. Third, if the perturbation is large, one of the vibrations will be specific in frequency, form, and atomic displacements with a frequency close to the corresponding benzene frequency and the second degenerate vibration will be entirely nonspecific. These effects will be discussed further for each type of substitution.

In the IR the type of ring substitution is usually determined by the characteristic out-of-plane CH deformation modes. However, the presence of

Table 13.2 Decomposition of Various Point Groups Under D_{6h} Symmetry

Wilson Vibration No.	Benzene[a)] Vibration	D_{6h} Species[b)]	D_{6h} Activity[c)]	D_{3h} Species	D_{3h} Activity	D_{2h} Species	D_{2h} Activity	C_{2v}^{y} Species	C_{2v}^{y} Activity	C_{2v}^{x} Species	C_{2v}^{x} Activity	C_s Species	C_s Activity
1	ν (Ring)	a_{1g}	R	a_1'	R	a_g	R	a_1	IR,R	a_1	IR,R	a'	IR,R
2	ν (CH)	a_{1g}	R	a_1'	R	a_g	R	a_1	IR,R	a_1	IR,R	a'	IR,R
3	δ (CH)	a_{2g}	-	a_2'	-	b_{3g}	R	b_2	IR,R	b_2	IR,R	a'	IR,R
4	γ (Ring)	b_{2g}	-	a_2''	IR	b_{2g}	R	b_1	IR,R	a_2	R	a"	IR,R
5	γ (CH)	b_{2g}	-	a_2''	IR	b_{2g}	R	b_1	IR,R	a_2	R	a"	IR,R
6a	δ (Ring)	e_{2g}^{a}	R	e_a'	IR,R	a_g	R	a_1	IR,R	a_1	IR,R	a'	IR,R
6b		e_{2g}^{b}		e_b'		b_{3g}	R	b_2	IR,R	b_2	IR,R	a'	IR,R
7a	ν (CH)	e_{2g}^{a}	R	e_a'	IR,R	a_g	R	a_1	IR,R	a_1	IR,R	a'	IR,R
7b		e_{2g}^{b}		e_b'		b_{3g}	R	b_2	IR,R	b_2	IR,R	a'	IR,R
8a	ν (Ring)	e_{2g}^{a}	R	e_a'	IR,R	a_g	R	a_1	IR,R	a_1	IR,R	a'	IR,R
8b		e_{2g}^{b}		e_b'		b_{3g}	R	b_2	IR,R	b_2	IR,R	a'	IR,R
9a	δ (CH)	e_{2g}^{a}	R	e_a'	IR,R	a_g	R	a_1	IR,R	a_1	IR,R	a'	IR,R
9b		e_{2g}^{b}		e_b'		b_{3g}	R	b_2	IR,R	b_2	IR,R	a'	IR,R
10a	γ (CH)	e_{1g}^{a}	R	e_a''	R	b_{1g}	R	a_2	R	b_1	IR,R	a"	IR,R
10b		e_{1g}^{b}		e_b''		b_{2g}	R	b_1	IR,R	a_2	R	a"	IR,R
11	γ (CH)	a_{2u}	IR	a_2''	IR	b_{3u}	IR	b_1	IR,R	b_1	IR,R	a"	IR,R

12	δ (Ring)	b_{1u}	-	a_1'	R	b_{1u}	IR	a_1	IR,R	b_2	IR,R	a'	IR,R
13	ν (CH)	b_{1u}	-	a_1'	R	b_{1u}	IR	a_1	IR,R	b_2	IR,R	a'	IR,R
14	ν (Ring)	b_{2u}	-	a_2'	-	b_{2u}	IR	b_2	IR,R	a_1	IR,R	a'	IR,R
15	δ (CH)	b_{2u}	-	a_2'	-	b_{2u}	IR	b_2	IR,R	a_1	IR,R	a'	IR,R
16a	γ (Ring)	e_{2u}^a	-	e_a''	R	a_u	-	a_2	R	b_1	IR,R	a"	IR,R
16b		e_{2u}^b		e_b''		b_{3u}	IR	b_1	IR,R	a_2	R	a"	IR,R
17a	γ (CH)	e_{2u}^a	-	e_a''	R	a_u	-	a_2	R	b_1	IR,R	a"	IR,R
17b		e_{2u}^b		e_b''		b_{3u}	IR	b_1	IR,R	a_2	R	a"	IR,R
18a	δ (CH)	e_{1u}^a	IR	e_a'	IR,R	b_{1u}	IR	a_1	IR,R	b_2	IR,R	a'	IR,R
18b		e_{1u}^b		e_b'		b_{2u}	IR	b_2	IR,R	a_1	IR,R	a'	IR,R
19a	ν (Ring)	e_{1u}^a	IR	e_a'	IR,R	b_{1u}	IR	a_1	IR,R	b_2	IR,R	a'	IR,R
19b		e_{1u}^b		e_b'		b_{2u}	IR	b_2	IR,R	a_1	IR,R	a'	IR,R
20a	ν (CH)	e_{1u}^a	IR	e_a'	IR,R	b_{1u}	IR	a_1	IR,R	b_2	IR,R	a'	IR,R
20b		e_{1u}^b		e_b'		b_{2u}	IR	b_2	IR,R	a_1	IR,R	a'	IR,R

a) ν = stretch; δ = in-plane deformation; γ = out-of-plane deformation.

b) Symmetry axes for each point group are shown in Figure 13.3.

c) IR = infrared active, R = Raman active, - = inactive in both IR and Raman.

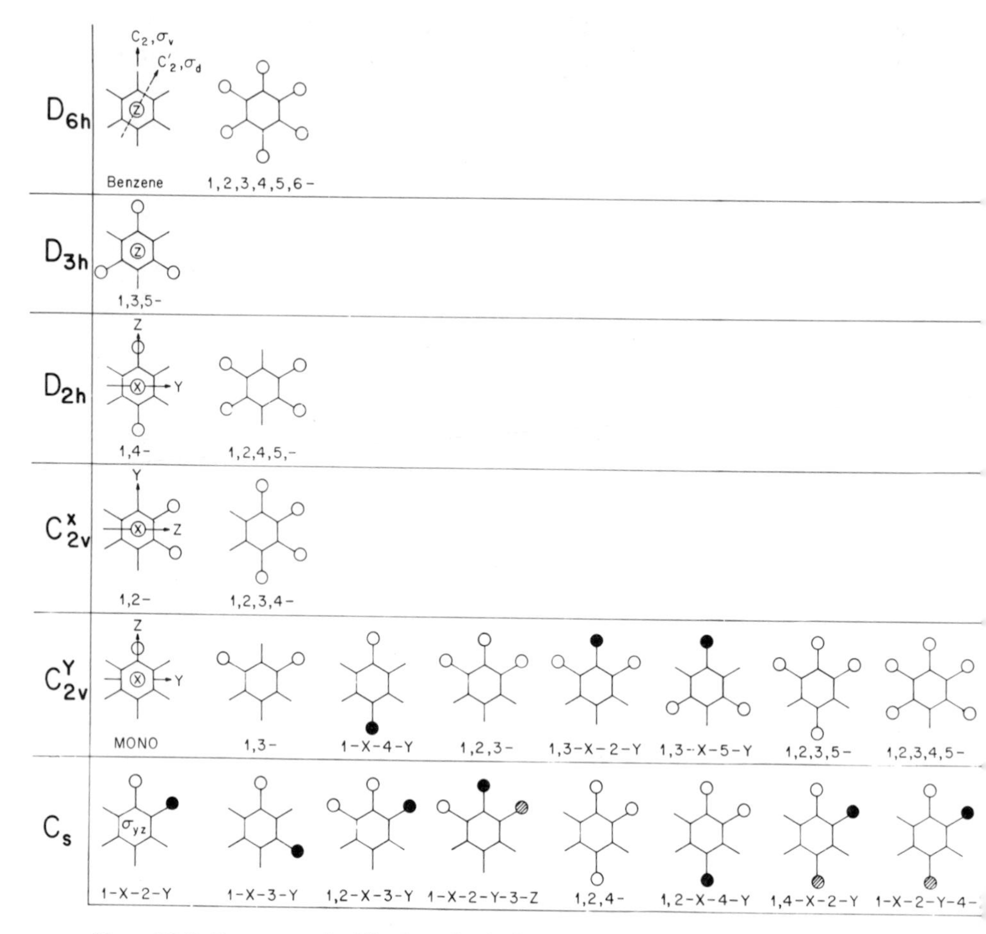

Figure 13.3 Symmetry classification of substituted benzenes.

electron-withdrawing groups on the ring can significantly perturb both the frequency and intensity of these bands so as to obscure these correlations. Nevertheless, the characteristic Raman ring modes described in the following sections are not adversely affected by such substituents.

13.3 MONOSUBSTITUTED BENZENES

If the substituent is monoatomic and lies in the plane of the ring, the monosubstituted benzene has C^{y}_{2v} symmetry. The 30 fundamental vibrations are distributed as follows: $11a_1$ (IR,R) + $10b_2$ (IR,R) + $3a_2$ (R) + $6b_1$ (IR,R). For the monosubstituted phenyl ring 24 vibrations are essentially independent of the substituent attached to the ring and the other 6 are "X-sensitive" vibrations, that is, in these modes the substituent moves with appreciable amplitude. The Cartesian displacements of the atoms during these vibrations and the frequency ranges for these modes are illustrated in Figure 13.4.

There are five CH stretching vibrations in monosubstituted benzenes whose frequencies fall in the range 3100–3000 cm^{-1}. In the IR three bands are generally found in this region. In the Raman only one strong band is usually observed in the region 3070–3030 cm^{-1}. Goubeau and Köhler (61) reported the average frequency of the Raman band for the aromatic CH stretch in monoalkylbenzenes to be 3057 cm^{-1}

The benzene e_{2g} degenerate ring stretch (No. 8a,8b) is not greatly perturbed upon substitution so that the two resulting frequencies in the Raman lie close to that for the benzene vibration (1596 cm^{-1}). This behavior also occurs for other types of substitution, therefore, this pair of vibrations are indicative of the presence of an aromatic ring but are not good group frequencies for distinguishing substitution type. The e_{1u} degenerate vibration of benzene at 1486 cm^{-1} also splits into two bands (No. 19a,19b) that are close to the original benzene frequency; however, these bands have little intensity in the Raman and are often masked by the antisymmetric deformation of attached methyl groups (1458–1436 cm^{-1}). The 1310 cm^{-1} b_{2u} benzene ring stretch (No. 14) is only slightly shifted upon substitution but occurs as a weak Raman band in the range obscured by the methylene twist and wag of attached alkyl substituents. The remaining ring stretch derived from the a_{1g} benzene vibration No. 1 (992 cm^{-1}) is substituent sensitive and according to Varsanyi (62) falls in the range 1100–1060 cm^{-1} for heavy substituents (mass above 25 amu) and in the range 830–620 cm^{-1} for light substituents (mass below 25 amu). For *n*-alkylbenzenes this vibration occurs at 780–730 cm^{-1} in the Raman spectra.

The in-plane ring deformation or trigonal ring "breathing" vibration derived from the b_{1u} benzene vibration No. 12 (1010 cm^{-1}) gives rise to an intense Raman band in monosubstituted benzenes at 1010–990 cm^{-1}. In benzene this vibration is essentially a contraction (expansion) of the ring at the triangle formed by the 1,3,5-carbon atoms, accompanied by an expansion (contraction) of the ring at the triangle formed by the 2,4,6-carbon atoms. This mode is termed the "Star of David" vibration. Upon monosubstitution the motion of the two triangles is decoupled and the vibrational amplitudes at positions 1, 3, and 5 are damped. Further substitution in the 3 (meta) or 3,5, (1,3,5-tri) position(s) does not alter the vibrational frequency range 1010-990 cm^{-1}. Therefore, an intense Raman band in this range is characteristic of mono-, meta-, and 1,3,5-trisubstituted benzenes. In the IR this band appears

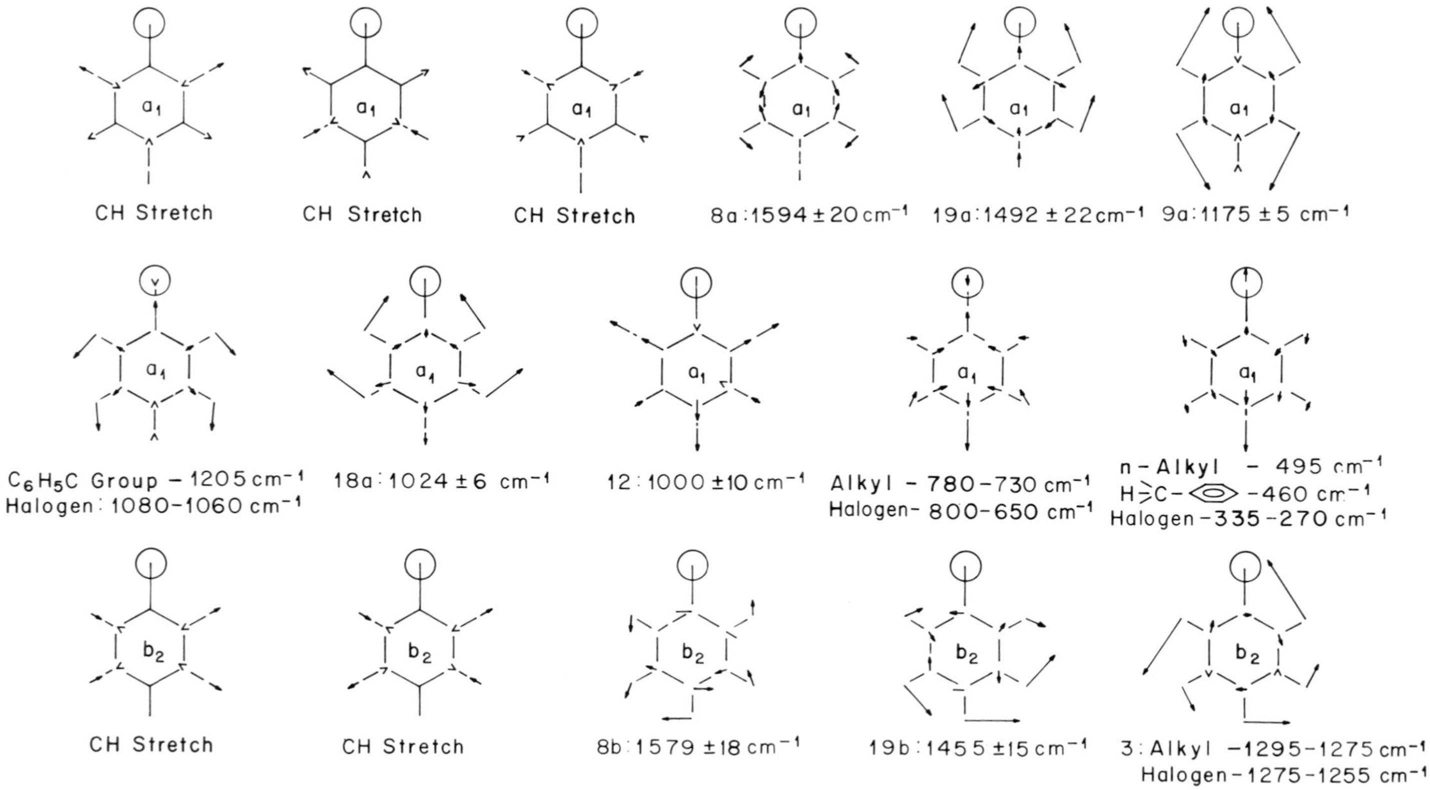

a_1 CH Stretch

a_1 CH Stretch

a_1 CH Stretch

a_1 8a: $1594 \pm 20\ cm^{-1}$

a_1 19a: $1492 \pm 22\ cm^{-1}$

a_1 9a: $1175 \pm 5\ cm^{-1}$

a_1 C_6H_5C Group – $1205\ cm^{-1}$
Halogen: $1080–1060\ cm^{-1}$

a_1 18a: $1024 \pm 6\ cm^{-1}$

a_1 12: $1000 \pm 10\ cm^{-1}$

a_1 Alkyl – $780–730\ cm^{-1}$
Halogen – $800–650\ cm^{-1}$

a_1 n-Alkyl – $495\ cm^{-1}$
H>C–(benzene ring) – $460\ cm^{-1}$
Halogen – $335–270\ cm^{-1}$

b_2 CH Stretch

b_2 CH Stretch

b_2 8b: $1579 \pm 18\ cm^{-1}$

b_2 19b: $1455 \pm 15\ cm^{-1}$

b_2 3: Alkyl – $1295–1275\ cm^{-1}$
Halogen – $1275–1255\ cm^{-1}$

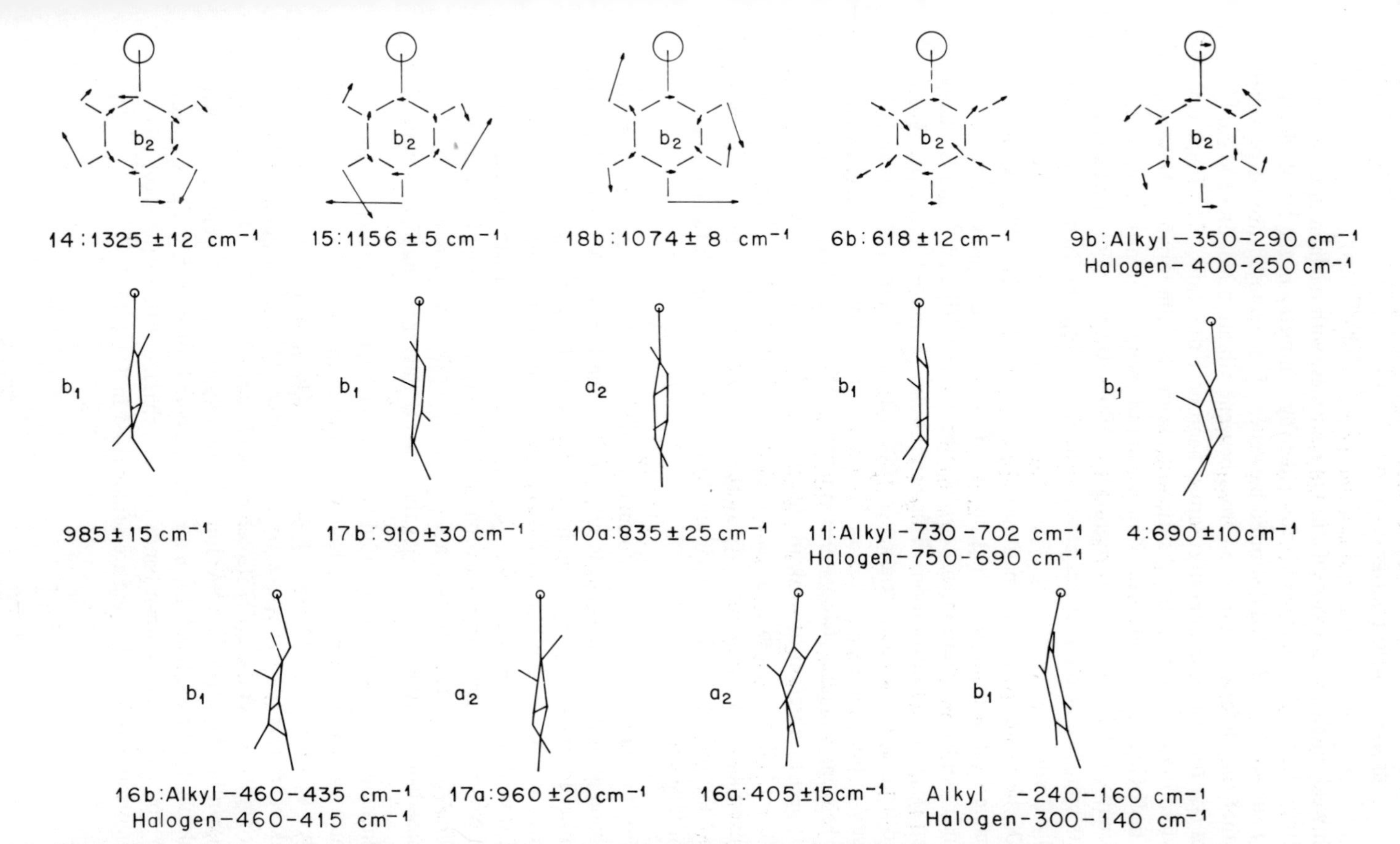

Figure 13.4 Vibrations of monosubstituted benzenes. Cartesian displacements adapted from refs. 29, 30.

with weak intensity in monosubstituted benzenes, with medium intensity in *m*-disubstituted benzenes, and is forbidden by symmetry in the IR of identically substituted 1,3,5-trisubstituted benzenes. For monosubstituted benzenes another ring deformation derived from benzene vibration No. 6b (606 cm^{-1}) gives rise to a medium intensity Raman band near 618 cm^{-1}. The frequency of vibration No. 6a in monosubstituted benzenes is quite variable. However, for methylbenzene (toluene) a Raman band of medium intensity occurs at 521 cm^{-1} and for higher *n*-alkyl groups at 500–490 cm^{-1}. For alkyl substituents branched at the point of attachment to the ring, this frequency falls to 460 ± 2 cm^{-1}.

Of the in-plane CH deformations (No. 3, 9a, 9b, 15, 18a, 18b) the most valuable for identification is that derived from 18a, which gives rise to a Raman band of medium intensity in the range 1030–1018 cm^{-1} and is characteristic of monosubstitution. This band is also prominent in the IR. The band from No. 18b falls in the range 1082–1065 cm^{-1} but is very weak in intensity in both the IR and Raman. Upon monosubstitution the degenerate e_{2g} benzene vibration (No. 9a, 9b) at 1178 cm^{-1} is perturbed and gives rise to a specific vibration of weak to medium intensity in the Raman at 1180–1170 cm^{-1} (No. 9a) and to a nonspecific band at 410–160 cm^{-1} (No. 9b). A Raman band of medium intensity at 1160–1150 cm^{-1} is due to vibration No. 15. Vibration No. 3 (1330–1250 cm^{-1}) is very weak in both the IR and Raman spectra.

For the out-of-plane ring deformations vibration No. 4 occurs as an intense IR band at 700–680 cm^{-1} but is very weak in the Raman. Vibration No. 16a is either forbidden or very weak in the IR and is also a very weak band in the Raman at 420–390 cm^{-1}; vibration No. 16b occurs in the range 560–430 cm^{-1} and is weak in the Raman. For the out-of-plane CH deformations only vibration No. 5 gives rise to a Raman band of sufficient intensity. It occurs at 1000–970 cm^{-1} and appears as a shoulder on the low frequency side of the intense band at 1000 cm^{-1} due to the trigonal ring "breathing" vibration.

In summary, a monosubstituted benzene ring can be identified in the Raman by the presence of the following bands: an intense band at 1000 cm^{-1} with a shoulder near 995 cm^{-1}, a medium to strong band near 1028 cm^{-1}, a weak band near 618 cm^{-1}, two weak bands near 1170 cm^{-1} and a doublet near 1600 cm^{-1}, and the CH stretching band near 3060 cm^{-1}. Alkyl substitution can be identified by a medium to strong band near 1205 cm^{-1}, another weak band near 495 cm^{-1} for *n*-alkyl groups, and for branched alkyls, a band near 460 cm^{-1}. The Raman bands that are characteristic of the alkyl substituents attached to the ring are summarized in Table 13.3

Table 13.3 Characteristic Raman Frequencies for Alkyl Substituents

Frequency (cm^{-1})	Vibration
1460–1440	CH_3 symmetric deformation
1390–1370	CH_3 antisymmetric deformation
1360–1330	Tertiary CH deformation
1310–1300	CH_2 twist and wag
1260–1250, 1220–1200	*tert*-Butyl deformation
1175–1165, 1150–1140	Isopropyl deformation
1141–1132	CH_3 in *n*-alkyl substituent
1075	*n*-Alkyl substituent
960–950, 920, 835–795	Isopropyl C–C vibration
930–925, 750–710	*tert*-Butyl C–C vibration
890–810	*n*-Alkyl C–C vibration
465–430, 370–255	Isopropyl C–C deformation
360–280	*tert*-Butyl C–C deformation

13.4 DISUBSTITUTED BENZENES

The Cartesian displacements for the specific vibrations in the IR and Raman spectra of ortho-, meta-, and para-disubstituted benzenes are illustrated in Figures 13.5, 13.6, and 13.7, respectively. In the Raman spectra of disubstituted benzenes a strong band caused by the CH stretch occurs in the range 3100–3050 cm^{-1}. A doublet centered about 1600 cm^{-1} (No. 8a, 8b) also occurs in the Raman spectra for all of these derivatives.

Ortho-disubstituted benzenes (Spectra 78–80) exhibit a strong Raman band between 1060 and 1020 cm^{-1} which is accompanied by two weak bands at 1295–1250 and 1170–1150 cm^{-1}. Ortho-dialkylbenzenes also exhibit prominent Raman bands specific for the $C_6H_4C_2$ group at 1230–1215, 740–715, and 600–560 cm^{-1}. For the ortho-C_6H_4 CX group (X = Cl, Br, I) the latter two strong Raman bands occur at 680–650 and 560–540 cm^{-1}.

Meta-disubstituted benzenes (Spectra 81–83) can easily be identified in the Raman by an intense polarized band near 1000 cm^{-1} due to the trigonal ring "breathing" vibration (No. 12) together with weak Raman bands at 1260–1210, 1180–1160, and 1080–1060 cm^{-1}. Meta-dialkylbenzenes can further be identified by prominent Raman bands at 740–700 and 540–520 cm^{-1}. Other meta-substituted benzenes have a strong band in the 700–640 cm^{-1} range.

Para-disubstituted benzenes (Spectra 84–86) can be identified in the Raman

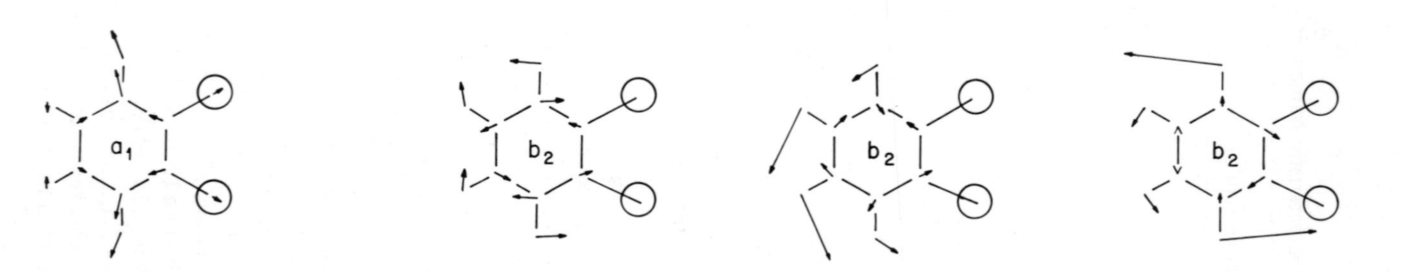
a1
a1
a1
a1
a1
8a : 1587 ± 22cm⁻¹
19b : 1485 ± 35cm⁻¹
14 : 1300 ± 50cm⁻¹
15 : 1159 ± 9cm⁻¹
18b : 1034 ± 22 cm⁻¹

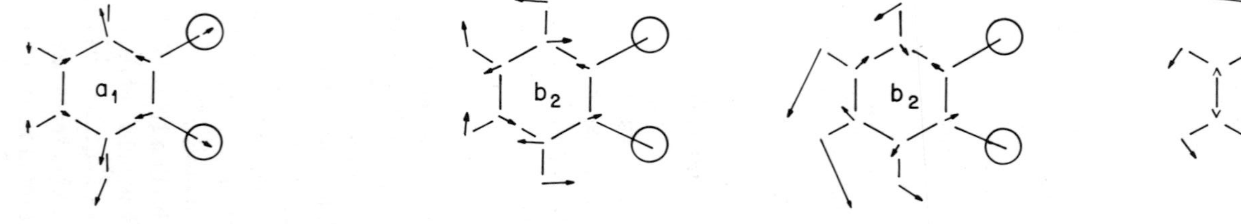
a1
b2
b2
b2
Alkyl - 705 ± 75cm⁻¹
8b : 1605 ± 20 cm⁻¹
19a : 1450 ± 20cm⁻¹
3 : 1273 ± 21 cm⁻¹

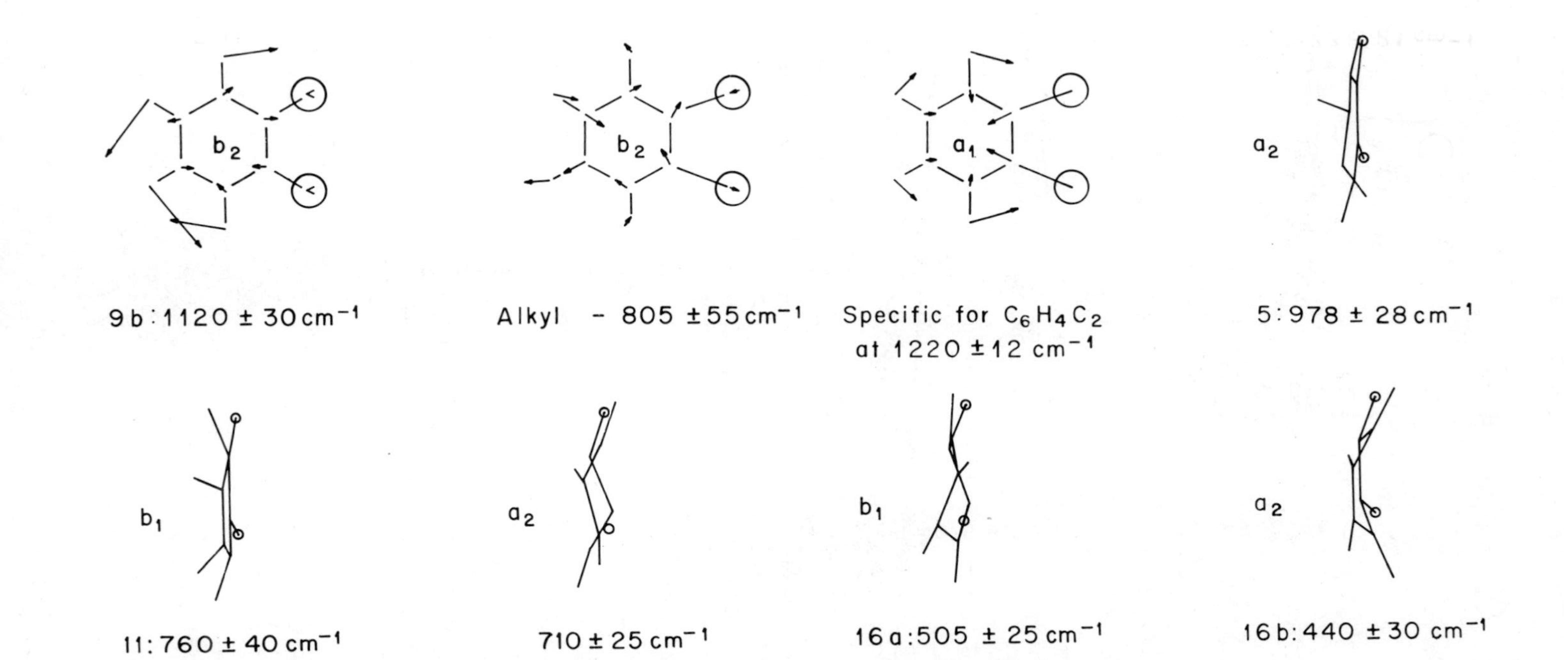

Figure 13.5 Specific vibrations of ortho-disubstituted benzenes. Cartesian displacements adapted from refs. 29, 30.

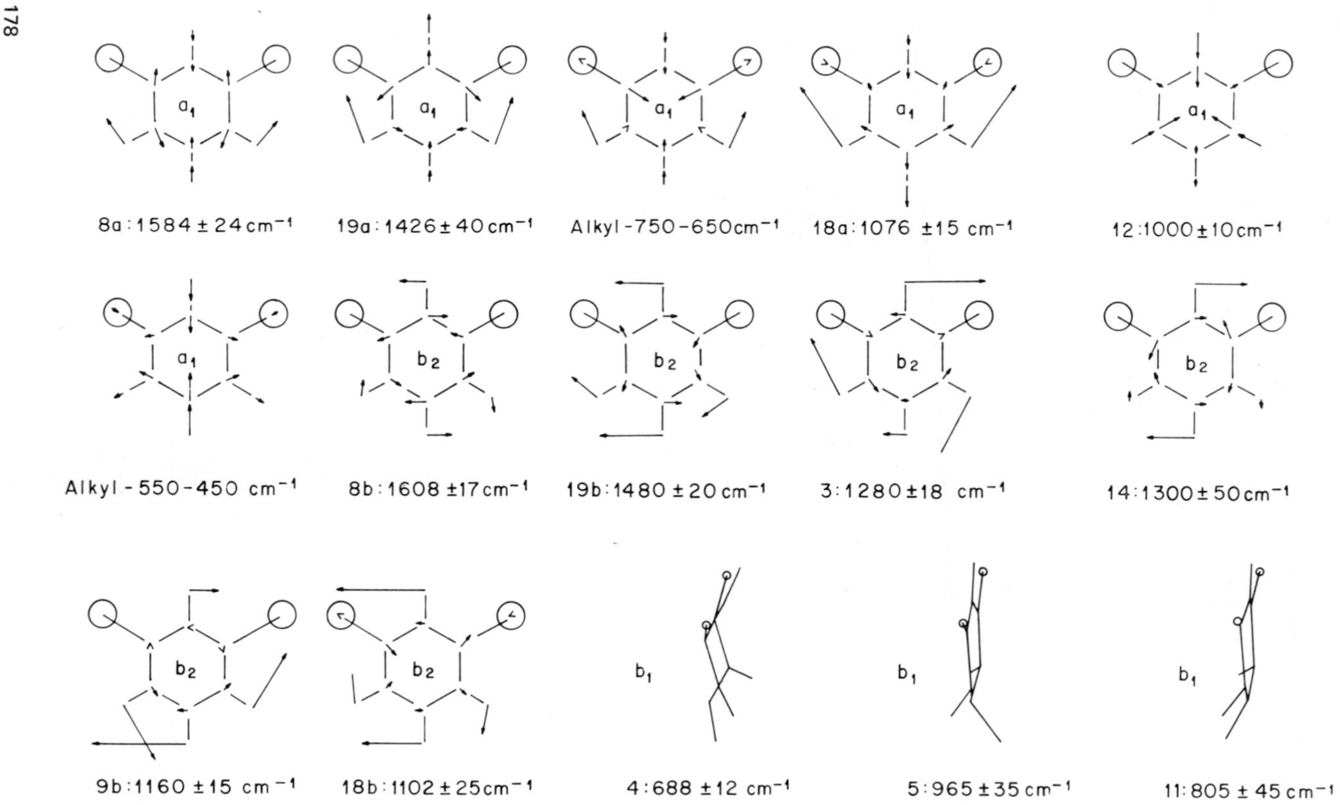

Figure 13.6 Specific vibrations of meta-disubstituted benzenes. Cartesian displacements adapted from refs. 29, 30.

by the presence of two bands at 1180–1150 cm^{-1} (m) and at 650–630 cm^{-1} (m). A very strong Raman band also occurs in the range 830–720 cm^{-1} for these derivatives. The frequency of this band for para-dialkylbenzenes falls in the range 830–780 cm^{-1}. A band specific for the p-$C_6H_4C_2$ moiety occurs at 1230–1200 cm^{-1} (νs-s).

13.5 TRISUBSTITUTED BENZENES

In the Raman spectra of 1,2,3-trisubstituted benzenes (Spectra 87,88) an intense polarized band is located in the range 670–500 cm^{-1}. For 1,2,3-trialkyl derivatives the band falls near 650 cm^{-1}. The dependence of this frequency on the mass and position of the substituents is illustrated for the methyl- (light substituent) and chloro- (heavy substituent) series in the following table:

Position	1	2	3	Frequency (cm^{-1})
	CH_3	CH_3	CH_3	658
	CH_3	Cl	CH_3	589
	CH_3	Cl	Cl	568
	Cl	CH_3	Cl	589
	Cl	Cl	Cl	516

Other characteristic Raman bands for 1,2,3-trisubstitution are 1100–1050 (medium), 490–470 (weak), and two bands of medium to weak intensity between 300 and 200 cm^{-1}.

In the Raman spectra of 1,2,4-trisubstituted benzenes (Spectra 89,90) one of the most intense bands (polarized) occurs in the range 750–650 cm^{-1}. For alkyl substituents this band appears near 740 cm^{-1} and for 1,2,4-trichlorobenzene at 677 cm^{-1}. Other characteristic Raman bands for 1,2,4-trisubstitution occur at 1280–1200 (medium), 580–540 (variable intensity), and 500–450 cm^{-1} (variable intensity).

1,3,5-Trisubstituted benzenes (Spectra 91,92) have an intense polarized Raman band near 1000 cm^{-1} due to the trigonal ring "breathing" vibration; this is also the case for mono- and meta-disubstituted benzenes. In those derivatives with identical substituents (D_{3h} symmetry) the vibrational pair No. 8a, 8b is still degenerate and only one band appears in the Raman near 1600 cm^{-1}. For 1,3,5-trialkylbenzenes other prominent Raman bands occur at 580–510, 520–480, and 280–250 cm^{-1}.

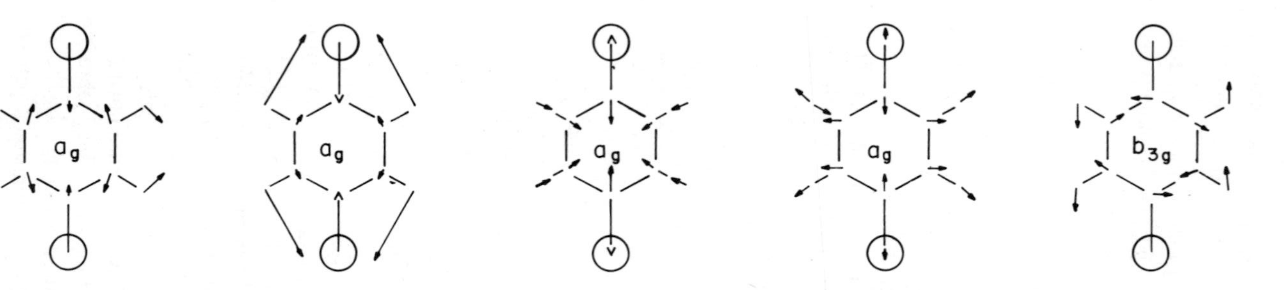
a_g
8a:1600 ± 29 cm^{-1}
a_g
9a:1166 ± 24 cm^{-1}
a_g
1210 ± 10 cm^{-1}
Specific for $C_6H_4C_2$
a_g
870–770 cm^{-1}
$C_6H_4C_2$
b_{3g}
8b:1578 ± 26 cm^{-1}

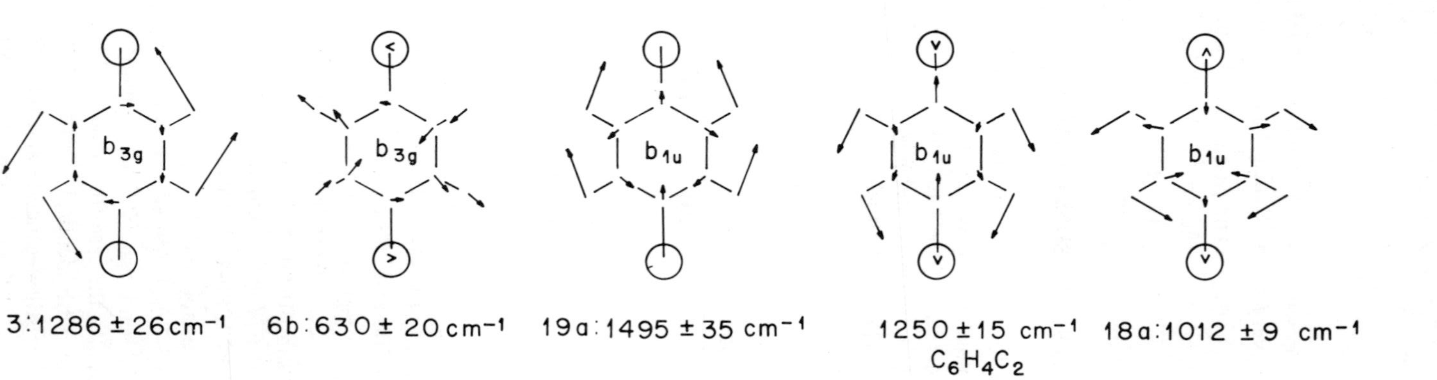
b_{3g}
3:1286 ± 26 cm^{-1}
b_{3g}
6b:630 ± 20 cm^{-1}
b_{1u}
19a:1495 ± 35 cm^{-1}
b_{1u}
1250 ± 15 cm^{-1}
$C_6H_4C_2$
b_{1u}
18a:1012 ± 9 cm^{-1}

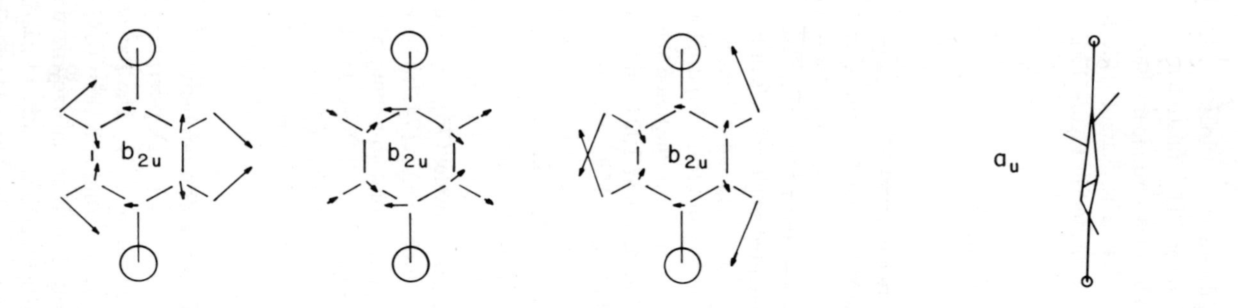

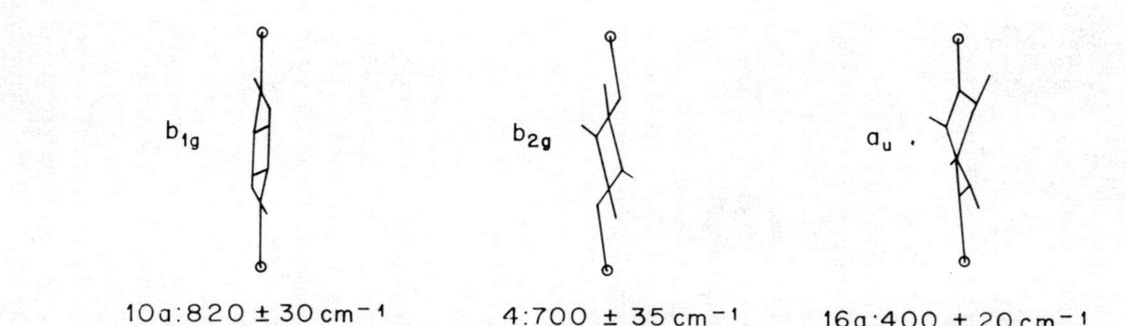

Figure 13.7 Specific vibrations of para-disubstituted benzenes. Cartesian displacements adapted from refs. 29, 30.

13.6 POLYSUBSTITUTED BENZENES

A limited number of Raman spectra are available for tetra- (Spectra 93,94), penta- (Spectrum 95), and hexasubstituted benzenes. The substituents in most cases were methyl, fluoro, or chloro groups. The most prominent Raman bands for these derivatives are shown in Table 13.4. The characteristic Raman frequencies for all the substituted benzenes are collected in Figure 13.8

13.7 BIPHENYLS AND HIGHER POLYPHENYLS

Zerbi and Sandroni (71) have collected and critically reviewed the large number of papers concerning the assignment of the vibrational spectra of biphenyl and its deuterated species. More recently, the IR and Raman spectra of biphenyl have been reexamined by Pasquier (72,73) and Barrett and Steele (74). A set of polarized Raman spectra from oriented single crystals of biphenyl and biphenyl-d_{10} has been obtained by Bree, Pang, and Rabeneck (75). The liquid and solid state Raman spectra of biphenyl (Spectrum 96) display the characteristic frequencies of monosubstituted benzenes. For the liquid state, the intense band from the ring "breathing" vibration occurs at 1004 cm^{-1}. The other typical ring stretching bands are located at 1612, 1595, and 1509 cm^{-1} and a strong band arising from an in-plane C-H deformation is found at 1031 cm^{-1}. An intense polarized Raman band occurs at 1285 cm^{-1} which is assigned to the C–C stretch of the carbon atoms linking the two rings.

The Raman spectra of various mono- and dialkyl, and halo derivatives of biphenyl have been obtained by Bonino and Manzoni-Ansidei (76). 2,2′- and 3,3′-dimethylbiphenyl (77) and the 4,4′-dihalogenobiphenyls (74,78) have also been the subject of Raman studies. The C–C bridge bond stretching vibration occurs as a strong band in the range 1300–1280 cm^{-1} in all these derivatives. Bands characteristic of the substitution pattern on each of the benzene rings are also observed. Steele, Nanney, and Lippincott (79) have recorded the Raman spectrum of decafluorobiphenyl and found that the most intense bands occurred at 1662, 1472, and 390 cm^{-1}.

Mukerji et al. have investigated the Raman spectra of *o*-terphenyl (1,2-diphenylbenzene) (80), *m*-terphenyl (1,3-diphenylbenzene) (81), *p*-terphenyl (1,4-diphenylbenzene) (82), and 1,3,5-triphenylbenzene (83). The Raman spectra of *p*-polyphenyls have been discussed by Sandroni and Geiss (84) and Ting (85). The Raman spectra of all these polyphenyls are characterized by a doublet near 1600 cm^{-1} that may only be partially resolved, a band near 1290 cm^{-1} due to the C–C bridge bond stretching, and the ring "breathing" vibration close to 1000 cm^{-1}.

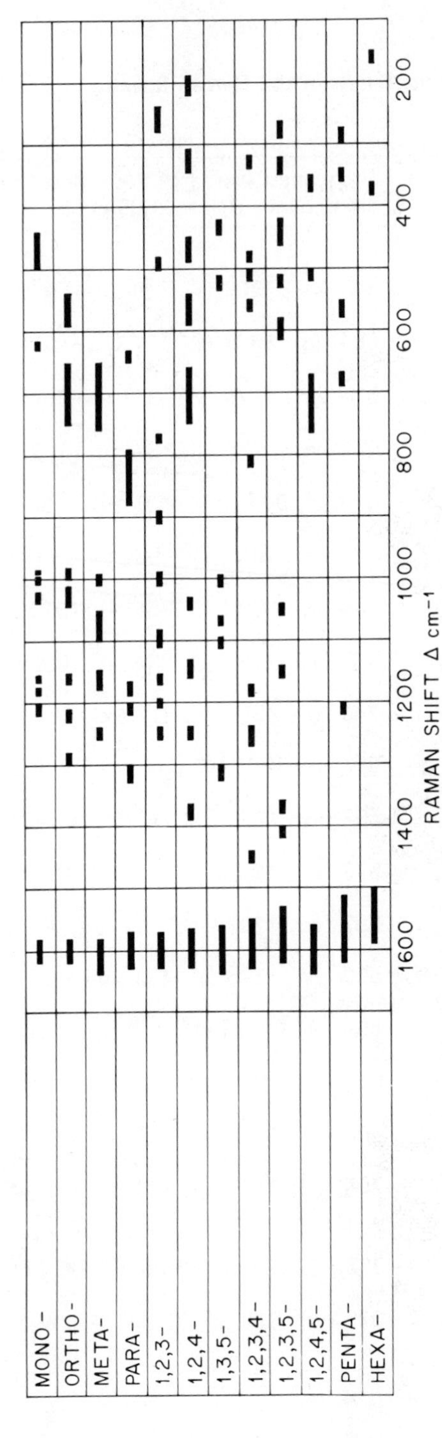

Figure 13.8 Characteristic frequencies in the Raman spectra of substituted benzenes.

Table 13.4 Prominent Bands in the Raman Spectra of Polysubstituted Benzenes

Benzene Derivative	Frequency (cm^{-1}) of Four Most Prominent Bands Below 2000 cm^{-1}	Reference
1,2,3,4-Tetramethyl-	648 (vs), 1260 (s), 1390 (m), 470 (m)	–
1,2,3,4-Tetrafluoro-	684 (vs), 1328 (s), 374 (ms), 489 (m)	63
1,2,3,4-Tetrachloro	515 (vs), 1175 (vs), 332 (vs), 1130 (s)	41
1,2,3,5-Tetramethyl-	574 (vs), 1289 (s), 1379 (m), 733 (m)	64
1,2,3,5-Tetrafluoro	578 (s), 994 (ms), 786 (ms), 1384 (m)	63
1,2,3,5-Tetrachloro-	381 (vs), 192 (vs), 1564 (vs), 1176 (vs)	41
1,2,4,5-Tetramethyl-	734 (vs), 500 (s), 1256 (s), 1379 (ms)	64
1,2,4,5-Tetraisopropyl-	719 (vs), 1264 (s), 884 (s), 1056 (m)	65
1,2,4,5-Tetrafluoro-	748 (vs), 487 (m), 417 (m), 295 (m)	66
1,2,4,5-Tetrachloro-	352 (vs, 190 (m), 684 (vs), 1160 (vs)	41
Pentamethyl-	567 (vs), 1289 (vs), 1379 (vs), 481 (s)	64
Pentafluoro-	578 (s), 719 (s), 1410 (ms), 1648 (ms)	67
Pentachloro-	387 (vs), 1208 (vs), 344 (vs), 200 (vs)	41
Hexamethyl-	550 (vs), 1296 (s), 450 (m), 361 (m)	64
Hexafluoro-	559 (s), 1490 (ms), 443 (ms), 1655 (m)	68
Pentafluoro-chloro-	516 (vs), 1643 (vs), 1595 (vs), 116 (vs)	69
Pentafluoro-bromo-	496 (vs), 114 (vs), 1639 (s), 583 (s)	69
Pentafluoro-iodo-	204 (vs), 489 (s), 1633 (s), 581 (s)	69
1,3,5-Trifluoro-2,4,6-trichloro-	382 (s), 579 (s), 184 (s), 1600 (m)	70
Hexachloro-	372 (vs), 323 (s), 219 (s), 1222 (m)	41

13.8 NAPHTHALENES

A considerable amount of experimental and theoretical work has been done on the vibrational assignments of naphthalene. Mitra and Bernstein (86) have reviewed the assignments in the IR and Raman spectra up to 1959. The Raman spectrum of powdered naphthalene has been reexamined by Stenman (87) and the polarized Raman spectra of a single crystal of naphthalene have been reported by Suzuki, Yokoyama, and Ito (88) and by Hanson and Gee (89). The vibrational spectra and assignments of naphthalene-d_8 have been discussed by Bree and Kydd (90). Raman data that have been recorded for other deuterated naphthalenes include naphthalene-α-d_4 (86) and 1,5-, 1,8-, and 2,6-deuterated naphthalenes (91). Several normal coordinate analyses for both the planar (92–95) and nonplanar (96) vibrations have been performed.

In the Raman spectrum of naphthalene (symmetry D_2h) the strongest bands arise from the planar a_g modes (polarized in liquid and single crystal). In the powder spectrum, these frequencies are: 3056, 1576, 1464, 1383, 1147, 1021, 764, and 513 cm^{-1} (Spectrum 98). Gockel (97) has recorded the Raman spectra of a series of α- and β-monosubstituted naphthalenes (substituents -OH, CH_3, CN, SH, Cl, Br) and observed the following characteristic frequencies:

α-Substituted naphthalene: 1580–1560, 1387–1373, 1025–1012, and 535–512 cm^{-1} (Spectrum 99)

β-Substituted naphthalene: 1585–1570, 1388–1383, 1026–1012, 767–762, and 519–512 cm^{-1}.

Luther has examined the Raman spectra of a series of monoalkyl naphthalenes (98) and of dimethyl- and trimethyl-substituted naphthalenes (99). The dimethyl derivatives could be distinguished on the basis of the strong bands occurring in the following ranges:

Naphthalene	1640–1570 cm^{-1} Range	800–550 cm^{-1} Range
1,2-Dimethyl	1591, 1577	744, 689, 635, 569, 537
1,3-Dimethyl-	1631, 1577	744, 726, 608, 574
1,4-Dimethyl	1585	705, 665, 638
1,5-Dimethyl-	1586	636
1,6-Dimethyl-	1634, 1579	574, 719
1,7-Dimethyl-	1631, 1580	716, 706, 539
1,8-Dimethyl-	1599, 1570	668, 634, 544
2,3-Dimethyl-	1628, 1574	737
2,6-Dimethyl-	1638, 1570	755
2,7-Dimethyl-	1632, 1570	773

13.9 ANTHRACENE AND OTHER POLYCYCLIC AROMATIC HYDROCARBONS

The Raman spectrum of anthracene has been examined in solution (100,101) and in the crystalline form (88,101–103). The polarized Raman spectra of anthracene-d_{10} single crystals have been reported by Bree and Kydd (104). A normal coordinate treatment of the nonplanar vibrations of anthracene has been carried out by Evans and Scully (105) and that for the planar vibrations by Neto, Scrocco, and Califano (95). The most intense bands in the Raman spectrum of the polycrystalline solid occur at 1403, 748, and 396 cm^{-1}.

Manzoni-Ansidei (106) has reported the Raman spectra of 1-chloro and 2-chloro-anthracene. The IR and Raman spectra of 9-monosubstituted and 9,10-disubstituted anthracenes in the solid state and in solution have been investigated by Brigodiot and Lebas (107,108). All these derivatives exhibit intense Raman bands at 1414–1384 and 426–390 cm^{-1}.

Raman spectra have also been reported and vibrational assignments made for the following polycyclic aromatic hydrocarbons: acenaphthene (96,99, 109,110); phenanthrene (106,111,112); 1-methyl-7-isopropylphenanthrene (106); pyrene (113,114); and triphenylene (115,116).

REFERENCES

1. W. R. Angus, C. R. Bailey, C. K. Ingold, and C. L. Wilson, *J. Chem. Soc., 1936,* 912.
2. C. K. Ingold, C. R. Raisin, and C. L. Wilson, *J. Chem. Soc., 1936,* 915.
3. W. R. Angus, C. K. Ingold, and A. H. Leckie, *J. Chem. Soc., 1936,* 925.
4. C. R. Bailey, J. B. Hale, C. K. Ingold, and J. W. Tompson, *J. Chem. Soc., 1936,* 931.
5. C. K. Ingold and C. L. Wilson, *J. Chem. Soc., 1936,* 941.
6. C. K. Ingold and C. L. Wilson, *J. Chem. Soc., 1936,* 955.
7. W. R. Angus, C. R. Bailey, J. B. Hale, and C. K. Ingold, *J. Chem. Soc., 1936,* 960.
8. A. H. Leckie, C. R. Raisin, J. W. Tompson, and C. L. Wilson, *J. Chem. Soc., 1936,* 966.
9. C. Wilson, *J. Chem. Soc., 1936,* 1210.
10. R. C. Lord and E. Teller, *J. Chem. Soc., 1937,* 1728.
11. C. R. Bailey, C. K. Ingold, H. G. Poole, and C. L. Wilson, *J. Chem. Soc., 1946,* 222.
12. L. H. P. Weldon and C. L. Wilson, *J. Chem. Soc., 1946,* 235.
13. A. P. Best and C. L. Wilson, *J. Chem. Soc., 1946,* 239.
14. L. H. P. Weldon and C. L. Wilson, *J. Chem. Soc., 1946,* 244.

15. H. G. Poole, *J. Chem. Soc., 1946,* 245.
16. C. R. Bailey, S. C. Carson, and C. K. Ingold, *J. Chem. Soc., 1946,* 252.
17. C. R. Bailey, J. B. Hale, N. Herzfeld, C. K. Ingold, A. H. Leckie, and H. G. Poole, *J. Chem. Soc., 1946,* 255.
18. N. Herzfeld, J. W. Hobden, and H. G. Poole, *J. Chem. Soc., 1946,* 272.
19. C. R. Bailey, S. C. Carson, R. R. Gordon, and C. K. Ingold, *J. Chem. Soc., 1946,* 288.
20. C. R. Bailey, R. R. Gordon, J. B. Hale, N. Herzfeld, C. K. Ingold, and H. G. Poole, *J. Chem. Soc., 1946,* 299.
21. N. Herzfeld, C. K. Ingold, and H. G. Poole, *J. Chem. Soc., 1946,* 316.
22. R. D. Mair and D. F. Hornig, *J. Chem. Phys., 17,* 1236 (1949).
23. A. Langseth and R. C. Lord, Jr., *Kgl. Dan. Vidensk. Selsk. Math-fys. Medd., 16,* No. 6, 1–85 (1938).
24. K. W. F. Kohlrausch, *Ramanspektren,* Akademische Verlagsges., Leipzig 1943.
25. G. Varsanyi, *Vibrational Spectra of Benzene Derivatives,* Academic Press, New York, 1969.
26. E. B. Wilson, Jr., *Phys. Rev., 45,* 706 (1934).
27. G. Herzberg, *Infrared and Raman Spectra of Polyatomic Molecules,* Van Nostrand, Princeton, 1945, p. 118.
28. R. R. Randle and D. H. Whiffen, in *Molecular Spectroscopy,* G. Sell, Ed., Institute of Petroleum, London, 1955, p. 111–128.
29. J. R. Scherer, *Planar Vibrations of Chlorinated Benzenes,* Dow Chemical Co., Midland, Michigan, 1963.
30. J. R. Scherer, *Spectrochim. Acta, 24A,* 747 (1968).
31. R. N. Jones and C. Sandorfy, in *Chemical Applications of Spectroscopy,* W. West, Ed., Interscience, New York, 1956.
32. N. B. Colthup, L. H. Daly, and S. E. Wiberley, *Introduction to Infrared and Raman Spectroscopy,* Academic Press, New York, 1964.
33. A. M. Bogomolov, *Opt. Spectrosc., 9,* 162 (1960).
34. A. M. Bogomolov, *Opt. Spectrosc., 10,* 162 (1961).
35. A. M. Bogomolov, *Opt. Spectrosc., 12,* 99 (1961).
36. A. M. Bogomolov, *Opt. Spectrosc., 13,* 90 (1962).
37. A. M. Bogomolov, *Opt. Spectrosc., 13,* 183 (1962).
38. J. R. Scherer, J. C. Evans, W. W. Muelder, and J. Overend, *Spectrochim. Acta, 18,* 57 (1962).
39. J. R. Scherer, J. C. Evans, and W. W. Muelder, *Spectrochim. Acta, 18,* 1579 (1962).
40. J. R. Scherer, *Spectrochim. Acta, 19,* 601 (1963).
41. J. R. Scherer and J. C. Evans, *Spectrochim. Acta, 19,* 1739 (1963).
42. J. R. Scherer, *Spectrochim. Acta, 21,* 321 (1965).

43. J. R. Scherer, *Spectrochim. Acta, 22,* 1179 (1966).
44. J. R. Scherer, *Spectrochim. Acta, 23A,* 1489 (1967).
45. J. H. S. Green, W. Kynaston, and A. S. Lindsey, *Spectrochim. Acta, 17,* 486 (1961).
46. J. H. S. Green, *Spectrochim. Acta, 17,* 607 (1961).
47. J. H. S. Green, *Spectrochim. Acta, 18,* 39 (1962).
48. J. H. S. Green, *Spectrochim. Acta, 24A,* 1627 (1968).
49. J. H. S. Green and W. Kynaston, *Spectrochim. Acta, 25A,* 1677 (1969).
50. J. H. S. Green, *Spectrochim. Acta, 26A,* 1503 (1970).
51. J. H. S. Green, D. J. Harrison, W. Kynaston, and D. W. Scott, *Spectrochim. Acta, 26A,* 1515 (1970).
52. J. H. S. Green, *Spectrochim. Acta, 26A,* 1523 (1970).
53. J. H. S. Green, *Spectrochim. Acta, 26A,* 1913 (1970).
54. J. H. S. Green and D. J. Harrison, *Spectrochim. Acta, 26A,* 1925 (1970).
55. J. H. S. Green, D. J. Harrison, and W. Kynaston, *Spectrochim. Acta, 27A,* 793 (1971).
56. J. H. S. Green, D. J. Harrison, and W. Kynaston, *Spectrochim. Acta, 27A,* 807 (1971).
57. J. H. S. Green and H. A. Louwers, *Spectrochim. Acta, 27A,* 817 (1971).
58. J. H. S. Green, D. J. Harrison, and W. Kynaston, *Spectrochim. Acta, 27A,* 2199 (1971).
59. J. H. S. Green, D. J. Harrison, and W. Kynaston, *Spectrochim. Acta, 28A,* 33 (1972).
60. *J. Chem. Phys., 23,* 1997 (1955).
61. J. Goubeau and E. Köhler, *Chem. Ber., 75,* 65 (1942).
62. G. Varsanyi, *Acta. Chim. Acad. Sci. Hung., 50,* 225 (1966).
63. D. Steele, *Spectrochim. Acta, 18,* 915 (1962).
64. G. Durocher, *J. Chim. Phys., 66,* 988 (1969).
65. D. E. Nicholson, *Anal. Chem., 32,* 1634 (1960).
66. E. E. Ferguson, R. L. Hudson, J. R. Nielsen, and D. C. Smith, *J. Chem. Phys., 21,* 1464 (1953).
67. D. Steele and D. H. Whiffen, *Spectrochim. Acta, 16,* 368 (1960).
68. D. Steele and D. H. Whiffen, *Trans. Faraday Soc., 55,* 369 (1959).
69. I. J. Hyams, E. R. Lippincott, and R. T. Bailey, *Spectrochim. Acta, 22,* 695 (1966).
70. J. R. Nielsen and H. D. Brandt, *J. Mol. Spectrosc., 17,* 334 (1965).
71. G. Zerbi and S. Sandroni, *Spectrochim. Acta, 24A,* 483, 511 (1968).
72. B. Pasquier and J. M. Lebas, *J. Chim. Phys., 64,* 765 (1967).
73. B. Pasquier, *Mol. Cryst. Liq. Cryst., 11,* 35 (1970).

74. R. M. Barrett and D. Steele, *J. Mol. Struct.*, *11*, 105 (1972).
75. A. Bree, C. Y. Pang, and L. Rabeneck, *Spectrochim. Acta*, *27A*, 1293 (1971).
76. G. B. Bonino and R. Manzoni-Ansidei, *Mem. Acad. Sci. Inst. Bologna, Cl. Sci. Fis.* [9], *1*, Sept. 7pp. (Feb. 18, 1934).
77. G. Kortüm and H. Maier, *Z. Phys. Chem. (Frankfurt)* [*N.F.*], *7*, 207 (1956).
78. P. Nanni, F. Viani, and V. Lorenzelli, *J. Mol. Struct.*, *6*, 133 (1970).
79. D. Steele, T. R. Nanney, and E. R. Lippincott, *Spectrochim. Acta*, *22*, 849 (1966).
80. S. K. Mukerji and S. A. Aziz, *Nature*, *142*, 477 (1938).
81. S. K. Mukerji and S. A. Aziz, *Phil. Mag.*, *31*, 231 (1941).
82. S. K. Mukerji and L. Singh, *Phil. Mag.*, *37*, 874 (1946).
83. S. K. Mukerji and L. Singh, *Nature*, *150*, 347 (1942).
84. S. Sandroni and F. Geiss, *Z. Anal. Chem.*, *220*, 321 (1966).
85. Chen-Hanson Ting, *J. Chin. Chem. Soc. (Taipei)*, *16*, 137 (1969).
86. S. S. Mitra and H. J. Bernstein, *Can. J. Chem.*, *37*, 553 (1959).
87. F. Stenman, *J. Chem. Phys.*, *54*, 4217 (1971).
88. M. Suzuki, T. Yokoyama, and M. Ito, *Spectrochim. Acta*, *24A*, 1091 (1968).
89. D. M. Hanson and A. E. Gee, *J. Chem. Phys.*, *51*, 5052 (1969).
90. A. Bree and R. A. Kydd, *Spectrochim. Acta*, *26A*, 1791 (1970).
91. P. Dizabo, H. E. Gatica, and N. LeCalve, *J. Chim. Phys.*, *66*, 1497 (1969).
92. D. E. Freeman and I. G. Ross, *Spectrochim. Acta*, *16*, 1393 (1960).
93. D. B. Scully and D. H. Whiffen, *Spectrochim. Acta*, *16*, 1409 (1960).
94. J. R. Scherer, *J. Chem. Phys.*, *36*, 3308 (1962).
95. N. Neto, M. Scrocco, and S. Califano, *Spectrochim. Acta*, *22*, 1981 (1966).
96. D. B. Scully and D. H. Whiffen, *J. Mol. Spectrosc.*, *1*, 257 (1957).
97. H. Gockel, *Z. Phys. Chem.*, *29B*, 79 (1935).
98. H. Luther, *Z. Elektrochem.*, *52*, 210 (1948).
99. H. Luther and C. Reichel, *Z. Phys. Chem.*, *195*, 103 (1950).
100. N. Abasbegovic, N. Vukotic, and L. Colombo, *J. Chem. Phys.*, *41*, 2575 (1964).
101. M. Brigodiot and J. M. Lebas, *Compt. Rend.*, *268B*, 51 (1969).
102. L. Colombo and J. P. Mathieu, *Bull. Soc. Fr. Mineral. Crist.*, *83*, 250 (1960).
103. Chen-Hanson Ting, *J. Chin. Chem. Soc. (Taipei)*, *16*, 123 (1969).
104. A. Bree and R. A. Kydd, *Chem. Phys. Lett.*, *3*, 357 (1969).

105. D. J. Evans and D. B. Scully, *Spectrochim. Acta, 20,* 891 (1964).
106. R. Manzoni-Ansidei, *Gazz. Chim. Ital., 67,* 790 (1937).
107. M. Brigodiot and J. M. Lebas, *Spectrochim. Acta, 27A,* 1315 (1971).
108. M. Brigodiot and J. M. Lebas, *Spectrochim. Acta, 27A,* 1325 (1971).
109. J. P. Mathieu, M. Ecollan, and J. F. Ecollan, *J. Chim. Phys., 50,* 250 (1954).
110. A. Bree, R. A. Kydd, and T. N. Misra, *Spectrochim. Acta, 25A,* 1815 (1969).
111. V. Schettino, N. Neto, and S. Califano, *J. Chem. Phys., 44,* 2724 (1966).
112. V. Schettino, *J. Chem. Phys., 46,* 302 (1967).
113. R. Mecke and W. E. Klee, *Z. Elektrochem., 65,* 327 (1961).
114. N. Neto and C. DiLauro, *Spectrochim. Acta, 26A,* 1175 (1970).
115. R. Mecke and K. Witt, *Z. Naturforsch., 21A,* 1899 (1966).
116. V. Schettino, *J. Mol. Spectrosc., 34,* 78 (1970).

CHAPTER FOURTEEN

HETEROCYCLIC COMPOUNDS — THREE-AND FOUR MEMBERED RINGS

14.1 THREE-MEMBERED RINGS

Detailed assignments of the fundamental vibrations have been made for ethylene oxide (oxirane) (1-5, 47, 48), ethylene imine (aziridine) (3,5-7), and ethylene sulfide (thiirane) (1,2,8). In the Raman spectra of these three-membered ring heterocyclic compounds, the most intense bands arise from the carbon-hydrogen stretching vibrations and the ring stretching modes. The frequencies of these vibrations are listed in Table 14.1 along with those for cyclopropane. Because of the presence of ring strain in these small rings, the C–H stretches are displaced from the ranges normally found for open-chain methylene stretches (2929–2912 and 2861–2849 cm^{-1} for the antisymmetric and symmetric stretches, respectively) to the region usually identified with the C–H stretches of olefins and aromatic compounds. The most intense vibration in the Raman spectra is that arising from the symmetric ring "breathing" vibration in which each of the three bonds expands or contracts in phase. For the heterocycles there are two other distinct in-plane ring stretchings: a symmetric ring vibration in which the two carbon-heteroatom bonds expand (contract) and the carbon-carbon bond contracts (expands) and an antisymmetric ring vibration in which the carbon-carbon bond remains relatively unchanged and one of the carbon-heteroatom bonds expands (contracts) and the other carbon-heteroatom bond contracts (expands). The intensity of the symmetric in-plane stretch in the Raman is greater than that of the antisymmetric ring vibration.

Table 14.1 Vibrations of Three-Membered Rings

	Vibration Frequency cm^{-1}				
Compound	Methylene Antisymmetric C–H Stretch	Methylene Symmetric C–H Stretch	Ring "Breathing" Mode	Symmetric Ring Vibration	Antisymmetric Ring Vibration
Cyclopropane	3090	3039	1188	740[a]	740[a]
Ethylene oxide	3063	3005	1266	877	892
Ethylene imine	3059	2999	1212	857	904[b]
Ethylene sulfide	3085	3010	1112	625	660

[a]These two vibrations are degenerate in cyclopropane.
[b]Observed only in the IR.

Of the substituted heterocyclic three-membered rings, the most extensively studied both by IR and Raman spectroscopy is the epoxy (oxirane) ring. Lespieau and Gredy (9) recorded the Raman spectra of some ethylene and acetylene oxides; Canals, Mousseron, Souche, and Peyrot (10) reported that of substituted epoxycyclopentanes, and Mousseron (11) that of 1,2-epoxycyclanes. Ballaus and Wagner (12) studied the Raman spectra of epoxypropane and several of its halogenated derivatives. Spectra-structure correlations in the IR spectra of oxirane compounds were reported by Patterson (13), Bomstein (14), and Kirchner (15). Evseeva and Sverdlov (26) investigated both the IR and Raman spectra of several epoxypropanes and epoxybutanes. Rambaud and Vessiere (49) examined the Raman spectra of several epoxyesters of butyric acid. From the Raman spectra Batuev, Akhrem, and Matveeva (50) were able to differentiate the erythro- and threo-isomers of *cis*- and *trans*-2-methyl-1-hydroxycyclohexylethylene oxide.

The characteristic bands in the Raman spectra of compounds containing the epoxy (oxirane) ring (e.g., Spectrum 100, Appendix 2) are listed in Table 14.2. The most intense band in the Raman spectra for these compounds is the ring "breathing" vibration that occurs in the region 1280-1240 cm^{-1}. Derkosch et al. (51) found that in di-, tri-, and tetra-halogen substituted ethylene oxides, the ring "breathing" vibration occurred in the range 1360–1320 cm^{-1}. A strong band, due to the antisymmetric ring vibration, is located at 920–880 cm^{-1}. The other symmetric ring vibration is somewhat substituent dependent; for monosubstituted rings, this band is found at 850–820 cm^{-1} and for 1,1-and 1,2-disubstituted rings at 820–770 cm^{-1}. The type of substitution can also often be checked by the presence or absence of the absorptions arising from the stretchings of methylene or methine groups attached to the ring as shown in Table 14.2.

The physical and chemical properties of ethylene imine (aziridine) and its derivatives have been the subject of a recent book by Dermer and Ham (16). The IR absorption spectra of these compounds have been studied by Tempe (17) (*N*-substituted derivatives), Lattes, Martino, and Mathis-Noel (18) (aziridines substituted at the ring carbons), and Spell (19) (*N*-substituted aziridines). Aside from the spectra of the parent compound, Raman data have been reported for propylene imine by Durig, Bush, and Harris (20), for phenyl-substituted ethylene imines by Aleksanyan et al. (21), for *N*-chloro- and *N*-bromoaziridines by Russell, Bishop, and Limberg (22), and for several alkyl substituted ethylene imines by Sheinker, Peresleni, and Braz (52).

The characteristic frequencies in the Raman spectra of aziridine rings are given in Table 14.3. Spell (19) has found that for *N*-substituted aziridines the characteristic IR absorption frequencies are: 3090–3040 cm^{-1}, antisymmetric ring CH_2 stretch; 3005–2979 cm^{-1}, symmetric ring CH_2 stretch; 1330–1250 cm^{-1}, symmetric ring "breathing" vibration; 1170–1150 cm^{-1}, ring CH_2 wagging; 860–745 cm^{-1}, ring deformation and CH_2 rocking modes; and 1365–1315 cm^{-1}, ring CH_2 twisting. For nonactivated aziridines in which the ring nitrogen is basic, the symmetric ring "breathing" band falls in the range 1278–1254 cm^{-1}, but in activated aziridines in which the group attached to the nitrogen can transmit a partial positive charge to the nitrogen this ring vibration occurs at 1360–1315 cm^{-1}. In the Raman spectra of *N*-substituted aziridines the most prominent bands are due to the ring "breathing" vibration, the symmetric in-plane ring deformation and the ring methylene stretches. In those aziridines in which a hydrogen is attached to the ring nitrogen, two bands can generally be found due to the nitrogen-hydrogen stretch in the range 3350–3250 cm^{-1}. The higher frequency arises from molecules that are unassociated and the lower frequency from molecules associated by hydrogen bonding. Also, the symmetric N–H bend can be found in the Raman at 1300–1270 cm^{-1}.

Shagidullin and Grechkin (23) have investigated the Raman spectra of a wide variety of organophosphorus derivatives of ethylene imine as well as several organoarsenic and organosilicon derivatives. The ring "breathing" vibration observed at 1340–1235 cm^{-1} proved to be the most characteristic and intense band. Other characteristic bands included the carbon-hydrogen stretch of the ring at about 3050 cm^{-1} and the antisymmetric vibration of the ring located at 935 cm^{-1}. A preliminary analysis of the IR and Raman spectra of 1,1′-bisaziridyl (C_2H_4N-NC_2H_4) has been reported by Rademacher and Lüttke (27). Bjork et al. (53) examined the vibrational spectra of CF_2N_2 and concluded from the evidence in the IR and Raman spectra that the compound possessed a diazirine structure. A very strong Raman band at 1560 cm^{-1} was assigned to the N=N stretch. Mitchell and Merritt (54,55)

Table 14.2 Characteristic Frequencies in the Raman Spectra of Epoxy Compounds

		Vibration Frequency (cm^{-1})					
Compound	Synonym	Antisymmetric Ring CH_2 Stretch	Symmetric Ring CH_2 Stretch	Methine Ring CH Stretch	Ring "Breathing"	Symmetric In-plane Ring Deformation	Antisymmetric In-Plane Ring Deformation
A. Monosubstituted Ring							
Oxirane	Ethylene oxide	3063	3005	—	1266	877	892
1,2-Epoxypropane	Propylene oxide	3056	3005	2964	1264	828	894
1,2-Epoxybutane	1,2-butylene oxide	3052	2993	2969	1259	830	903
2,3-Epoxy-1-propanol	Glycidol	3070	3005	—	1258	832	903
1-Chloro-2,3-epoxypropane	Epichloro-hydrin	3068	3010	2964	1250	845	904
1-Bromo-2,3-epoxypropane	Epibromo-hydrin	3071	3020	2968	1255	847	915
1-Iodo-2,3-epoxypropane	Epiiodo-hydrin	3054	3000	2960	1253	837	912
1,2-Epoxy-3-butene	—	3075	3000	—	1245	818	916

1,2-Epoxyethylene-benzene	Styrene oxide	3065	2998	—	1256	815	917
1,2-Epoxy-3-phenoxypropane	—	3058	3009	—	1255	815	913
B. Disubstituted Ring							
1,2-Epoxy-2-methyl-propane	Isobutylene oxide	3048	2992	—	1274	795	902
cis-2,3-Epoxy-butane	*cis*-2,3-Butyl-ene oxide	—	—	2992	1277	779	886
trans-2,3-Epoxy-butane	*trans*-2,3-Butyl-ene oxide	—	—	2983	1255	812	887
Methyl epoxy-2,3-butanoate	—	—	—	2979	1235	780	864
Ethyl epoxy-2,3-butanoate	—	—	—	2977	1237	782	865
Propyl epoxy-2,3-butanoate	—	—	—	2990	1240	782	863
Relative Raman intensity		Strong-medium	Very strong-strong	Strong-medium	Very strong	Strong	Strong

Table 14.3 Characteristic Frequencies in the Raman Spectra of Aziridines

		Vibration Frequency (cm^{-1})					
Compound	Synonym	N-H Stretch[a]	Antisymmetric Ring CH_2 Stretch	Symmetric Ring CH_2 Stretch	Ring "Breathing"	Symmetric In-Plane Ring Deformation	Antisymmetric In-Plane Ring Deformation
Aziridine	Ethylene imine	3302,3251	3059	2999	1212	855	1297
2-Methylaziridine	Propylene imine	3305,3257	3057	2997	1243	827	1271
2-Phenylaziridine	—	3310	3062	3000	1230	831	1270
N-Methylaziridine	—	—	3060	2982	1201	818	—
N-Chloroaziridine	—	—	3082	3008	1210	868	—
N-Bromoaziridine	—	—	3088	3002	1209	802	—
Relative Raman intensity		Medium	Strong	Very strong	Very strong	Strong	Medium

[a]The first frequency corresponds to the N–H stretch in unassociated molecules and the second to that of the hydrogen-bonded species.

have studied the IR of 3,3-dimethyldiazirine as well as the methylchloro- and methylbromodiazirines. The N=N stretch in these compounds was located at 1590–1570 cm^{-1}.

The physical and chemical properties of the thirranes (olefin sulfides) have been reviewed by Sander (24) and by Goodman and Reist (25). However, aside from the parent compound, little IR and Raman work has been done on this class of compounds.

14.2 FOUR-MEMBERED RINGS

The most prominent bands in the Raman spectra of the parent compounds for several four-membered heterocyclic rings are summarized in Table 14.4. The vibrations of trimethylene oxide (oxetane) have been discussed by Kohlrausch and Reitz (28), Barrow and Searles (29), and Zürcher and Günthard (30,31). The spectra of trimethylene imine (azetidine) and trimethylene sulfide (thietane) have been assigned by Lippert and Prigge (32) and that of thietane by Scott et al. (33). More recently, the vibrational spectra of silacyclobutane have been investigated by Laane (34) and those of trimethylene selenide by Harvey, Durig, and Morrissey (35).

In these four-membered rings there are six vibrational modes that can be assigned to the ring. Four of these can be designated as stretching of the ring bonds, although there can be mixing with other vibrations of the same symmetry species. In molecules with heteroatoms similar in mass and bond force constant to carbon, the most intense band in the Raman occurs for the ring "breathing" vibration. In cyclobutane this vibration occurs at 1001 cm^{-1} whereas for trimethylene imine and trimethylene oxide (Spectrum 101) it is located at 1026 and 1029 cm^{-1}, respectively. When the heteroatom is much heavier than carbon (e.g., Si, S, and Se), the carbon-carbon stretches decouple from the carbon-heteroatom stretches to a greater extent and each can be considered separately. In general, the in-phase and out-of-phase *C*-heteroatom stretches give rise to strong–very strong Raman bands (see Table 14.4). Besides these four ring stretching modes, there are two ring deformation vibrations. The first can approximately be described as an in-plane ring deformation that often occurs as a strong Raman band, especially in the heterocyclic four-membered rings containing heavier heteroatoms. The second ring deformation is the out-of-plane or ring-puckering vibration that occurs below 200 cm^{-1}. This vibration is generally not found in the Raman spectra with any appreciable intensity, if at all, and it is not a good characteristic frequency. As in the three-membered ring heterocycles, the carbon-hydrogen stretching fundamentals are found at higher frequencies than those in the corresponding straight chain molecules. Of the six carbon-hydrogen stretches of the ring

Table 14.4 Prominent Bands in the Raman Spectra of Four-Membered Heterocyclics

Compound	Frequency (cm^{-1})	Relative Raman Intensity	Approximate Description of Vibration
1. Trimethylene oxide (oxetane)	1140	Medium	Symmetric ring mode
	1029	Very strong	Ring "breathing" vibration
	931	Medium	Antisymmetric ring vibration
2. Trimethylene imine (azetedine)	1026	Very strong	Ring "breathing" vibration
	989	Medium	Antisymmetric ring vibration
3. Trimethylene sulfide (thietane)	991	Medium	Out-of-phase C–C stretch
	932	Strong	In-phase C–C stretch
	699	Strong	In-phase C–S stretch
	670	Very strong	Out-of-phase C–S stretch
	528	Strong	In-phase ring deformation
4. Silacyclobutane	932	Strong	Out-of-phase C–C stretch
	903	Strong	In-phase C–C stretch
	876	Very strong	In-phase C–Si stretch
	652	Very strong	Out-of-phase C–Si stretch
	539	Very strong	Ring deformation
5. Trimethylene selenide	937	Medium	In-phase ring deformation
	650	Strong	In-phase C–C stretch
	563	Strong	Out-of-phase C–Se stretch
	416	Very strong	In-phase C–Se stretch

methylene groups, the easiest to identify is that due to the antisymmetric CH stretching on the α-CH_2 groups that occurs in the range 3000–2980 cm^{-1} for four-membered rings. The other five stretching frequencies generally fall within the region 2908–2860 cm^{-1} but are difficult to sort out because of overlapping and Fermi resonance interactions.

A few oxetane derivatives have been studied by Raman spectroscopy. Fonteyne et al. (36,37) investigated the Raman spectra of 3,3-dimethyloxetane and 3,3-bis(cyanomethyl) oxetane. As in oxetane, the ring "breathing" vibration is quite intense and falls in the range 1030–1010 cm^{-1}. Barrow and Searles (29) in an IR study of oxetane compounds found that the most prominent IR bands occurred at 1250–1200 cm^{-1} (CH_2 wag) and at 980–970 and 900 cm^{-1} (ring vibrations). Table 14.5 lists some of the prominent bands in oxetane derivatives containing carbonyl and/or vinyl groups exo to the ring. Taufen and Murray (38) investigated the Raman spectra of ketene dimer and β-butyrolactone. Durig et al. examined the IR and Raman spectra of 3-methyleneoxetane (39), β-propiolactone (40), β-butyrolactone (41), and diketene (42). In those derivatives in which a carbonyl group is exocyclic to the strained four-membered ring, the carbonyl stretch appears at a considerably higher frequency (1900–1820 cm^{-1}) than in open-chained, unconjugated ketones (1720 cm^{-1}).

Little Raman work has been done on azetidine derivatives. In an IR examination of some *N*-substituted azetidines, Leary and Topsom (43) found the following characteristic frequencies: 2940–2810 cm^{-1}, N–H stretches; 1350–1330 cm^{-1} (*N*-arylazetidines); 1238–1233 cm^{-1}; 1211–1196 cm^{-1} (*N*-alkylazetidines); 1178–1152 cm^{-1}; and 1126–1124 cm^{-1}.

Downs and Haas (44) have examined the IR and Raman spectra of trifluoromethanesulphenyl isocyanate dimer $(CF_3SNCO)_2$ and phenyl isocyanate dimer $(C_6H_5NCO)_2$, both of which contain a cyclic uretidine-1,3-dione structure (I). They have assigned the eighteen fundamental modes of this framework in both molecules. The totally symmetric ring stretch occurs about 820 cm^{-1}, the in-phase carbonyl stretch at 1960–1930 cm^{-1}, and the out-of-phase carbonyl stretch at 1825–1780 cm^{-1}.

The Raman spectrum of the thietane derivative, α-methyltrimethylene sulfide, was reported by Akishin et al. (45). They found that upon substitution, the in-phase C–S stretch shifted to 700 cm^{-1}, the out-of-phase C–S stretch shifted to 628 cm^{-1}, and the ring deformation shifted to 516 cm^{-1} as compared with the same vibrations in the parent compound.

The IR and Raman spectra of tetrafluoro-1,3-dithietane (II), a dimer of thiocarbonyl fluoride, have been reported by Durig and Lord (46). The most intense band in the Raman occurs at 516 cm^{-1} and is attributed to a symmetric ring expansion involving mainly the sulfur atoms. Other strong bands at 668 and 833 cm^{-1} are assigned to ring expansions involving the CF_2 groups

Table 14.5 Characteristic Raman Bands in Oxetane Derivatives Containing Carbonyl and/or Vinyl Groups

		Vibration Frequency (cm^{-1})				
Compound	Structure	Antisymmetric CH_2 Stretch	Ring "Breathing" Mode	Other Ring Deformations	C=O Stretch	C=C Stretch
Oxetane		3001	1029	1140, 931	–	–
3-Methylene oxetane		2935	966	855, 695	–	1693
β-Propiolactone		3023	1007	906,876, 757	1830	–
β-Butyrolactone		3017	711	825	1818	–
Diketene		3127	673	868,807	1895	1693

and the in-plane ring deformation, respectively.

I

II

REFERENCES

1. K. Venkateswarlu and P. A. Joseph, *J. Mol. Struct., 6,* 145 (1970).
2. T. Henshall and J. M. Freeman, *Can. J. Chem., 46,* 2135 (1968).
3. W. J. Potts, *Spectrochim. Acta, 21,* 511 (1965).
4. R. C. Lord and B. Nolin, *J. Chem. Phys., 24,* 656 (1956).
5. H. W. Thompson and W. T. Cave, *Trans. Faraday Soc., 47,* 946, 951 (1951).
6. H. T. Hoffman, Jr., G. E. Evans, and G. Glockler, *J. Am. Chem. Soc., 73,* 3028 (1951).
7. R. W. Mitchell, J. C. Burr, Jr., and J. A. Merritt, *Spectrochim. Acta, 23A,* 195 (1967).
8. G. B. Guthrie, Jr., D. W. Scott, and G. Waddington, *J. Am. Chem. Soc., 74,* 2795 (1952).
9. R. Lespieau and B. Gredy, *Compt. Rend., 196,* 399 (1933).
10. E. Canals, M. Mousseron, L. Souche, and P. Peyrot, *Compt. Rend., 202,* 1989 (1936).
11. M. Mousseron, *Bull. Soc. Chim. Fr., 1946,* 629.
12. O. Ballaus and J. Wagner, *Z. Phys. Chem., 45B,* 272 (1940).
13. W. A. Patterson, *Anal. Chem., 26,* 823 (1954).
14. J. Bomstein, *Anal. Chem., 30,* 544 (1958).
15. H. H. Kirchner, *Z. Phys. Chem. Neue Folge, 39,* 273 (1963).
16. O. C. Dermer and G. E. Ham, *Ethylenimine and Other Aziridines–Chemistry and Applications,* Academic Press, New York, 1969.
17. J. Tempe, *Compt. Rend., 259,* 1717 (1964).
18. A. Lattes, R. Martino, and R. Mathis-Noel, *Compt. Rend., 263C,* 49 (1966).
19. H. Spell, *Anal. Chem., 39,* 185 (1967).
20. J. R. Durig, S. F. Bush, and W. C. Harris, *J. Chem. Phys., 50,* 2851 (1969).
21. V. T. Aleksanyan, M. Y. Lukina, S. V. Zotova, and G. V. Loza, *Dokl. Akad. Nauk. USSR, 171,* 1027 (1966) (English translation).
22. J. W. Russell, M. Bishop, and J. Limburg, *Spectrochim. Acta, 25A,* 1929 (1969).
23. R. R. Shagidullin and N. P. Grechkin, *J. Gen. Chem. USSR, 38,* 148 (1967).
24. M. Sander, *Chem. Rev., 66,* 297 (1966).
25. L. Goodman and E. J. Reist, in Chapter 4, *The Chemistry of Organic Sulfur Compounds,* Vol. 2, N. Kharasch, and C. Y. Myers, Eds., Pergamon Press, Oxford, 1966.
26. L. A. Evseeva and L. M. Sverdlov, *Russ. J. Phys. Chem., 43,* 468 (1969).

27. P. Rademacher and W. Lüttke, *Angew. Chem., Intl. Ed., 9,* 245 (1970).
28. K. W. F. Kohlrausch and A. W. Reitz, *Z. Phys. Chem., 45B,* 249 (1940).
29. G. M. Barrow and S. Searles, *J. Am. Chem. Soc., 75,* 1175 (1953).
30. R. F. Zürcher and H. H. Günthard, *Helv. Chim. Acta, 38,* 849 (1955).
31. R. F. Zürcher and H. H. Günthard, *Helv. Chim. Acta, 40,* 89 (1957).
32. E. Lippert and H. Prigge, *Ber. Bunsenges. Phys. Chem., 67,* 554 (1968).
33. D. W. Scott, H. L. Finke, W. N. Hubbard, J. P. McCullough, C. Katz, M. E. Gross, J. F. Messerly, R. E. Pennington, and G. Waddington, *J. Am. Chem. Soc., 75,* 2795 (1953).
34. J. Laane, *Spectrochim. Acta, 26A,* 517 (1970).
35. A. B. Harvey, J. R. Durig, and A. C. Morrissey, *J. Chem. Phys., 50,* 4949 (1969).
36. R. Fonteyne, P. Cornand, and M. Ticket, *Natuurev. Tijdschr. Ned. Indii, 25,* 67 (1943).
37. R. Fonteyne and M. Ticket, *Natuurev. Tijdschr. Ned. Indii, 25,* 49 (1943).
38. H. J. Taufen and M. J. Murray, *J. Am. Chem. Soc., 67,* 754 (1945).
39. J. R. Durig and A. C. Morrissey, *J. Chem. Phys., 45,* 1269 (1966).
40. J. R. Durig, *Spectrochim. Acta, 19,* 1225 (1963).
41. J. R. Durig and A. C. Morrissey, *J. Mol. Struct., 2,* 377 (1968),
42. J. R. Durig and J. N. Willis, Jr., *Spectrochim. Acta, 22,* 1299 (1966).
43. G. J. Leary and R. D. Topsom, *Spectrochim. Acta, 21,* 1161 (1965).
44. A. J. Downs and A. Haas, *Spectrochim. Acta, 23A,* 1023 (1967).
45. P. A. Akishen, N. G. Rambidi, K. Yu. Novitskii, and Yu. K. Yur'ev, *Vestnik Mosdov. Univ. 9,* No. 3, *Ser. Fiz. Mat. i Estestven. Nauk,* No. 2, 77–80 (1954); from *Chem. Abstr. 48,*10436d.
46. J. R. Durig and R. C. Lord, *Spectrochim. Acta, 19,* 769 (1963).
47. J. H. Wray, *J. Phys. B. (Proc. Phys. Soc.), 1,* 485 (1968).
48. L. A. Evseeva and L. M. Sverdlov, *Izv. Vyssh. Ucheb. Zavd., Fiz., 11,* 118 (1968).
49. R. Rambaud and M. Vessiere, *Bull. Soc. Chim. Fr., 1960,* 1114.
50. M. I. Batuev, A.A. Akhrem, and A. D. Matveeva,*Dokl, Akad. Nauk/SSSR, 137,* 1113 (1961).
51. J. Derkosch, E. Ernstbrunner, E. G. Hoffman, F. Osterreicher, and E. Ziegler, *Monatsh. Chem., 98,* 956 (1967).
52. Yu. N. Sheinker, E. M. Peresleni, and G. I. Braz, *Zh. Fiz. Khim., 29,* 518 (1955).
53. C. W. Bjork, N. C. Craig, R. A. Metsch, and J. Overend, *J. Am. Chem. Soc., 87,* 1186 (1965).
54. R. W. Mitchell and J. A. Merritt, *J. Mol. Spectrosc., 22,* 165 (1967).
55. R. W. Mitchell and J. A. Merritt, *J. Mol. Spectrosc., 27,* 197 (1968).

CHAPTER FIFTEEN

HETEROCYCLIC COMPOUNDS — NONAROMATIC FIVE-MEMBERED RINGS

15.1 RINGS CONTAINING ONE HETEROATOM

Detailed vibrational assignments have been made for tetrahydrofuran (1-5), pyrrolidine (1,3,6-7), tetrahydrothiophene (3,8) (Spectrum 102, Appendix 2), silacyclopentane (9), germylcyclopentane (10), and tetrahydroselenophene (11). The molecular structures of these compounds are similar to that of cyclopentane. The cyclopentane ring is puckered and one of the internal degrees of freedom, which in a planar ring would be vibrational, must be treated as a free pseudorotation (12-16). This pseudorotation can be thought of as a bending motion in which successive members of the ring are out of the plane of the ring. In cyclopentane, the potential barrier to rotation is zero and the molecule can be regarded as a free pseudorotator. In tetrahydrofuran and pyrrolidine the insertion of the oxygen or nitrogen heteroatom introduces a small barrier (< 0.5 kcal/mole for tetrahydrofuran) to pseudorotation but these molecules may be considered as almost free pseudorotators. However, the introduction of heavier heteroatoms (S, Se, Si, Ge) into the ring leads to considerable barriers to pseudorotation (2-6 kcal/mole). Therefore, these molecules exist in a number of different conformations that are in equilibrium and correspond to minima in the pseudorotation potential. Some of the basic conformations of these rings are the half-chair (twisted form), which can be represented as I or II and the envelope form III. The half-chair form has C_2 symmetry whereas the envelope form has C_S. Buys, Altona, and Havinga (17) have shown that 3-substituted tetrahydrofurans can exist in 20 energetically different C_2 and C_S forms.

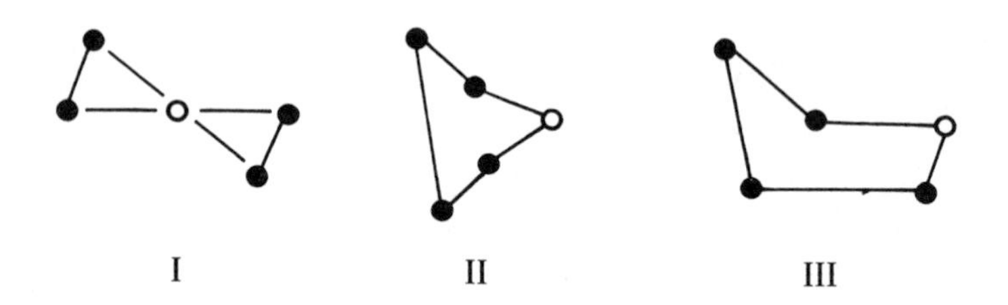

In describing the normal modes of these molecules, it is convenient to picture them as modified forms of the modes that would occur in a planar configuration with C_{2v} symmetry. In these five-membered heterocyclic rings, nine of the modes can be characterized as ring modes. Seven of these can be considered "in-plane" modes and two as "out-of-plane" modes. One of these latter modes is the pseudorotation. In Table 15.1 the vibrational frequencies of these ring modes are listed along with their relative intensities in the Raman spectra. For the molecules containing the lighter heteroatoms, the most intense vibration is the totally symmetric ring stretching mode or ring "breathing" vibration. For heavier heteroatoms, this vibration is better described as the symmetric C–X–C stretch.

The following derivatives of tetrahydrofuran have been investigated in the IR and Raman: 2-mono- and 2,5-disubstituted compounds (4,18); hydrofuranols (19), and 3-halogeno- and *trans*-3,4-dihalogenotetrahydrofurans (17). In these compounds the ring "breathing" vibration found at 914 cm^{-1} in tetrahydrofuran is split into several bands (generally two or three) in the range 1000–890 cm^{-1}. The most intense of these bands in the Raman spectra lies between 950 and 900 cm^{-1}. Another ring stretching vibration appears as a medium to strong band in the region 1080–1040 cm^{-1}.

Nahum (20,21) has investigated the Raman spectra of 2,3- and 2,5-dihydro-1 furan. Akishin et al. (22) have also reported the Raman spectrum of the latter compound. Veda and Shimanouchi (23) have shown that 2,5-dihydrofuran possesses a planar ring skeleton and, therefore, has C_{2v} symmetry. Other studies of derivatives of 2,5-dihydrofuran have been made by Belogorodskii et al. (24) and by Eiduss et al. (25). The Raman spectrum of 2,3-dihydrobenzofuran (coumaran) has been studied by Kohlrausch and Seka (26). 2,5-Dihydrofuran is characterized by a strong Raman band at 982 cm^{-1} and very strong bands at 1099, 1481, and 1622 cm^{-1}. On the other hand, 2,3-dihydrofuran has strong Raman bands at 925 and 1135 cm^{-1} and a very strong band at 1619 cm^{-1}. The bands around 1600 cm^{-1} arise from the C=C stretch in these compounds.

Table 15.1 Ring Modes in the Vibrational Spectra of Saturated Five-Membered Rings

Type	Symmetry Species			Tetrahydrofuran		Pyrrolidine		Tetrahydro-thiophene		Germylcyclopentane	
	C_{2v}	C_s	C_2	Raman cm^{-1}	Intensity	Raman cm^{-1}	Intensity	Raman cm^{-1}	Intensity	Raman cm^{-1}	Intensity
Ring stretches	a_1	a'	a	1174	Medium	980	Medium	960	Strong	848	Strong
	a_1	a''	a	1028	Strong	—	—	822	Strong	761	Medium
	b_1	a''	b	—	—	925	Medium	1036	Medium	946	Medium
	b_1	a''	b	1071	Medium	872	—	884	Strong	588	Medium
Ring "breathing"	a_1	a'	a	914	Very strong	899	Very strong	688	Very strong	635	Very strong
In-plane ring deformations	a_1	a'	a	596	Weak	593	Weak	520	Medium	481	Weak
	b_1	a''	b	—	—	349	Weak	471	Strong	345	Strong
Out-of-plane ring deformations	a_2	a''	a	276	Weak	300	Very weak	297	Weak	273	Weak
	b_2	a'	b	215	Very weak	219	Very weak	—	—	—	—

Derivatives of pyrrolidine that have been studied by Raman spectroscopy include *N*-methylpyrrolidine (3), *N*-ethyl- and *N*-phenylpyrrolidine (1), proline and hydroxyproline (27). In pyrrolidine, the monomeric N–H stretching frequency is at 3350 cm^{-1} and the frequency for the associated form (hydrogen-bonded) occurs at 3305 cm^{-1}. In isoelectric proline and hydroxyproline and in their hydrochloride salts, Raman bands appear near 3000 and 3025 cm^{-1}. These bands are not present in the Raman spectra of their sodium salts. In *N*-methyl- and *N*-ethylpyrrolidine, the ring "breathing" vibration occurs as a very strong Raman band at 899 and 904 cm^{-1}, respectively. For proline it is located at 902, 905, and 904 cm^{-1} for the cation, dipolar ion, and anion forms, respectively. For hydroxyproline strong bands are found in the range 880–850 cm^{-1} depending upon the ionic form of the molecule.

Detailed vibrational assignments have been made for the tetrahydrothiophene derivatives, tetramethylene sulfoxide (28), sulfolane (tetramethylene sulfone) (37), and 2,5-dihydrothiophene (29). From the vibrational data, it is probable that tetramethylene sulfoxide is a puckered ring with C_2 symmetry. The in-phase C–S–C stretch occurs as a strong band in the Raman at 667 cm^{-1} whereas the out-of-phase C–S–C stretch appears at 636 cm^{-1}. The S=O stretch of the sulfoxide gives rise to a very strong Raman band at 1030 cm^{-1}. The S=O bend at 330 cm^{-1} is a strong Raman band. In sulfolane, the symmetric C–S–C stretch is assigned a frequency of 675 cm^{-1} and the antisymmetric C–S–C stretch is at 734 cm^{-1}. The O=S=O antisymmetric stretch is located at 1306 cm^{-1} and the symmetric O=S=O stretch at 1144 cm^{-1}. 2,5-Dihydrothiophene possesses a planar structure with C_{2v} symmetry. The in-phase C–S–C stretch occurs at 711 cm^{-1} as an intense Raman band and the out-of-phase C–S–C stretch at 641 cm^{-1}. The frequency of the C=C bond is at 1637 cm^{-1} in this compound.

Raman spectra of derivatives of these five-membered heterocyclic rings in which one or more carbonyl groups are attached to the ring include γ-butyrolactone (30), unsaturated lactones (38), ascorbic and tetronic acids (31), cyclic five-membered ring anhydrides (32–34), 2-pyrrolidone (35,36), and succinimide (32). Raman data for γ-lactones and cyclic anhydrides are collected in Tables 15.2 and 15.3, respectively. Saturated five-membered ring γ-lactones generally exhibit one carbonyl band at 1795–1750 cm^{-1}. The analogous open-chain esters have their carbonyl stretch near 1740 cm^{-1}. The increase in the C=O stretching frequency has been attributed to the change in the hybridization of the carbon atom in the carbonyl group upon ring closure (39). In those γ-lactones in which a double bond is conjugated with the carbonyl group, two bands are observed in the carbonyl stretching region, the first at 1790–1777 cm^{-1} and the second at 1765–1740 cm^{-1}. The relative intensities of these two bands are extremely sensitive to changes in the polarity of the solvent and to temperature, but are independent of concentration. It

Table 15.2 Characteristic Raman Frequencies for γ-Lactones

Name	Structure	Carbonyl Frequencies (cm^{-1})	Other Strong Identifying Raman Bands (cm^{-1})
γ-Butyrolactone	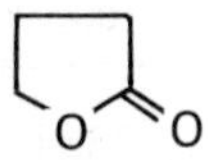	1770	1166,1037, 929 (Ring vibrations)
$\Delta^{\alpha\beta}$-Butenolide	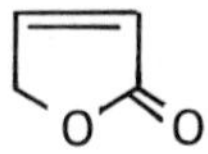	1778,1745 (in $CHCl_3$) 1784,1742 (in CCl_4)	1618,1601 (C=C)
β-*n*-Butyl-$\Delta^{\alpha\beta}$-butenolide	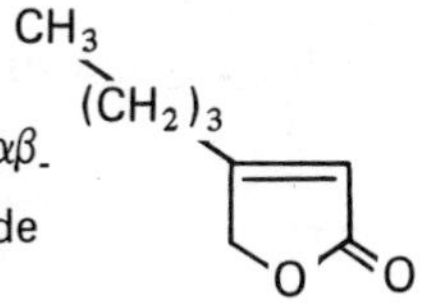 CH_3 $(CH_2)_3$	1779,1747 (in $CHCl_3$) 1781,1747 (in CCl_4)	1636 (C=C)
β-Phenyl-$\Delta^{\alpha\beta}$-butenolide	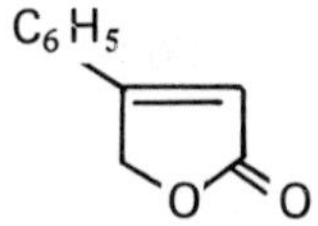 C_6H_5	1789,1751 (in $CHCl_3$) 1786,1758 (in CCl_4)	1694,1628 (C=C) 1712,1626 (C=C)
Phthalide	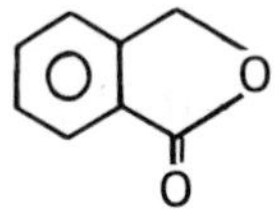	1753	

was assumed (38) that an intramolecular vibrational effect is involved which requires an accidental degeneracy between the carbonyl stretch and an overtone or combination band (Fermi resonance). In the five-membered ring anhydrides, two carbonyl frequencies are observed due to mechanical coupling effects (40). In the IR, the lower frequency band is more intense whereas in the Raman the higher frequency band is appreciably stronger. In acyclic anhydrides, the bands have about the same intensity. In the saturated five-membered cyclic anhydrides the symmetric C=O mode occurs at 1870–1840 cm^{-1} and the antisymmetric C=O mode at 1790–1770 cm^{-1}. When the carbonyl group is conjugated with a double bond, the frequencies are 1850–1830 and 1780–1750 cm^{-1}, respectively, for the two coupled carbonyl stretches.

Of the γ-lactams only 2-pyrrolidone has been extensively studied. In the Raman spectrum of the liquid, there are three bands in the carbonyl stretching region. The band at 1707 cm^{-1} is attributed to the monomeric form. The other two frequencies arise from coupled carbonyl stretchings in the cyclic

Table 15.3 Characteristic Raman Frequencies for Cyclic Anhydrides with Five-Membered Heterocyclic Rings

Name	Structure	Antisymmetric C=O Stretch (cm^{-1})	Symmetric C=O Stretch (cm^{-1})	Other Strong Identifying Raman Bands (cm^{-1})
Succinic anhydride	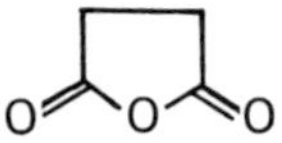	1782	1854	616 (Ring)
Methylsuccinic anhydride	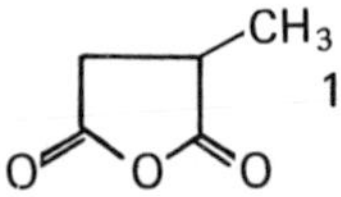	1774	1843	629 (Ring)
cis-Hexahydrophthalic anhydride	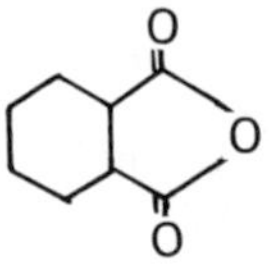	1776	1850	637 (Ring)
trans-Hexahydrophthalic anhydride	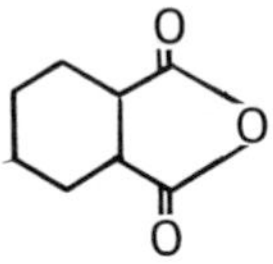	1782	1860	642 (Ring)
Maleic anhydride	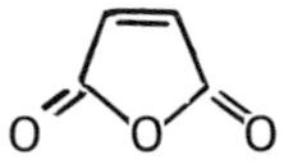	1775	1844	631 (Ring), 1587, 1616 (C=C)
Citraconic anhydride	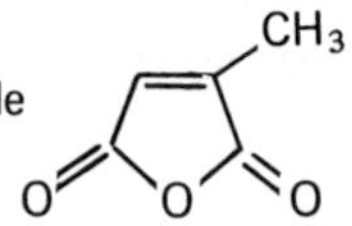	1765	1835	655 (Ring), 1646 (C=C)
2,3-Dimethylmaleic anhydride	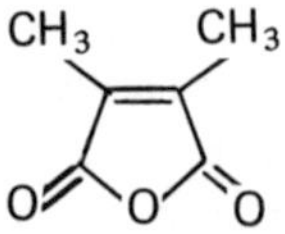	1757	1835	638 (Ring), 1679 (C=C)
Phthalic anhydride	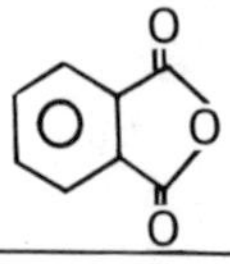	1775	1841	634 (Ring)

dimer formed by hydrogen bonding. This behavior is characteristic of the cyclic amides. The band at 1659 cm^{-1} corresponds to the symmetric C=O mode of the dimer and the 1695 cm^{-1} band corresponds to the antisymmetric vibration. The most intense Raman band occurs at 888 cm^{-1} and is attributed to the ring "breathing" vibration. In thiopyrrolidone (30) the frequency of the C=S stretch in the Raman spectrum is located at 1109 cm^{-1} and the ring "breathing" vibration at 880 cm^{-1}. Recently, the chemistry and physical properties (including IR spectra) of cyclic carboxylic monoimides have been reviewed (41). However, in the Raman only succinimide has been extensively studied (32). In the molten state it displays two bands in the carbonyl stretching region, a strong band at 1760 cm^{-1} and a weak band at 1712 cm^{-1}.

15.2 RINGS CONTAINING TWO OXYGEN ATOMS

The IR and Raman spectra of 1,3-dioxolane and several methyl- and hydroxymethyl-substituted dioxolanes have been investigated by Barker et al. (42,43). The vibrational assignments for these molecules are summarized in Table 15.4. In 1,3-dioxolane, the ring "breathing" vibration gives rise to a very strong highly polarized band at 939 cm^{-1}. In 2-substituted dioxolanes this band is also quite prominent but in the 4-substituted compounds the Raman intensity is quite variable and often weak. Other ring stretching vibrations with frequencies near 1158 and 1087 cm^{-1} frequently occur in the spectra of substituted dioxolanes, as well as the in-plane ring deformations located near 690 and 640 cm^{-1}. The frequency of the methylene scissoring vibration of the 2-methylene group occurs at 1509 cm^{-1} in 1,3-dioxolane and at 1506 cm^{-1} in 4-methyl-dioxolane. This band does not occur in any of the 2-substituted dioxolanes. The frequencies of the in-phase and out-of-phase scissoring vibrations of the 4- and 5-methylene groups both occur near 1480 cm^{-1}. The intensity of this band decreases when one or both of these positions are substituted.

The Raman spectra of several methylenedioxy derivatives in which the Δ^4-dioxoline ring (IV) is fused to a benzene ring have been studied. The compounds include piperonal (44), safrole (44,45), isosafrole (44,45), and dihydrosafrole (45). Characteristic frequencies in the Raman for the 3,4-methylene-

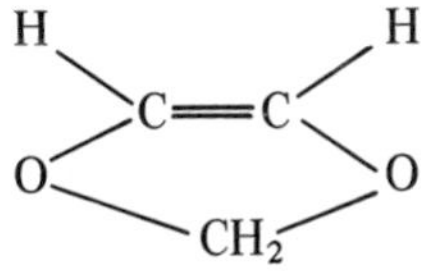

H H
C=C
O O
CH_2

IV

Table 15.4 Characteristic Frequencies in the Raman Spectra of Substituted 1,3-Dioxolanes[a,b]

Vibrational Assignment	Frequency (cm^{-1})						
	Unsub-stituted	2-Methyl-	4-Methyl-	*cis*-2,4-Dimethyl	*trans*-2,4-Dimethyl	2,2-Di-methyl	2,2,4-Trimethyl
CH_2 (C_2) antisymmetric C–H stretch	—	2997 vs	2983 vs	2982 vs	2984 vs	2991 vs	2986 vs
CH_2 (C_4, C_5) symmetric C–H stretch	2894 vs	2885 s	2876 s	2874 vs	2872 vs	2875 vs	2868 s
CH_2 scissors (C_2)	1509 s	—	1506 s	—	—	—	—
CH_2 scissors (C_4,C_5)	1481 s	1480 s	1480 m	1481 m	1478 m	1482 s	1483 w
CH_2 wag (C_2)	1397 w–m	—	1397 m	—	—	—	—
Symmetric CH_2 wag (C_4,C_5)	1352 w	1356 w	—	—	—	1347 w	—
CH_2 wag (C_5)	—	—	1335 w	1342 w	1334 w–m	—	1314 w
Antisymmetric CH_2 wag (C_4,C_5)	1329 w	1320 w–m	1309 m	1300 w	1301 m	1347 w	—
CH_2 twist (C_2)	1246 m	—	1250 w	—	—	—	—
Antisymmetric CH_2 twist (C_4,C_5)	1210 m	1219 w–m	1216 w–m	1218 w	1225 m	1220 m	—
Antisymmetric ring stretch	1158 (IR)	1151 w	1149 w	1164, w–m 1144 w–m	1153 m	—	—
Symmetric ring stretch	1088 w–m	—	1077 w	1085 w	1083 m	1064 w	1095 vw
Symmetric ring stretch	1038 vw	1026 m	1008 w–m	1021 w–m	1035 m	1040 vw	1056 vw
Antisymmetric ring stretch	962 w–m	—	—	947 w	942 w	955 m	979 w–m
Ring "breathing"	939 vs	947 vs	938 m–s	932 w	923 w	946 vs	947 w

CH_2 rock (C_2)	725 w–m	–	726 m	–	–	–	–
In-plane ring deformation	671 vw	695 m	–	695 w–m	694 w–m	675 vw	692 vw
In-plane ring deformation	658 vw	637 vw	619 vw	622 w	639 w	638 s	640 s

[a]Assignments given in refs. 42 and 43.
[b]Ring numbering system:

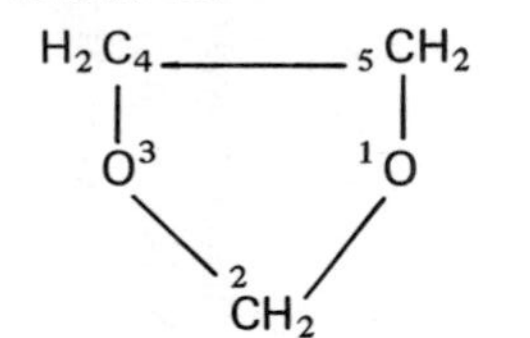

dioxybenzene group, as well as certain characteristic substituent bands are presented in Table 15.5.

The normal vibrations of ethylene carbonate (V) (1,3-dioxolan-2-one) have been investigated by Angell (46) and by Mecke et al. (47). The carbonyl stretching frequency in the Raman spectrum occurs at 1795 cm^{-1} and strong bands at 1071, 971, 894, and 716 cm^{-1} have been assigned to skeletal modes. Vinylene carbonate (VI) has been the subject of an IR and Raman study by Dorris et al. (48). The C=O stretch appears as a band in the Raman spectrum at 1835 cm^{-1}. The skeletal "breathing" vibration occurs as a strong polarized band at 907 cm^{-1}, and the frequencies of other skeletal stretches are located at 1627, 1085, and 1040 cm^{-1}. Oeda (49) has examined several 1,3-dioxolanes in which the carbonyl group is attached to the 4-position. In acetone lactic acid (2,2,5-trimethyl-1,3-dioxolan-4-one) and in acetone leucic acid (2,2-dimethyl-5-isobutyl-1,3-dioxolan-4-one) the C=O stretching frequency in the

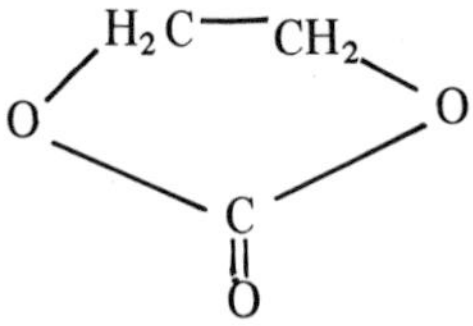

V 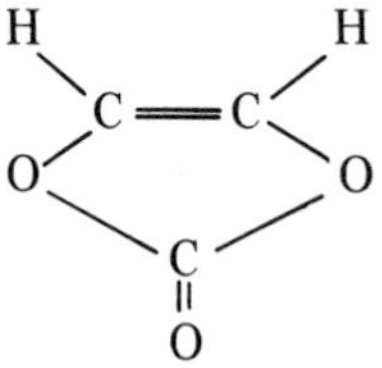 VI

Raman is located near 1790 cm^{-1}. Two other strong Raman bands occurred about 623 and 700 cm^{-1} for both compounds.

15.3 RINGS CONTAINING TWO NITROGEN ATOMS

Durig et al. (50) have assigned the vibrational spectra of 1-pyrazoline (VII). Some prominent Raman bands are: 1548 cm^{-1} (N=N stretch); 1027 cm^{-1} (C-C-C antisymmetric stretch); 941 cm^{-1} (C-C-C symmetric stretch); and 880 cm^{-1} (C-N in-phase stretch).

Other compounds that have been investigated (51-53) in which the nitrogens are adjacent in the ring include antipyrene (2,3-dimethyl-1-phenyl-3-pyrazolin-5-one) and pyramidone (4-dimethylamino-2,3-dimethyl-1-phenyl-3-pyrazolin-5-one). In the solid state, both of these compounds exhibit strong Raman bands near 1640, 1590, 1320, 1310, 1165, and 997 cm^{-1}.

Spectra of heterocyclic rings containing nitrogen atoms in the 1,3-positions include ethylene urea (2-imidazolidinone) (VIII) (47) and ethylene thiourea (2-imidazolidinethione) (54). In the spectrum of ethylene urea amide-type bands occur at 1660 cm^{-1} (C=O stretch and Amide I), 1502 cm^{-1} (Amide II),

Table 15.5 Characteristic Bands in the Raman Spectra of Some Methylenedioxy Compounds

Name		Structure	Frequency (cm^{-1})	
			Bands Characteristic of the 3,4-methylene-dioxybenzene Moiety	Substituent Vibrations
Piperonal	3,4-Methylenedioxy-benzaldehyde	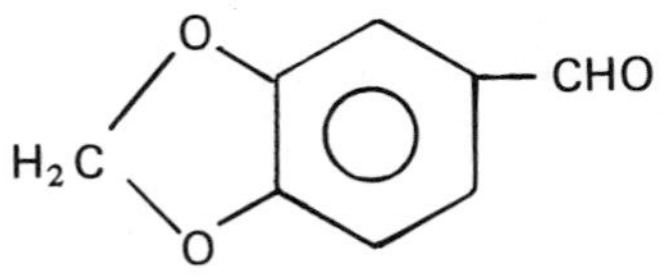 O, H_2C, O, CHO	1606,1499,1444,1348, 1290,1245,1185,1036, 782,715,649,372,222	2820 Aldehyde C–H 1684 Aldehyde C=O stretch
Safrole	1-Allyl-3,4-methylene-dioxybenzene	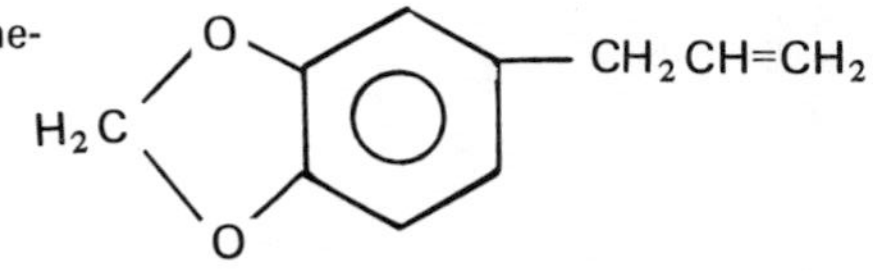 O, H_2C, O, $CH_2CH=CH_2$	1608,1496,1442,1352, 1292,1236,1187,1032, 778,717,649,370,216	1632 C=C stretch
Isosafrole	1-Propenyl-3,4-methyl-enedioxybenzene	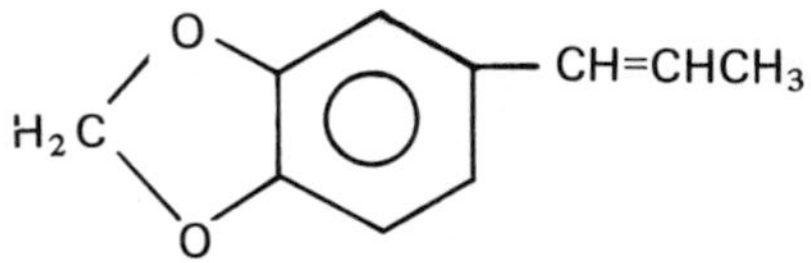 O, H_2C, O, $CH=CHCH_3$	1618,1492,1448,1356, 1289,1250,1184,1039, 790,718,649,369,222	1656 C=C stretch
Dihydrosafrole	1-Propyl-3,4-methyl-enedioxybenzene	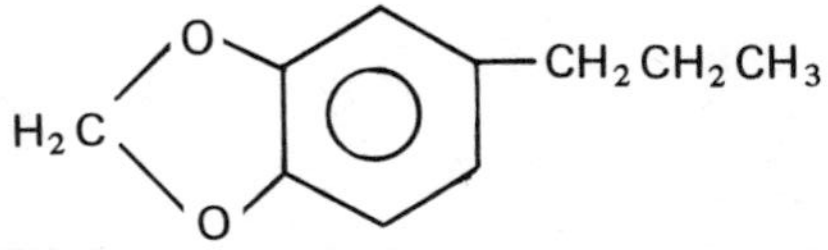 O, H_2C, O, $CH_2CH_2CH_3$	1610,1503,1444,1312, 1247,1187,1033,789, 712,634,365,211	—

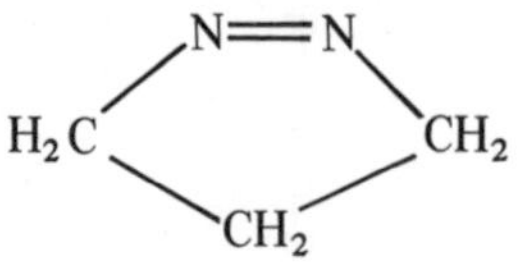

N═N
H_2C CH_2
CH_2

VII

H_2C—CH_2
HN NH
C
‖
O

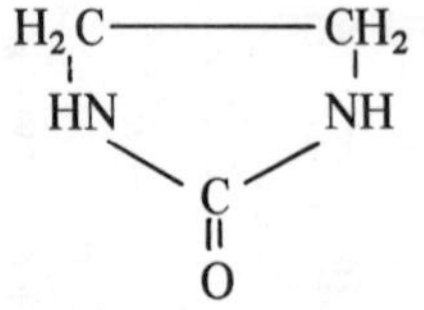

VIII

and 1274 cm^{-1} (Amide III). For ethylene thiourea, these Raman bands occur at 1210 cm^{-1} (C=S stretch), 1518 cm^{-1} (Amide II), and 1275 cm^{-1} (Amide III). The ring "breathing" vibration has been assigned to a very strong band at 507 cm^{-1}.

15.4 RINGS CONTAINING TWO SULFUR ATOMS

Assignments of the fundamental modes have been reported for ethylene dithiocarbonate (1,3-dithiolane-2-one) (IX) (47) and for ethylene trithiocarbonate (1,3-dithiolane-2-thione) (47,54). In the spectrum of ethylene dithiocarbonate, the C=O stretching frequency is located at 1638 cm^{-1}. Bands arising from various skeletal modes occur at 939, 888, 826, and 467 cm^{-1}. In the Raman spectrum of ethylene trithiocarbonate, a very strong band at 1065 cm^{-1} is assigned to the C=S stretch. Bands at 947, 883, 832, and 673 cm^{-1} are attributed to ring stretches, whereas a band at 474 cm^{-1} is due to a ring deformation. The ring "breathing" vibration is located at 505 cm^{-1}; two strong bands at 383 and 247 cm^{-1} are assigned to deformations involving the C=S bond.

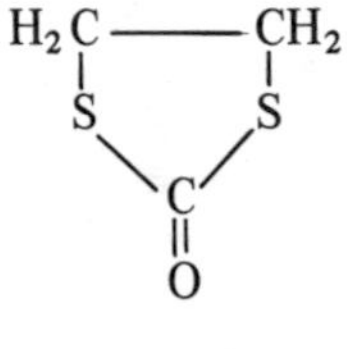

H_2C—CH_2
S S
C
‖
O

IX

H_2C—CH_2
HN S
C
‖
S

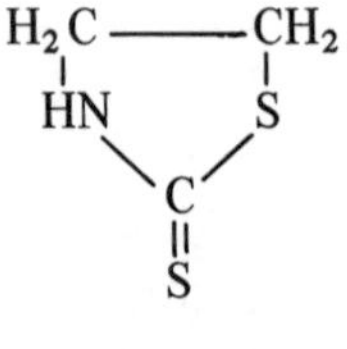

X

15.5 RINGS CONTAINING TWO DIFFERENT HETEROATOMS

The Raman spectrum of thiazolidine-2-thione (IX) has been studied by Mecke et al. (47). A very strong band arising from the C=S stretch occurs in the Raman at 1047 cm^{-1}. Strong bands attributed to various skeletal modes are located at 1204, 1080, 927, and 580 cm^{-1}. Weak bands originating from deformations involving the N-H group occur at 1515 and 650 cm^{-1}.

REFERENCES

1. K. W. F. Kohlrausch and A. W. Reitz, *Z. Phys. Chem., 45B,* 249 (1940).
2. H. Luther, F. Lampe, J. Goubeau, and W. Rodewald, *Z. Naturforsch., 5,* 34 (1950).
3. H. Tschamler and H. Voetter, *Monatsh. Chem., 83,* 302 (1952).
4. G. M. Barrow and S. Searles, *J. Am. Chem. Soc., 75,* 1175 (1953).
5. A. Palm and E. R. Bessell, *Spectrochim. Acta, 16,* 459 (1960).
6. J. C. Evans and J. C. Wahr, *J. Chem. Phys., 31,* 655 (1959).
7. J. P. McCullough, D. R. Douslin, W. N. Hubbard, S. S. Todd, J. F. Messerly, I. A. Hossenlopp, F. R. Frow, J. P. Dawson, and G. Waddington, *J. Am. Chem. Soc., 81,* 5884 (1959).
8. W. N. Hubbard, H. L. Finke, D. W. Scott, J. P. McCullough, C. Katz, M. E. Gross, J. F. Messerly, R. E. Pennington, and G. Waddington, *J. Am. Chem. Soc., 74,* 6025 (1952).
9. J. R. Durig and J. N. Willis, Jr., *J. Mol. Spectrosc., 31,* 320 (1969).
10. J. R. Durig and J. N. Willis, Jr., *J. Chem. Phys., 52,* 6108 (1970).
11. W. H. Green and A. Harvey, *24th Symposium on Molecular Structure and Spectroscopy,* Ohio State University, Columbus, Ohio, 1969.
12. J. E. Kilpatrick, K. S. Pitzer, and R. Spitzer, *J. Am. Chem. Soc., 69,* 2483 (1947).
13. J. P. McCullough, *J. Chem. Phys., 29,* 966 (1958).
14. F. A. Miller and R. G. Inskeep, *J. Chem. Phys., 18,* 1519 (1950).
15. K. S. Pitzer and W. E. Donath, *J. Am. Chem. Soc., 81,* 3213 (1959).
16. W. J. Lafferty, D. W. Robinson, R. V. St. Louis, J. W. Russell, and H. L. Strauss, *J. Chem. Phys., 42,* 2915 (1965).
17. H. R. Buys, C. Altona, and E. Havinga, *Tetrahedron, 24,* 3019 (1968).
18. A. P. Kilimov, M. A. Svechnikova, V. I. Shevchenko, V. V. Smirnov, F. V. Kvasnyuk-Mudryl, and S. B. Zotov, *Khim. Geterotsikl. Soedin., 1967,* 579.
19. N. Baggett, S. A. Barker, A. B. Foster, R. H. Moore, and D. H. Whiffen, *J. Chem. Soc., 1960,* 4565.
20. R. Nahum, *Compt. Rend., 240,* 1898 (1955).
21. R. Nahum, *Ann. Chim. (Paris), 3,* 108 (1958).
22. P. A. Akishin, N. G. Rambidi, I. K. Korobitsyna,G. Ya. Kondrat'eva, and Yu. K. Yur'ev, *Vestn. Mosk. Univ. 10,* No. 12, *Ser. Fiz. Mat. i Estestv. Nauk.,* No. 8, 103 (1955); from *Chem. Abstr. 50,* 8329i (1956).
23. T. Veda and T. Shimanouchi, *J. Chem. Phys., 47,* 4042 (1967).
24. V. V. Belogorodskii, E. G. Savich, L. A. Pavlova, and E. D. Venus-Danilova, *J. Gen. Chem. (USSR), 36,* 2061 (1966).
25. J. Eiduss, D. O. Lolya, K. Venters, and A. Grinvalde, *Latv. PSR Zinat, Akad. Vestis, 1970,* 18; from *Chem. Abstr. 73,* 19965x (1970).

26. K. W. F. Kohlrausch and R. Seka, *Chem. Ber., 71,* 1563 (1938).
27. D. Garfinkel, *J. Am. Chem. Soc., 80,* 3827 (1958).
28. P. Klaeboe, *Acta Chem. Scand., 22,* 369 (1968).
29. W. H. Green and A. B. Harvey, *Spectrochim. Acta, 25A,* 723 (1969).
30. R. Mecke, R. Mecke, and A. Luttringhaus, *Chem. Ber., 90,* 975 (1957).
31. J. T. Edsall and E. L. Sazall, *J. Am. Chem. Soc., 65,* 1312 (1943).
32. K. W. F. Kohlrausch, A. Pongratz, and R. Seka, *Chem. Ber., 66,* 1 (1933).
33. L. Kahovec and K. W. F. Kohlrausch, *Z. Elektrochem., 43,* 285 (1937).
34. L. Kahovec and S. Mardascheiv, *Z. Elektrochem., 43,* 288 (1937).
35. A. E. Parsons, *J. Mol. Spectrosc., 6,* 201 (1961).
36. M. Rey-Lafon, M. Forel, and J. Lascombe, *J. Chim. Phys., 64,* 1435 (1967).
37. J. E. Katon and W. R. Feairheller, Jr., *Spectrochim. Acta, 21,* 199 (1965).
38. R. N. Jones, C. L. Angell, T. Ito, and R. J. D. Smith, *Can. J. Chem., 37,* 2007 (1959).
39. H. K. Hall, Jr. and R. Zbinden, *J. Am. Chem. Soc., 80,* 6428 (1958).
40. L. J. Bellamy, B. R. Connelly, A. R. Philpotts, and R. L. Williams, *Z. Elektrochem., 64,* 563 (1960).
41. M. K. Hargreaves, J. G. Pritchard, and H. R. Dave, *Chem. Rev., 70,* 439 (1970).
42. S. A. Barker, E. J. Bourne, R. M. Pinkard, and D. H. Whiffen, *J. Chem. Soc., 1959,* 802.
43. S. A. Barker, E. J. Bourne, R. M. Pinkard, and D. H. Whiffen, *J. Chem. Soc., 1959,* 807.
44. B. Susz, E. Perrottet, and E. Briner, *Helv. Chim. Acta, 19,* 548 (1936).
45. G. Dupont and R. Dulou, *Bull. Soc. Chim. Fr., 3, (5),* 1639 (1936).
46. C. L. Angell, *Trans. Faraday Soc., 52,* 1178 (1956).
47. R. Mecke, R. Mecke, and A. Luttringhaus, *Chem. Ber., 90,* 975 (1957).
48. K. L. Dorris, J. E. Boggs, A. Danti, and L. L. Altpeter, Jr., *J. Chem. Phys., 46,* 1191 (1967).
49. H. Oeda, *Bull. Chem. Soc. Jap., 10,* 187 (1935).
50. J. R. Durig, J. M. Karriker, and W. C. Harris, *J. Chem. Phys., 52,* 6096 (1970).
51. G. B. Bonino and R. Manzoni-Ansidei, *Mem. R. Accad. Sci. Ist., Bologna, 9,* 3 (1934).
52. A. W. Reitz, *Z. Phys. Chem., 46B,* 165 (1940).
53. E. Canals and P. Peyrot, *Compt. Rend., 206,* 1179 (1958).
54. P. Klaboe, *Acta Chem. Scand., 22,* 1532 (1968).

CHAPTER SIXTEEN

HETEROCYCLIC COMPOUNDS — AROMATIC FIVE-MEMBERED RINGS

16.1 RINGS CONTAINING ONE HETEROATOM

The assignment of the fundamental vibrations of furan has been discussed by Reitz (1), Thompson and Temple (2), Lecomte (3), Guthrie et al. (4), Bak et al. (5), and Rico, Barrachina, and Orza (6). For pyrrole the basic vibrational assignment was made by Lord and Miller (7) with some subsequent modifications by Mirone (8), Morcillo and Orza (9), and Prima (10). These assignments have been reviewed by Jones (26). An assignment of the fundamental frequencies of thiophene (Spectrum 103, Appendix 2) and its deuterated derivatives has been given by Rico, Orza, and Morcillo (11) along with a review of earlier studies. More recent work on thiophene includes that of Bolotina and Sverdlov (12) and Aleksanyan et al. (13–15). Data on the IR and Raman spectra of selenophene have been reviewed by Magdesieva (16). Complete assignments for the vibrational spectra of selenophene have been reported by Gerding et al. (17), Trombetti and Zauli (18), Milazzo (19), and Aleksanyan et al. (13,15,20). Valence force field calculations for thiophene, furan, and pyrrole and their deuterium and methyl derivatives have been performed by Scott (27,28).

Microwave investigations of pyrrole (21), furan (22), thiophene (23), and selenophene (24–25) have confirmed the planarity of these molecules. For

C_{2v} symmetry, the 21 fundamental vibrations of furan, thiophene, and selenophene are distributed among the symmetry species as follows:

$$\Gamma = 8a_1 + 3a_2 + 7b_1 + 3b_2$$

For pyrrole, the distribution is

$$\Gamma = 9a_1 + 3a_2 + 8b_1 + 4b_2$$

where the three additional vibrations are due to the N-H stretching (a_1), the N-H in-plane deformation (b_1), and the N-H out-of-plane deformation (b_2). The vibrational assignments of these molecules are given in Table 16.1. The numbering of the vibrational modes of these heterocyclic molecules that will be used is that of Rico, Orza, and Morcillo (11). The earlier nomenclature of Lord and Miller (7) for the vibrations of the pyrrole molecule is also given in Table 16.1.

The four carbon-hydrogen stretching vibrations (Nos. 1, 2, 12, and 13) occur in the range 3200–3000 cm^{-1}, which is characteristic of aromatic C-H stretches. The N-H stretching vibration of pyrrole occurs as a sharp band at 3495 cm^{-1} for dilute solutions in nonpolar solvents, whereas for concentrated solutions in which the molecules are hydrogen-bonded it occurs as a broad band at 3400 cm^{-1}. Of the four C-H in-plane deformations (Nos. 6, 7, 15, and 16) only the two symmetric C-H deformations (Nos. 6 and 7) give rise to Raman bands of appreciable intensity. The C-H out-of-plane vibrations (Nos. 9, 10, 19, and 20) appear as very weak bands, if at all, in the Raman spectra. This observation is also true of the in-plane and out-of-plane deformations of the N-H bond in pyrrole. The nine remaining vibrations can be described as skeletal modes. Vibration 3, which is the ring "breathing" mode, occurs as a very strong to intense band in the Raman. Vibrations 4 and 17 can be regarded as the symmetric and antisymmetric carbon-heteroatom stretches, respectively. In a similar manner, vibrations 5 and 14 can be viewed as the C=C symmetric and antisymmetric stretches, respectively. In both cases, the symmetric stretches result in bands of much greater intensity in the Raman. The frequencies of the in-plane ring deformations (Nos. 8 and 18) and the out-of-plane ring deformations (Nos. 11 and 21) occur in the region 850–400 cm^{-1} with a variable Raman intensity.

The characteristic frequencies in the laser Raman spectra of monosubstituted pyrroles are listed in Table 16.2. The extensive references to the Raman spectra of substituted pyrroles have been catalogued in the review of Jones (26). The monosubstituted pyrroles can be distinguished by their characteristic bands in the range 1600–1400 cm^{-1}, which arise mainly from the ring stretching modes. Raman spectra of polysubstituted pyrroles have been studied by Bonino, Manzoni-Ansidei, and Pratesi (29,30) and by Stern and Thalmayer (31). For polyalkyl derivatives of pyrrole some characteristic bands arising from ring stretching and C-H in-plane deformation modes are listed on the top of the next page.

1,2-dialkyl pyrroles : 1543, 1505, 1482, 1404, 1258, 1085, 1050, 887 cm^{-1}
2,3-dialkyl pyrroles : 1589, 1514, 1470, 1403, 1278, 1061, 1035 cm^{-1}
2,4-dialkyl pyrroles : 1565, 1510, 1414, 1286, 1061, 941 cm^{-1}
2,5-dialkyl pyrroles : 1605, 1514, 1418, 1258, 1034, 999, 765 cm^{-1}
3,4-dialkyl pyrroles : 1577, 1520, 1062, 979 cm^{-1}
1,2,5-trialkyl pyrroles : 1566, 1549, 1394, 1035, 973, 754 cm^{-1}
1,3,4-trialkyl pyrroles : 1533, 1407, 1054, 934, 771 cm^{-1}
2,3,4-trialkyl pyrroles : 1595, 1525, 1300, 1060, 958, 780 cm^{-1}
2,3,5-trialkyl pyrroles : 1600, 1520, 1292, 999, 763 cm^{-1}
2,3,4,5-tetraalkyl pyrroles : 1607, 1528, 1061, 952, 764 cm^{-1}

Characteristic frequencies in the Raman spectra of substituted furans are listed in Table 16.3. Early work on the Raman spectra of 2-substituted-, 2,5- and 2,3-disubstituted furans was conducted by Han et al. (32–34) and by Bonino and Manzoni-Ansidei (35). They established that the furfuryl group (I) could be readily distinguished from the furfurylidine (II) and 2-furoyl (III) groups on the basis of their Raman spectra in the region of 1600–1460

O — CH_2— (I) O — C-H (II) O — C(=O)— (III)

cm^{-1}. Sobolev et al. (36) in a study of alkyl-substituted furans observed that the spectrum in the region of the stretching frequencies of the C–H bonds varied depending on the number and type of substituents. In furan there are three intense bands at 3157, 3122, and 3092 cm^{-1}. When the heterocyclic ring is substituted at the 2-position, the 3092 cm^{-1} band disappears and the relative intensity of the two other bands changes. For 2,5-disubstitution only one intense band at 3114 cm^{-1} remains. They also found that in the spectrum of 2-vinyl-furan all the characteristic bands in the range 1650–1500 cm^{-1} were split into two components, which was attributed to the existence of two rotational isomers. A similar effect was observed in furfural. Grigg, Sargent, and Knight (37) reported that in the spectra of a series of furans with carbonyl substituents at the 2-positions two carbonyl bands were present whose solvent and concentration dependence indicated the presence of rotational isomerism in these compounds. The Raman spectra of 2-propyl-, 2-propenyl-, and 2-cyclopropyl furan, as well as 1,2-di(fur-2′-yl)ethylene have been discussed by Treschchova et al. (38) who state that in these compounds when the number of conjugated double bonds increases by one the Raman intensity of the C=C

Table 16.1 Vibrational Assignments of Pyrrole, Furan, Thiophene, and Selenophene

				Pyrrole		Furan		Thiophene		Selenophene	
LM Number[a]	ROM Number[b]	Description of Mode[c]	Symmetry Species	Frequency (cm^{-1})	Raman Intensity	Frequency (cm^{-1})	Raman Intensity	Frequency (cm^{-1})	Raman Intensity	Frequency (cm^{-1})	Raman Intensity
8	1	ν(CH)	a_1	3133	90	3167	100	3107	100	3106	97
7	2	ν(CH)	a_1	3100	40	3140	65	3084	52	3071	74
6	5	ν(Ring)	a_1	1467	50	1491	85	1491	64	1423	100
5	4	ν(Ring)	a_1	1379	90	1384	36	1358	45	1345	23
4	6	δ(CH)	a_1	1237	5	1066	20	1081	44	1080	37
2	7	δ(CH)	a_1	1076	30	995	7	1035	50	1014	68
3	3	ν(Ring)	a_1	1144	100	1140	43	832	58	760	94
1	8	δ(Ring)	a_1	708	30	871	10	606	28	457	57
9	—	ν(NH)	a_1	3400	40	—	—	—	—	—	—
16	12	ν(CH)	b_1	3111	90	3161	100	3107	100	3106	97
17	13	ν(CH)	b_1	3133	40	3129	65	3076	27	3061	66
15	14	ν(Ring)	b_1	1530	5	1556	—	1502	—	1521	3
12	15	δ(CH)	b_1	1045	10	1267	—	1257	—	1244	1
11	16	δ(CH)	b_1	1015	5	1180	—	1085	—	1087	37
14	17	ν(Ring) + δ(Ring)	b_1	1418	—	1040	12	896	3	828	1
10	18	δ(Ring)	b_1	647	30	873	—	751	11	627	34

13	—	δ(NH)	b_1	1146	—	—	—	—	—	—	—
20	9	γ(CH)	a_2	868	10	863	—	898	—	881	1
19	10	γ(CH)	a_2	711	—	728	5	683	—	684	1
18	11	γ(Ring)	a_2	618	—	613	—	565	3	544	3
24	20	γ(CH)	b_2	841	20	838	—	867	—	794	8
22	19	γ(CH)	b_2	735	—	745	—	714	—	701	—
23	21	γ(Ring)	b_2	601	—	603	9	453	—	398	17
21	—	γ(NH)	b_2	464	—	—	—	—	—	—	—

[a]Numbering of modes according to Lord and Miller (7).

[b]Numbering of Modes according to Rico, Orza, and Morcillo (11).

[c]ν-Stretching vibration; δ-in-plane deformation; γ-out-of-plane deformation.

Table 16.2 Characteristic Frequencies in the Raman Spectra of Substituted Pyrroles

Type of Substitution	Frequency (cm^{-1})	Relative Raman Intensity	Approximate Description of Vibration
1-Substituted	ca. 3130	Very strong	C–H stretch
	ca. 3105	Very strong	C–H stretch
	1510–1490	Very strong	Ring stretch
	1390–1380	Very strong	Ring stretch
	1290–1280	Very strong	C–N stretch
	1095–1080	Strong	C–H in-plane deformation
	1065–1055	Strong	C–H in-plane deformation
	620–605	Medium	In-plane ring deformation
2-Substituted	3385–3380	Medium	Bonded N–H stretch (unassociated N–H stretch ca. 3530 cm^{-1})
	ca. 3120	Strong	C–H stretch
	ca. 3100	Strong	C–H stretch
	ca. 3045	Medium	C–H stretch
	1570–1560	Medium	Ring stretch
	1475–1460	Very strong	Ring stretch
	1420–1400	Strong	Ring stretch
	1120–1100	Weak	N–H in-plane deformation
	1090–1080	Strong	C–H in-plane deformation
3-Substituted	ca. 3400	Medium	Bonded N–H stretch
	ca. 3130	Strong	C–H stretch
	ca. 3100	Strong	C–H stretch
	ca. 3035	Medium	C–H stretch
	1570–1560	Medium	Ring stretch
	1490–1480	Strong	Ring stretch
	1430–1420	Medium	Ring stretch
	1080–1060	Strong	C–H in-plane deformation

bands increases approximately by a power of ten. Other Raman studies of substituted furans include that of Kimel'fel'd et al. (39) on bromo-, alkylthio-, and cyano-derivatives and that of Hillers and Berzina (40) on alkyl-substituted furans.

Table 16.4 contains the characteristic frequencies for several types of substitution of the thiophene ring. Peron, Saumagne, and Lebas (41,42) have reviewed the earlier Raman studies of 2-substituted thiophenes, presented the

IR and Raman spectra of 20 2-substituted thiophenes, and assigned the fundamental vibrations of the thiophenic ring. Horak, Hyams, and Lippincott (43) have assigned the vibrational spectra of 2-chloro-, 2-bromo-, and 2-iodothiophene. They found that the substituent-sensitive vibrations of the 2-halothiophenes were very similar to those found in the corresponding monohalogenobenzenes. Bolotina and Sverdlov (44) have reported normal coordinate analyses for 2-methyl- and 3-methylthiophene. The spectra of 2-cyano- and 3-cyanothiophenes as well as 3-bromothiophene were presented by Kimel'fel'd et al. (39). Polysubstituted halogeno-thiophenes were investigated by Bonino and Manzoni-Ansidei (35) (2,5-dibromo- and 2,3,5-trichlorothiophene) and by Kohlrausch and Schreiner (45) (tetrachloro- and tetrabromothiophene).

The characteristic frequencies in the Raman spectra of 2-substituted selenophenes are given in Table 16.5. Treshchova et al. (38,46) have studied the Raman spectra of 2-propyl-, 2-cyclopropyl-, 2-propenyl-, and 2-(2′-methylcyclopropyl) selenophene. The spectra of 2- and 3-methyl- and 2- and 3-cyanoselenophene were investigated by Kimel'fel'd et al. (39). The 2-ethyl-, 2-propyl-, 2-butyl-, 2,3,5-trimethyl-, and 2,3,4,5-tetramethyl-derivatives of selenophene were the subject of a vibrational study by Eiduks, Polko, and Yur'ev (47).

Raman spectra that have been reported for fused benzo-derivatives of these heterocyclic compounds include indole (benzo[b]pyrrole) (48), 2-methylindole (48), benzo[b]furan (coumarone) (48,49), benzo[b]thiophene(thionaphthene) (48,49), dibenzofuran (50–53), carbazole(dibenzopyrrole) (54), and dibenzothiophene (50).

16.2 RINGS CONTAINING HETEROATOMS IN THE 1,2-POSITIONS

The vibrational assignments in the IR and Raman spectra of pyrazole (IV), isoxazole (V), and isothiazole (VI) are summarized in Table 16.6. Recently,

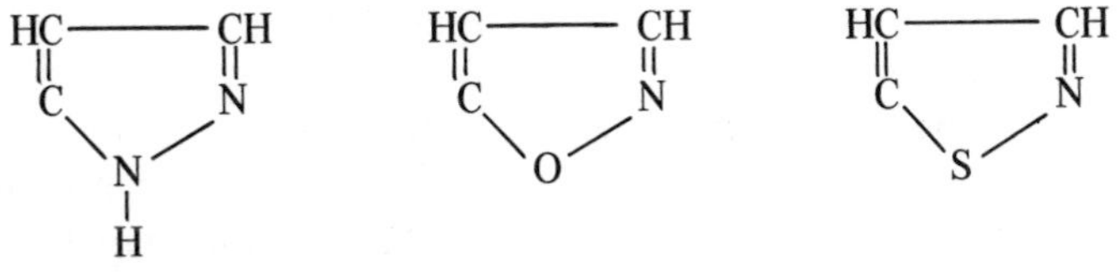

IV V VI

the vibrational assignments for pyrazole in the vapor, solution, and solid states have been discussed by Zecchina et al. (55). Earlier, the Raman

Table 16.3 Characteristic Frequencies in the Raman Spectra of Substituted Furans

Type of Substitution	Frequency (cm^{-1})	Relative Raman Intensity	Approximate Description of Vibration
2-Substituted	3157–3148	Very strong	C–H stretch
	3125–3119	Very strong	C–H stretch
	1605–1590	Strong	Ring stretch characteristic of 2-furfuryl group
	1585–1560	Strong	Ring stretch characteristic of 2-furfurylidene or 2-furoyl group
	1515–1490	Very strong	Ring stretch characteristic of 2-furfuryl group
	1480–1460	Very strong	Ring stretch characteristic of 2-furfurylidene or 2-furoyl group
	1390–1370	Strong	Ring stretch
	1230–1220	Medium	C–H in-plane deformation
	1160–1140	Medium	C–H in-plane deformation
	1080–1060	Medium	C–H in-plane deformation
	1020–992	Strong–medium	Ring "breathing" vibration
	930–920	Medium–weak	C–H out-of-plane deformation
	888–880	Medium–weak	C–H out-of-plane deformation
	635–605	Weak	Out-of-plane ring deformation
2,5-Disubstituted	3115	Very strong	C–H stretch
	ca. 1620	Weak	Ring stretch
	1600–1571	Very strong	Ring stretch
	1530–1500	Strong–weak	Weak intensity for alkyl substituents, strong for electron-accepting substituents

Table 16.4 Characteristic Frequencies in the Raman Spectra of Substituted Thiophenes

Type of Substitution	Frequency (cm^{-1})	Relative Raman Intensity	Approximate Description of Vibration
2-Substituted	3130–3109	Very strong–strong	C–H stretch
	3110–3084	Very strong–strong	C–H stretch
	3092–3078	Very strong–strong	C–H stretch
	1538–1504	Strong–medium	Ring stretch
	1443–1398	Very strong	Ring stretch
	1365–1335	Strong–medium	Ring stretch
	1240–1215	Medium	C–H in-plane deformation
	1085–1074	Medium	C–H in-plane deformation
	1052–1029	Medium	C–H in-plane deformation
	916–890	Weak	C–H out-of-plane deformation
	867–842	Strong	Ring in-plane deformation
	854–778	Weak	C–H out-of-plane deformation
	768–736	Medium	Ring in-plane deformation
	724–670	Weak	C–H out-of-plane deformation
	571–544	Weak	Ring out-of-plane deformation
3-Substituted	3110–3100	Very strong–strong	C–H stretch
	3090–3050	Very strong–strong	C–H stretch
	1542–1492	Medium	Antisymmetric C=C stretch
	1410–1380	Very strong–medium	Symmetric C=C stretch
	1376–1362	Medium	Ring stretch
	935–880	Strong–medium	Antisymmetric C–S stretch
	850–825	Very strong–strong	Symmetric C–S stretch

Table 16.5 Characteristic Frequencies in the Raman Spectra of Substituted Selenophenes

Type of Substitution	Frequency (cm^{-1})	Relative Raman Intensity	Approximate Description of Vibration
2-Substituted	3110–3100	Very strong	C–H stretch
	3090–3050	Very strong	C–H stretch
	1550–1530	Medium–weak	Ring stretch
	1460–1430	Strong	Ring stretch
	1345–1325	Medium	Ring stretch
	1100–1075	Medium	C–H in-plane deformation
	1040–1015	Medium	C–H in-plane deformation
	810–765	Medium	Ring "breathing" vibration
	635–615	Medium–weak	In-plane ring deformation

spectrum of the liquid state had been reported by Bonino and Manzoni-Ansidei (56). Anderson et al. (57) in an IR study of pyrazole in carbon tetrachloride solution found that in the concentration range 10^{-4}–1 *M* there existed an equilibria between the monomer, a cyclic dimer, and a cyclic trimer. Zecchina et al. (55) also confirmed that several bands disappeared at dilute concentrations. The IR and Raman spectra of isoxazole have been investigated by Borello (58) and by Califano, Piacenti, and Speroni (59). The vapor and liquid phase spectra of the deuterated derivative, isoxazole-d_3, were also measured by Adembri, Speroni, and Califano (60) to establish a final assignment. The assignment of the fundamental modes of isothiazole was performed by Califano, Piacenti, and Sbrana (61).

These three molecules possess C_s symmetry, which means that all of the fundamental frequencies are active in both the IR and Raman. There are nine modes that predominantly involve the C–H bonds (three C–H stretches, three in-plane and three out-of-plane deformations) and nine that involve the ring (five ring stretches including the ring "breathing" vibration, and two in-plane as well as two out-of-plane ring deformations). For pyrazole there are, in addition, three other modes—an N–H stretch, an in-plane and an out-of-plane N–H deformation. The C–H stretches occur in the range 3150–3050 cm^{-1}, which is typical of aromatic C–H bonds. Generally three typical ring stretching vibrations occur in the region 1600–1350 cm^{-1}.

Data that are available on the characteristic frequencies in the Raman

Table 16.6 Vibrational Assignments for Pyrazole, Isoxazole, and Isothiazole in the Vapor State

C_s Symmetry Species	Vibration Number	Approximate Description of Vibration	Pyrazole[a]	Isoxazole[b]	Isothiazole[c]
a'	1	C–H stretch	3140	3140	3120
	2	C–H stretch	–	3128	3099
	3	C–H stretch	3074	3086	3071
	4	Ring stretch	1530	1560	1489
	5	Ring stretch	1446	1432	1391
	6	Ring stretch	1394	1373	1342
	7	In-plane C–H deformation	1253	1217	1239
	8	In-plane C–H deformation	1359	1330	1071
	9	Ring "breathing"	1121	1128	1060
	10	In-plane C–H deformation	1057	1089	1041
	11	Ring stretch	1021	1021	872
	12	In-plane ring deformation	931	930	819
	13	In-plane ring deformation	910	856	755
		N–H stretch	3541	–	–
		In-plane N–H deformation	1121	–	–
a''	14	Out-of-plane C–H deformation	879	1033	910
	15	Out-of-plane C–H deformation	833	889	859
	16	Out-of-plane C–H deformation	744	764	726
	17	Out-of-plane ring deformation	668	632	590
	18	Out-of-plane ring deformation	612	595	478
		Out-of-plane N–H deformation	515	–	–

[a] Assignments from ref. 55.
[b] Assignments from refs. 59,60.
[c] Assignments from ref. 61.

spectra of substituted pyrazoles and oxazoles are collected in Table 16.7. The Raman spectra of some polysubstituted alkyl derivatives of pyrazole were reported by Bonino and Manzoni-Ansidei (62). Zerbi and Alberti (63,64) have assigned several characteristic frequencies in the IR for monoalkyl- and polyalkyl-substituted pyrazoles. Benzo [c] pyrazole(indazole) as well as the 1-methyl and 2-methyl derivatives were investigated by Kohlrausch and Seka (48,65). The Raman spectra of several monoalkyl, polyaklyl, and phenyl derivatives of oxazole have been reported by Milone and Tappi (66). Assignments in the IR and Raman spectra of substituted oxazoles have been discussed by Borello (58) and by Katritzky and Boulton (67). For substituted isothiazoles only the IR spectra have been investigated. These results are summarized in a review by Slack and Wooldridge (68) and are analogous with other five-membered heterocyclic ring systems.

16.3 RINGS CONTAINING HETEROATOMS IN THE 1,3-POSITIONS

The vibrational spectra of imidazole (VII) have been extensively studied.

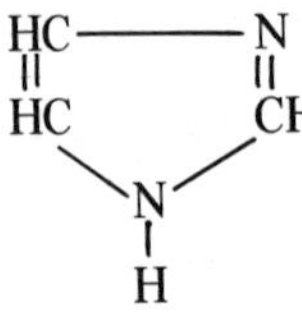

HC—N, HC=CH ring with N–H	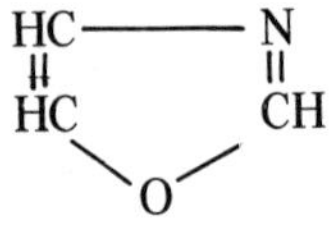 HC—N, HC=CH ring with O	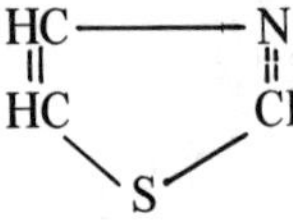 HC—N, HC=CH ring with S
VII	VIII	IX

Kohlrausch and Seka (69) first observed the Raman spectrum in the crystalline state. Garfinkel and Edsall (70) examined the Raman spectra of aqueous solutions of imidazole and the imidazolium ion. An IR study of the vapor, liquid, and solid states was conducted by Milone and Borello (71). A comprehensive IR and Raman investigation of imidazole, imidazole-1-d_1, imidazole-2,4,5-d_3, and imidazole-d_4 in the vapor, liquid, aqueous solution, and solid states was accomplished by Bellocq et al. (72) and by Perchard, Bellocq, and Novak (73). Vibrational assignments for the fundamental frequencies of imidazole in various physical states obtained from these studies are given in Table 16.8. Cordes and Walter (74) have also assigned the IR spectra of imidazole and its -d_4 derivative as well as performing a normal coordinate calculation using a Urey-Bradley force field. The low-frequency Raman spectra of imidazole single crystals have been reported by Colombo (75).

The most characteristic bands in the Raman spectra of imidazole and its

Table 16.7 Characteristic Frequencies in the Raman Spectra of Substituted Pyrazoles and Isoxazoles[a]

Compound	Type of Vibration and Frequency (cm^{-1})								
	Associated N–H Stretch	Aromatic C–H Stretch	Ring Stretches			Ring "Breathing"	C–H Deformation	Ring Deformation	
Pyrazoles									
3-Alkyl	3175	3125–3096	1580		1472	1020–1010	1050	940	830
4-Alkyl	3175	3125–3096	1575		1492	1010–990	1060–1040	950	855
N-Alkyl	—	3125–3096	1520		1397	1040–1030	1090–1060	970–950	850
3,5-Dialkyl	3154	3139	1580–1575		1477	1000–960	1050–1040	—	—
3,4-Dialkyl	3175	3129	1602–1582		1500–1462	—	1070–1065	—	—
3,4,5-Trialkyl	3175	—	1592–1587		1522–1508	1005–990	—	—	—
Isoxazoles									
3-Substituted	—	3150–3050	1670		1540	1015	1230–1220	920–910	—
5-Substituted	—	3150–3050	1670		1520	1001	1230–1210	940–900	—
3,5-Disubstituted	—	3150–3050	1675	1558	1501	1009	1205	915–890	—
3,4,5-Trisubstituted	—	—	1693	1568	1522	1026	1215	920–900	—

[a]Numbering of ring: 4, 3, 5, N 2, X 1 where X = N–H, O.

Table 16.8 Fundamental Vibrations of Imidazole in Various Physical States[a]

Symmetry Species	Vibration Number	Approximate Description of Vibration	Frequency (cm^{-1}) Vapor at 190°C	Aqueous Solution (ca. 5 *M*)	Liquid at 90°C	Solid
a′	1	N–H stretch	3518	–	2900	2800
	2	C–H stretch (4,5)	3160	3156	Masked	3145
	3	C–H stretch (4,5)	3135	3131	Masked	3125
	4	C–H stretch (2)	3135	3131	Masked	3125
	5	Ring stretch	1530	1532	1530	1541
	6	Ring stretch	1480	1485	1480	1485
	7	Ring stretch	1405	1428	1430	1446
	8	Ring stretch	1330	1328	1325	1324
	9	Ring stretch	1127	1135	1135	1142
	10	In-plane N–H deformation	–	1160	1235	1242
	11	In-plane C–H deformation	1260	1259	1255	1261
	12	In-plane C–H deformation	1074	1094	1090	1098
	13	In-plane C–H deformation	1055	1067	1060	1055
	14	In-plane ring deformation	890	828	930	936
	15	In-plane ring deformation	855	865	–	895
a″	1	Out-of-plane C–H deformation	930	914	915	925
	2	Out-of-plane C–H deformation	809	842	820	834
	3	Out-of-plane C–H deformation	723	757	740	746
	4	Out-of-plane ring deformation	668	666	660	660
	5	Out-of-plane ring deformation	626	620	615	619
	6	Out-of-plane N–H deformation	513	–	–	930

[a]Assignments taken from refs. 72,73.

substituted derivatives arise from the ring stretching vibrations (1600–1350 cm^{-1}), the in-plane C-H deformations (1300-1250 cm^{-1}), the in-plane ring deformations (930–820 cm^{-1}), and the out-of-plane ring deformations (690–620 cm^{-1}). Data available on the prominent characteristic bands in some imidazole derivatives are collected in Table 16.9. Kohlrausch and Seka (69) reported the Raman spectra of 1-methyl- and 2-methylimidazole. Perchard and Novak (76) have investigated the IR and Raman spectra of 1-methylimidazole and several of its deuterated analogues. Garfinkel and Edsall (70) have determined the Raman spectra in aqueous solution (ca. 25% solution) of the imidazolium ion, 4(5)-methylimidazole, 4-methylimidazolium chloride, and the sodium salt of histidine (α-amino-5-imidazolepropionic acid).

The Raman spectra of benzimidazole and its 1-methyl- and 2-methyl-derivatives have been recorded (69). A complete normal coordinate calculation has been carried out on benzimidazole by Cordes and Walter (77). Characteristic Raman frequencies for this fused ring system include the ring stretches (ca. 1460, 1415, 1320, and 1250 cm^{-1}), the N-H in-plane deformations (1590–1540 and 1150–1130 cm^{-1}), the C–H out-of-plane deformation near 850 cm^{-1}, and the in-plane ring deformations ca. 770 and 620 cm^{-1}. Vibrational assignments for oxazole (VIII) have been discussed by Borello, Zecchina, and Appiano (78) and by Sbrana, Castellucci, and Ginanneschi (79). These assignments are given in Table 16.10. In the Raman spectrum of the liquid, the most prominent bands include the ring stretches at 1544, 1502, 1330, and 1260 cm^{-1}, the in-plane C–H deformation at 1147 cm^{-1}, the ring "breathing" vibration at 1087 cm^{-1}, and the in-plane ring deformation at 1051 cm^{-1}. Kohlrausch and Seka (48) have reported the Raman spectrum of benzoxazole in the liquid state.

The Raman spectrum of thiazole (IX) as a liquid was first reported by Manzoni-Ansidei and Travagli (80). An IR study of thiazole in the vapor, liquid, and crystal states, as well as the Raman of the liquid was performed by Sbrana, Castellucci, and Ginanneschi (79). The assignment of the vibrational spectra made by these workers is given in Table 16.10. Davidovics et al. (81,82) have also studied the assignments in the IR and Raman spectra of thiazole and thiazole-2-*d*. Derivatives of thiazole that have been studied by Raman spectroscopy include methyl- and halogen-substituted (82,83) and benzothiazole (48). Monosubstituted thiazoles can be identified by the strong to very strong bands at 1520–1485, 1410–1380, and 1320–1295 cm^{-1} arising from ring stretching modes and a strong band ca. 870 cm^{-1} attributed to the ring "breathing" vibration. Bands of variable intensity arising from the in-plane ring deformations also appear near 750 and 600 cm^{-1}.

Table 16.9 Prominent Bands in the Raman Spectra of Substituted Imidazoles

		Frequency (cm^{-1})			
Compound	State	Ring Stretches	In-plane C–H Defor-mation	In-plane Ring Defor-mation	Out-of-plane Ring Defor-mation
Imidazole	Aqueous solution	1532, 1485, 1428, 1328, 1135	1259	865, 828	666, 620
Imidizolium ion	Aqueous solution	1533, 1449, 1405, 1125	—	874, 820	625
4 (5)-Methylimidazole	Aqueous solution	1568, 1525, 1486, 1420, 1144	1254	822	660, 621
4 (5)-Methylimidazolium-chloride	Aqueous solution	1561, 1480, 1391	1262	841	658, 633
Histidine, Na salt	Aqueous solution	1567, 1531, 1488, 1412, 1148	1265	866, 849	—
1-Methylimidazole	Liquid	1510, 1500, 1353, 1328, 1230	1288	904	665, 616
2-Methylimidazole	Solid	1483, 1413, 1368	1302	928	684

Table 16.10 Vibrational Assignments in the Raman Spectra of Oxazole and Thiazole (Liquid State)[a]

C_s Symmetry Species	Vibration Number	Approximate Description of Vibration	Frequency (cm^{-1})	
			Oxazole	Thiazole
a'	1	C–H stretch	3172	3119
	2	C–H stretch	3143	3085
	3	C–H stretch	3116	3067[c]
	4	Ring stretch	1544	1483
	5	Ring stretch	1502	1382
	6	Ring stretch	1330	1321
	7	In-plane C–H deformation	1260	1243
	8	In-plane C–H deformation	1147	1123
	9	In-plane C–H deformation	1096[b]	1045
	10	Ring "breathing"	1087	868
	11	Ring stretch	1051	888[b]
	12	In-plane ring deformation	910[b]	759
	13	In-plane ring deformation	858[b]	612
a''	14	Out-of-plane C–H deformation	902	849[b]
	15	Out-of-plane C–H deformation	832[b]	798[b]
	16	Out-of-plane C–H deformation	749[b]	717[b]
	17	Out-of-plane ring deformation	644[b]	603[b]
	18	Out-of-plane ring deformation	613	467

[a] Assignments from ref. 79.
[b] In IR spectrum of vapor.
[c] In IR spectrum of crystal.

16.4 RINGS CONTAINING THREE HETEROATOMS

The IR spectrum of 1,2,4-triazole (X) was investigated by Ottinge (84). The vibrational assignment of 1,2,3-triazole was discussed by Borello, Zecchina, and Guglielminotti (85) who found that the spectral data could be adequately interpreted in terms of the asymmetric structure (XI). In this triazole the ring stretching modes are assigned to frequencies at 1533, 1440, 1413, and 1383 cm^{-1} and the ring "breathing" mode to a frequency of 1164 cm^{-1}. Frequencies for the in-plane ring deformations are attributed to bands at 976 and 955 cm^{-1} and those of the out-of-plane ring deformations to 701 and 633 cm^{-1}. The

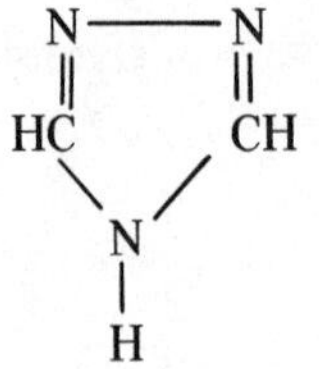

X 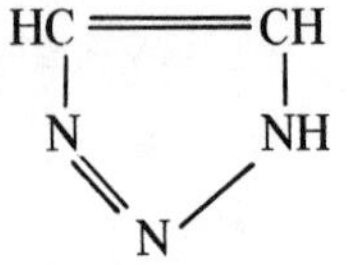 XI 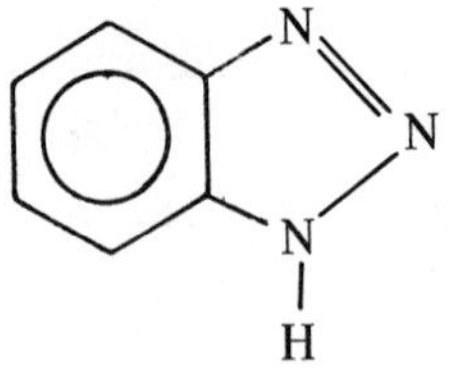 XII

Raman spectra of benzotriazole (IX) and its 1-methyl and 2-methyl derivatives have been recorded by Kohlrausch and Seka (48,65). Characteristic Raman bands in the solid state for these fused ring systems occur at 1625-1590, 1285-1260, 1010-1000, 860-845, and 790-780 cm^{-1}.

The normal vibrations of 1,2,5-oxadiazole (furazan) (XIII) have been discussed by Milone et al. (86). The vibrational assignments in the IR spectra of

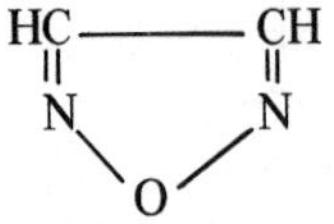

XIII

the vapor, liquid, and solution were first determined by Borello, Zecchina, and Guglielminotti (87). Sbrana, Ginanneschi, and Marzocchi (88) obtained the Raman spectrum of the liquid with polarization measurements as well as the IR spectra of the vapor, liquid, and solid phases. Except for minor discrepancies, the vibrational assignments for 1,2,5-oxadiazole in this paper were in agreement with those published about the same time by Christensen and Nielsen (89) who also examined both the IR and Raman spectra. These assignments for the Raman spectra are given in Table 16.11 along with those for 1,3,4-oxadiazole. The assignment of the fundamental vibrations of 1,3,4-oxadiazole was done by Christensen, Nielsen, and Nielsen (90,91) based on the IR spectra of the vapor, liquid, and solid and the Raman spectrum of the liquid.

The Raman spectra of dialkyl, alkyl aryl, and diaryl derivatives of 1,2,5-oxadiazole (furazan), 1,2,4-oxadiazole (azoxime), and 1,3,4-oxadiazole (oxbiazole) have been investigated by Milone and Muller (92-94). The vibrational assignments in these derivatives and their IR spectra are discussed by Milone and Borello (95,96). The characteristic frequencies in the Raman spectra of these oxadiazoles are summarized in Table 16.12.

Vibrational assignments for the fundamental frequencies found in the Raman

Table 16.11 Fundamental Frequencies in the Liquid Raman Spectra of Oxadiazoles

Symmetry Species of C_{2v}	Vibration Number	Approximate Description of Vibration	Frequency (cm^{-1})			
			1,2,5-Oxadiazole[a]	Relative Raman Intensity	1,3,4-Oxadiazole[b]	Relative Raman Intensity
a_1	1	C–H stretch	3149	Strong	3141	Strong
	2	Ring stretch	1420	Very strong	1536	Very strong
	3	Ring stretch	1313	Very strong	1495	Medium
	4	In-plane C–H deformation	1038	Weak	1272	Very strong
	5	In-plane ring deformation	996	Medium	1092	Strong
	6	In-plane ring deformation	860	Medium	951	Strong
a_2	7	Out-of-plane C–H deformation	820	Very weak	941	Weak
	8	Out-of-plane ring deformation	635	Very weak	655	Medium
b_1	9	C–H stretch	(3144)[c]	—	3141	Strong
	10	Ring stretch	1545	Very weak	(1592)[d]	—
	11	In-plane C–H deformation	1179	Very weak	1223	Strong
	12	In-plane ring deformation	949	Weak	(1078)[d]	—
	13	In-plane ring deformation	(888)[c]	—	925	Weak
b_2	14	Out-of-plane C–H deformation	(838)[c]	—	852	Weak
	15	In-plane ring deformation	621	Very weak	625	Medium

[a]Assignments given in refs. 88,89.
[b]Assignments given in refs. 90,91.
[c]Frequency taken from IR spectrum of vapor.
[d]Frequency taken from IR spectrum of the crystalline solid at –196°C.

Table 16.12 Characteristic Frequencies in the Raman Spectra of Oxadiazoles

Substituent	Frequency (cm^{-1})							
	Ring Stretching Modes				Other Characteristic Frequencies			
A. 1,2,5-Oxadiazoles								
Dialkyl substituted	–	–	1457	1402	1316	–	975	–
Alkyl aryl substituted	1546	1502	1451	–	1278	–	–	–
Diaryl substituted	1546	1489	1448	–	1293	–	–	–
B. 1,2,4-Oxadiazoles								
Alkyl aryl substituted	1550	1491	1451	–	1312	1060	1028	940
Diaryl substituted	1546	1495	1460	1348	1297	1055	1029	932
C. 1,3,4-Oxadiazoles								
Dialkyl substituted	1579	–	1438	–	–	1108	1043	923
Alkyl aryl substituted	1546	1482	1442	–	–	1102	1030	–
Diaryl substituted	1542	1479	1456	–	–	1115	1022	–

spectra of 1,2,5-thiadiazole and 1,3,4-thiadiazole are given in Table 16.13. The IR spectra of 1,2,5-thiadiazole in the vapor, liquid, solution, and crystalline phases as well as the Raman spectrum in the liquid phase have been examined by Benedetti, Sbrana, and Bertini (97,98). The complete vibrational assignments for 1,2,5-thiadiazole and 1,2,5-thiadiazole-d_2, based on the IR and Raman spectra, have been made by Soptrajanov and Ewing (99). Sbrana and Ginanneschi (100) have presented the vibrational assignment of 1,3,4-thiadiazole based on the IR spectra in the vapor and solid phases and in solution. Complete assignments of the fundamental vibration frequencies of 1,3,4-thiadiazole, 1,3,4-thiadiazole-2-d, 1,3,4-thiadiazole-2,5-d_2, 1,3,4-thiadiazole-2-^{13}C, and 1,3,4-thiadiazole-3-^{15}N were made by Christensen and Stroyer-Hansen (101) from the IR and Raman spectra of these compounds in various physical states.

Benedetti and Bertini (102) have proposed a vibrational assignment of the fundamentals of 1,2,5-selenadiazole based on the IR spectra recorded in the vapor, liquid, solution, and crystalline phases and the Raman spectrum recorded in the liquid state. The observed spectra offered evidence that the molecule is planar and possesses C_{2v} symmetry. The fundamental frequencies observed in this liquid Raman spectrum and their assignments are presented in Table 16.13.

Table 16.13 Vibrational Assignments of the Fundamental Modes in the Liquid Raman Spectra of Thiadiazoles and Selenadiazoles

Symmetry Species of C_{2v}	Vibration Number	Approximate Description of Vibration	Frequency (cm^{-1})					
			1,2,5-Thiadiazole[a]	Relative Raman Intensity	1,3,4-Thiadiazole[b]	Relative Raman Intensity	1,2,5-Selenadiazole[c]	Relative Raman Intensity
a_1	1	C–H stretch	3095	Very strong	3095	Strong	3056	Very strong
	2	Ring stretch	1350	Medium	(1403)[d]	—	1367	Medium
	3	Ring stretch	1250	Medium	1398	Strong	1292	Strong
	4	In-plane C–H deformation	1045	Medium	1228	Strong	1013	Medium
	5	In-plane ring deformation	805	Strong	957	Medium	727	Very strong
	6	In-plane ring deformation	685	Medium	896	Very strong	489	Strong
a_2	7	Out-of-plane C–H deformation	(908)[d]	—	836	Weak	(868)[d]	—
	8	Out-of-plane ring deformation	(500)[d]	—	622	Very strong	672	Weak
b_1	9	C–H stretch	3095	Very strong	3095	Strong	(3028)[d]	—
	10	Ring stretch	1455	Very weak	(1526)[d]	—	(1385)[d]	—
	11	In-plane C–H deformation	1215	Weak	1197	Medium	1237	Weak
	12	In-plane ring deformation	890	Very weak	906	Very weak	882	Weak
	13	In-plane ring deformation	770	Weak	752	Medium	586	Medium
b_2	14	Out-of-plane C–H deformation	(838)[d]	—	836	Weak	845	Very weak
	15	In-plane ring deformation	520	Very weak	483	Weak	439	Weak

[a] Assignments from refs. 98,99.
[b] Assignments from ref. 101.
[c] Assignments from ref. 102.
[d] Observed in IR spectrum only.

REFERENCES

1. A. W. Reitz, *Z. Phys. Chem., 33B,* 179 (1937).
2. H. W. Thompson and R. B. Temple, *Trans. Faraday Soc., 41,* 27 (1945).
3. J. Lecomte, *Bull. Soc. Chim. Fr., 1946,* 415.
4. G. B. Guthrie, D. W. Scott, W. N. Hubbard, C. Katz, J. P. McCullough, M. E. Gross, K. D. Williamson, and G. Waddington, *J. Am. Chem. Soc., 74,* 4662 (1952).
5. B. Bak, S. Brodersen, and L. Hansen, *Acta Chem. Scand., 9,* 749 (1955).
6. M. Rico, M. Barrachina, and J. M. Orza, *J. Mol. Spectrosc., 24,* 133 (1967).
7. R. C. Lord, Jr. and F. A. Miller, *J. Chem. Phys., 10,* 328 (1942).
8. P. Mirone, *Gazz. Chim. Ital., 86,* 165 (1956).
9. J. Morcillo and J. M. Orza, *An. Real Soc. Esp. Fes. y Quim (Mad.), 56B,* 231 (1960).
10. A. M. Prima, *Opt. Spektrosk., Akad. Nauk SSSR, Sb. Statei 3,* 157 (1967).
11. M. Rico, J. M. Orza, and J. Morcillo, *Spectrochim. Acta, 21,* 689 (1965).
12. E. N. Bolotina and L. M. Sverdlov, *Opt. Spectrosc., Suppl. 3, Mol. Spectrosc., Opt. Soc. Am.,* 75 (1968).
13. V. T. Aleksanyan, Ya. M. Kimelfeld, N. N. Magdesieva, and Yu. K. Yurev, *Opt. Spectrosc., 22,* 116 (1967).
14. V. T. Aleksanyan, Y. M. Kimelfeld, N. N. Magdesieva, and Yu. K. Yurev, *Opt. Spectrosc., Suppl. 3, Mol. Spectrosc., Opt. Soc. Am.,* 85 (1968).
15. V. T. Aleksanyan, Ya. M. Kimelfeld, and N. N. Magdesieva, *J. Struct. Chem. (USSR), 9,* 633 (1968).
16. N. N. Magdesieva, in *Advances in Heterocyclic Chemistry,* Vol. 12, Academic Press, New York, 1970, pp. 1–38.
17. H. Gerding, G. Milazzo, and H. H. Rossmark, *Rec. Trav. Chim., 72,* 957 (1953).
18. A. Trombetti and C. Zauli, *J. Chem. Soc. (A), 1967,* 1196.
19. G. Milazzo, *Gazz. Chim. Ital., 98,* 1511 (1968).
20. V. T. Aleksanyan, Ya. M. Kimelfeld, N. N. Magdesieva, and Yu. K. Yurev, *Opt. Spectrosc., Suppl. 3, Mol. Spectrosc., Opt. Soc. Am.,* 89 (1968).
21. B. Bak, D. Christensen, L. Hansen, and J. Rastrup-Andersen, *J. Chem. Phys., 24,* 720 (1956).
22. B. Bak, L. Hansen, and J. Rastrup-Andersen, *Disc. Faraday Soc., 19,* 30 (1955).
23. B. Bak, D. Christensen, L. Hansen-Nygaard, and J. Rastrup-Andersen, *J. Mol. Spectrosc., 7,* 58 (1961).

24. R. D. Brown, F. R. Burden, and P. D. Godfrey, *J. Mol. Spectrosc., 25,* 415 (1968).
25. N. M. Pozdeev, O. B. Akulinin, A. A. Shapkin, and N. N. Magdesieva, *Dokl. Akad. Nauk. (USSR), 185* 384 (1969).
26. R. A. Jones, in *Advances in Heterocyclic Chemistry,* Vol. 11, Academic Press, New York, 1970, pp. 443–459.
27. D. W. Scott, *J. Mol. Spectrosc., 31,* 451 (1969).
28. D. W. Scott, *J. Mol. Spectrosc., 37,* 77 (1971).
29. G. B. Bonino, R. Manzoni-Ansidei, and P. Pratesi, *Z. Phys. Chem., 22B,* 21 (1933).
30. G. B. Bonino, R. Manzoni-Ansidei, and P. Pratesi, *Z. Phys. Chem., 25B,* 348 (1934).
31. A. Stern and K. Thalmayer, *Z. Phys. Chem., 31B,* 403 (1936).
32. K. Matsuno and K. Han, *Bull. Chem. Soc. Jap., 9,* 327 (1934).
33. K. Han, *Bull. Chem. Soc. Jap., 11,* 701 (1936).
34. K. Matsuno and K. Han, *Bull. Chem. Soc. Jap., 12,* 155 (1937).
35. G. B. Bonino and R. Manzoni-Ansidei, *Z. Phys. Chem., 25B,* 327 (1934).
36. E. V. Sobolev, V. T. Aleksanyan, R. A. Karakhanov, I. F. Bel'skii, and V. A. Ovodova, *J. Struct. Chem. USSR, 4,* 330 (1963).
37. R. Grigg, M. V. Sargent, and J. A. Knight, *Tetrahedron Lett., 19,* 1381 (1965).
38. E. G. Treschova, D. Ekkhardt, and Yu. K. Yur'ev, *Russ. J. Phys. Chem., 38,* 159 (1964).
39. Ya. M. Kimel'fel'd, V. T. Aleksanyan, N. N. Magdesieva, and Yu. K. Yur'ev, *J. Struct. Chem. USSR, 7,* 42 (1966).
40. S. Hillers and A. Berzina, *Khim. Geterotsikl. Soeden., 1966,* 487.
41. J. J. Peron, P. Saumagne, and J. M. Lebas, *Spectrochim. Acta, 26A,* 1651 (1970).
42. J. J. Peron, P. Saumagne, J. M. Lebas, *Compt. Rend., 264B,* 797 (1967).
43. M. Horak, I. J. Hyams, and E. R. Lippincott, *Spectrochim. Acta, 22,* 1355 (1966).
44. E. N. Bolotina and L. M. Sverdlov, *Zh. Prikl. Spektrosk., 7,* 870 (1967).
45. K. W. F. Kohlrausch and H. Schreiner, *Acta Phys. Austriaca, 1,* 373 (1948).
46. Yu. K. Yur'ev, N. N. Mezentsova, and E. G. Treshchova, *Vestn. Mosk. Univ., Sec. II, Khim. 17,* 60 (1962); *Chem. Abstr. 57,* 294a.
47. J. Eiduks, T. Polko, and Yu. K. Yur'ev, *Latv. PSR Zinat. Akad. Vestis, 1963,* 63; *Chem. Abstr. 59,* 2302 h.
48. K. W. F. Kohlrausch and R. Seka, *Chem. Ber., 71,* 1563 (1938).
49. G. B. Bonino and R. Manzoni-Ansidei, *Ric. Sci., 8,* 225 (1937).
50. P. Donzelot and M. Chaix, *Compt. Rend., 202,* 851 (1936).

51. J. Behringer, *Z. Elektrochem., 62,* 544 (1958).
52. J. Behringer and J. Brandmüller, *Z. Angew. Phys., 14,* 674 (1962).
53. G. Davidovics, J. Chouteau, and H. Reymond, *Can. J. Chem., 44,* 3073 (1966).
54. A. Bree and R. Zwarich, *J. Chem. Phys., 49,* 3344 (1968).
55. A. Zecchina, L. Cerruti, S. Coluccia, and E. Borello, *J. Chem. Soc. (B), 1967,* 1363.
56. G. B. Bonino and R. Manzoni-Ansidei, *Rend. Accad. Lincei, 22,* 349 (1935).
57. D. M. W. Anderson, J. L. Duncan, and F. J. C. Rossotti, *J. Chem. Soc., 1961*, 140.
58. E. Borello, *Gazz. Chim. Ital., 89,* 1437 (1959).
59. S. Califano, F. Piacenti, and G. Speroni, *Spectrochim. Acta, 15,* 86 (1959).
60. G. Adembri, G. Speroni, and S. Califano, *Spectrochim. Acta, 19,* 1145 (1963).
61. S. Califano, F. Piacenti, and G. Sbrana, *Spectrochim. Acta, 20,* 339 (1964).
62. G. B. Bonino and R. Manzoni-Ansidei, *Accad. Naz. Lincei Rend. Sc. Fis. Mat. Nat., 22,* 438 (1935).
63. G. Zerbi and C. Alberti, *Spectrochim. Acta, 18,* 407 (1962).
64. G. Zerbi and C. Alberti, *Spectrochim. Acta, 19,* 1261 (1963).
65. K. W. F. Kohlrausch and R. Seka, *Chem. Ber., 73,* 162 (1940).
66. M. Milone and G. Tappi, *Gazz. Chim. Ital., 70,* 359 (1940).
67. A. R. Katritzky and A. J. Boulton, *Spectrochim. Acta, 17,* 238 (1961).
68. R. Slack and K. R. H. Wooldridge, *Adv. Heterocycl. Chem., 4,* 107–120 (1965).
69. K. W. F. Kohlrausch and R. Seka, *Chem. Ber., 71,* 985 (1938).
70. D. Garfinkel and J. T. Edsall, *J. Am. Chem. Soc., 80,* 3807 (1958).
71. M. Milone and E. Borello, *Advances in Molecular Spectroscopy,* Vol. 2, A. Mangini, Ed., Macmillan, New York, 1962, p. 885.
72. A. Bellocq, C. Perchard, A. Novak, and M. Josien, *J. Chim. Phys., 62,* 1334 (1965).
73. C. Perchard, A. Bellocq, and A. Novak, *J. Chim. Phys., 62,* 1344 (1965).
74. M. Cordes and J. L. Walter, *Spectrochim. Acta, 24A,* 237 (1968).
75. L. Colombo, *J. Chem. Phys., 49,* 4688 (1968).
76. C. Perchard and A. Novak, *Spectrochim. Acta, 23A,* 1953 (1967).
77. M. M. Cordes and J. L. Walter, *Spectrochim. Acta, 24A,* 1421 (1968).
78. E. Borello, A. Zecchina, and A. Appiano, *Spectrochim. Acta, 22,* 977 (1966).

79. G. Sbrana, E. Castellucci, and M. Ginanneschi, *Spectrochim. Acta, 23A*, 751 (1967).
80. R. Manzoni-Ansidei and G. Travagli, *Gazz. Chim. Ital., 71*, 677 (1941).
81. G. Davidovics, C. Garrigou-Lagrange, J. Chouteau, and J. Metzger, *Spectrochim. Acta, 23A*, 1477 (1967).
82. G. Davidovics, P. Roepstorff, J. Chouteau, and C. Garrigou-Lagrange, *Spectrochim. Acta, 23A*, 2669 (1967).
83. R. Manzoni-Ansidei and G. Travagli, *Gazz. Chim. Ital., 71*, 680 (1941).
84. W. Otting, *Chem. Ber., 89*, 2887 (1956).
85. E. Borello, A. Zecchina, and E. Guglielminotti, *J. Chem. Soc. (B), 1969*, 307.
86. M. Milone, E. Borello, and S. Nocilla, *Gazz. Chim. Ital., 81*, 896 (1951).
87. E. Borello, A. Zecchina, and E. Guglielminotti, *Gazz. Chim. Ital., 96*, 852 (1966).
88. G. Sbrana, M. Ginanneschi, and M. P. Marzocchi, *Spectrochim. Acta, 23A*, 1757 (1967).
89. D. H. Christensen and O. F. Nielsen, *J. Mol. Spectrosc., 24*, 477 (1967).
90. D. H. Christensen, J. T. Nielsen, and O. F. Nielsen, *J. Mol. Spectrosc., 24*, 225 (1967).
91. D. H. Christensen, J. T. Nielsen, and O. F. Nielsen, *J. Mol. Spectrosc., 25*, 197 (1968).
92. M. Milone and G. Müller, *Gazz. Chim. Ital., 63*, 334 (1933).
93. M. Milone, *Gazz. Chim. Ital., 63*, 456 (1933).
94. M. Milone and G. Müller, *Gazz. Chim. Ital., 65*, 241 (1935).
95. M. Milone and E. Borello, *Gazz. Chim. Ital., 81*, 368 (1951).
96. M. Milone and E. Borello, *Gazz. Chim. Ital., 81*, 677 (1951).
97. E. Benedetti, G. Sbrana, and V. Bertini, *Chim. Ind. Milan, 48*, 761 (1966).
98. E. Benedetti, G. Sbrana, and V. Bertini, *Gazz. Chim. Ital., 97*, 379 (1967).
99. B. Soptrajanov and G. E. Ewing, *Spectrochim. Acta, 22*, 1417 (1966).
100. G. Sbrana and M. Ginanneschi, *Spectrochim. Acta, 22*, 517 (1966).
101. D. H. Christensen and T. Stroyer-Hansen, *Spectrochim. Acta, 26A*, 2057 (1970).
102. E. Benedetti and V. Bertini, *Spectrochim. Acta, 24A*, 57 (1968).

CHAPTER SEVENTEEN

HETEROCYCLIC COMPOUNDS — NONAROMATIC SIX-MEMBERED RINGS

17.1 RINGS CONTAINING ONE HETEROATOM

The vibrational assignments for the bands observed in the Raman spectra of piperidine, tetrahydropyran (pentamethylene oxide), and tetrahydrothiopyran (pentamethylene sulfide) are given in Table 17.1. These molecules have been found to exist in the "chair" form, which possesses C_S symmetry; therefore, all fundamental vibrations are active in both the IR and Raman (a' vibrations are polarized).

Kahovec and Kohlrausch (1) have reviewed the early Raman spectra of piperidine (2-5). Voetter and Tschamler (6) have examined the Raman spectra of piperidine and *N*-methylpiperidine along with those of other six-membered carbocyclic and heterocyclic rings. The vibrational spectra of piperidine, *N*-deutero-, *N*-methyl-, and *N*-*tert*-butyl piperidine have been discussed by Anisimova et al. (7-10).

IR (11) and Raman measurements in the N-H stretching region of piperidine in dilute solutions of nonpolar solvents show that there is a band at 3343 cm^{-1} with a shoulder at 3315 cm^{-1}. The predominant high frequency band is assigned to the "chair" conformation of the piperidine molecule in which the hydrogen attached to the nitrogen atom is in the equatorial position and the lone pair orbital of the nitrogen is in the axial position. This form comprises about 60% of the total number of molecules. The low frequency band is assigned to the conformation in which the hydrogen attached to the nitrogen is in the axial form. In the C-H stretching region piperidines with β- and γ-methylene groups exhibit an antisymmetric stretch near 2933 cm^{-1} and a

Vibrational Assignment	Frequency (cm^{-1}) Piperidine	Relative Raman Intensity	Tetrahydro-pyran	Relative Raman Intensity	Symmetry Species	Tetrahydro-thiopyran	Relative Raman Intensity
1. N–H stretch	3339	Strong	—	—	—	—	—
2. C–H stretches	2936	Very strong	2957	Very strong	—	2958	Strong
	2890	Very strong	2929	Very strong		2895	Strong
	2803	Very strong	2900	Weak	—	2857	Strong
	2730	Strong	2849	Very strong	—		
3. CH_2 scissor	1451	Weak	1457	Strong	a'	1443	Strong
	1440	Strong	1437	Strong	a'	1425	Very strong
4. CH_2 wag	1283	Medium	1349	Weak	a''	1306	Weak
	1264	Medium	1300	Medium	a'	1264	Medium
5. CH_2 twist	1166	Medium	1274	Medium	a'	1143	Medium
	1146	Medium	1260	Medium	a'	1092	Medium
			1197	Medium	a''	1064	Medium
6. CH_2 rock	1050	Medium	1158	Medium	a'	933	Medium
	1021	Medium	1011	Strong	a'	816	Medium
			855	Weak	a'		
7. Skeletal stretching	947	Very weak	1100	Very weak	—	1014	Very strong
	858	Medium	1048	Strong	a'	965	Weak
	815	Very strong	874	Weak	a'	903	Medium
	754	Medium	818	Very strong	a'	692	Strong
8. Skeletal defor-						659	Very strong
mation	441	Medium	458	Weak	a'	400	Medium
	401	Weak	436	Weak	a''	348	Medium
	270	Very weak	403	Weak	a'	252	Very weak
			252	Very weak	a'	196	Weak

[a]Vibrational assignments are for the chair form in (C_s symmetry) in the liquid state. Assignments for piperidine are from ref. (1,10), for tetrahydropyran from ref. (22), and for tetrahydrothiopyran from ref. (6,34).

symmetric stretch ca. 2850 cm^{-1}. In conformations that have an axial lone pair the C–H stretching modes of the two α–CH_2 groups are coupled with the two components of the antisymmetric stretch occurring at 2915 and 2895 cm^{-1} and those of the symmetric stretch at 2803 and 2730 cm^{-1}. *cis*-2,6-Dimethylpiperidine, in which the methyl groups are in equatorial positions, also exhibit these perturbed bands. *trans*-2,6-Dimethylpiperidine has only one axial tertiary α–CH and the spectrum has only one absorption below 2840 cm^{-1}. 2,2,6,6-Tetramethylpiperidine exhibits the normal β- and γ-methylene group stretching vibrations. Also, for this molecule a single band due to the N–H stretch is located at 3312 cm^{-1}. In *N*-substituted compounds, electron-donating substituents (methyl-, *t*-butyl, amino) enhance the perturbed in-phase and out-of-phase α–CH_2 bands whereas electron-withdrawing substituents (benzoyl, *p*-tolyl) decrease this effect.

In the Raman spectra of *N*-methylpiperidine the strongest bands arising from ring vibrations are located at 860, 774, 570, 511, and 371 cm^{-1}. The intense band at 774 cm^{-1} is derived from the ring "breathing" vibration of piperidine (815 cm^{-1}). The Raman spectra of a series of cyclic enamines (β-piperidino-styrenes) were investigated by Dulou, Elkik, and Veillard (12) who found bands near 1657 and 1303 cm^{-1} characteristic of the trans configuration about the carbon-carbon double bond.

Rey-Lafon and Forel (13) have recorded the Raman spectra of δ-valerolactam (2-piperidone) (I) and several of its deuterated derivatives as liquids, as solids, and in solution. Other derivatives studied in the Raman include α-chloro-δ-valerolactam (3-chloro-2-piperidone) (14) and 1-vinyl-2-piperidone (15). δ-Valerolactam in solution exists principally as a cyclic centrosymmetric dimer. In the liquid, the dimer persists together with oliogomer chains. In the solid state, the molecule exists as the dimer and the low-frequency bands

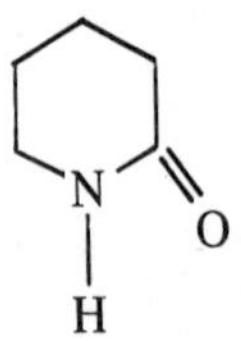

I 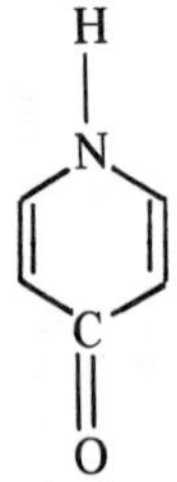 II 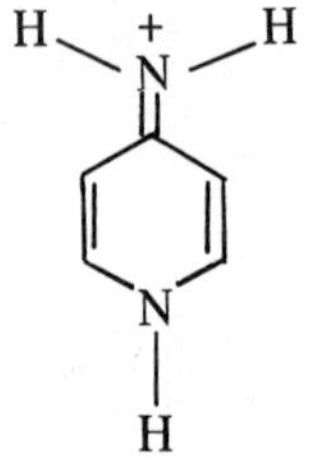 III

in the IR and Raman spectra can be interpreted in terms of hydrogen bonding of the type (N-H. . .O). In the gas phase, the carbonyl stretching frequency of the monomer occurs at 1715 cm^{-1} in the Raman spectrum; in the solid state a band at 1663 cm^{-1} with a shoulder at 1630 cm^{-1} is found. In the liquid spectrum, the two carbonyl frequencies of the centrosymmetric dimer are found at 1668 and 1627 cm^{-1} together with a band at 1645 cm^{-1} which is attributed to another associated form. The frequency of the totally symmetric ring stretching vibration in δ-valerolactam is located at 770–765 cm^{-1}. In the α-chloro derivative, this vibration couples with the carbon-chlorine stretch and is split into two components at 764 and 754 cm^{-1}.

Polycyclic derivatives of piperidine that have been investigated by Raman spectroscopy include 1,2,3,4-tetrahydroquinoline (16), 1,2,3,4-tetrahydroisoquinoline (16), and quinuclidine (1-azabicyclo [2.2.2] octane) (17).

Batts and Spinner (18) have investigated the Raman spectra of 4-pyridone (1,4-dihydro-4-oxopyridine) (II) and various deutero and 1-methyl derivatives in aqueous solution. It was shown that there is extensive mixing of the C=O and C=C stretches which gives rise to composite vibrations near 1640 and 1520 cm^{-1}. The lower frequency is quite sensitive to environmental effects and varies in an irregular manner. Below 1450 cm^{-1} the Raman spectra of 4-pyridone and its *N*-methyl derivative are quite similar and show no marked environmental effect. Both possess prominent polarized bands near 1030 and 850 cm^{-1}. It was also shown (19) from an IR and Raman study that the structure of the 4-aminopyridine cation can best be represented by the iminium form (III), since there is a strong resemblance to the spectrum of the corresponding 4-pyridone.

The Raman spectrum of tetrahydropyran (pentamethylene oxide) was first reported by Kahovec and Kohlrausch (1). More recent Raman studies include those of Akishin et al. (20), Baggett et al. (21), and Snyder and Zerbi (22). The latter authors have also derived a valence force field calculation for the assignment of the vibrational spectra. Voetter and Tschamler (6) and Burket and Badger (23) have also discussed the assignments of the fundamental frequencies. The ring bending modes of this molecule as well as those of related oxanes has been reexamined by Pickett and Strauss (24). The vibrational assignments for this molecule are given in Table 17.1. The strongest Raman band below 1500 cm^{-1} is the symmetric ring "breathing" vibration at 818 cm^{-1}. Other skeletal stretching modes give rise to bands at 1158, 1033, and 874 cm^{-1}. The ring deformation modes, which are to a good approximation separable from the other vibrational modes, result in weak bands in the Raman at 458, 436, 403, and 252 cm^{-1}.

Baggett et al. (21) have also investigated the spectra of 2-, 3-, and 4-hydroxytetrahydropyran and Bel'skii et al. (25) have studied the Raman spectra of some 2,6-dialkytetrahydropyrans. In all of these compounds, a prominent Raman band occurs in the region 820–800 cm^{-1} arising from the ring "breathing" vibration. Altona, Hageman, and Havinga (26) have investigated the IR and Raman spectra of *trans*-2,3 and *trans*-3,4-dihalogenotetrahydropyran. When the halogens are in the anti (diaxial) position, the two carbon-halogen stretching frequencies occur characteristically as a strongly coupled pair. The symmetric stretch (ν_{sym}) occurs as an intense band in the Raman and the antisymmetric stretch (ν_{asym}) gives rise to a strong IR-active band at lower frequency. The positions of the bands and their frequency difference are functions of the structure in the vicinity of the carbon-halogen band. Some representative frequencies are:

Compound	ν_{asym}	ν_{sym}
trans-2,3-Dichlorotetrahydropyran	645	738
trans-3,4-Dichlorotetrahydropyran	647	718
trans-2-Chloro-3-bromotetrahydropyran	591	701
trans-2,3-Dibromotetrahydropyran	556	666
trans-3,4-Dibromotetrahydropyran	555	665

In the diequatorial conformation, the carbon-halogen stretching vibrations are weakly coupled. The Raman spectrum of Δ^2-dihydropyran (IV) was studied by Akishin et al. (20) who found that a very strong band due to the C=C stretch at 1651 cm^{-1}. Other strong Raman bands occurred at 1245, 836, 816, 492, and 440 cm^{-1}. The tetrasubstituted derivative 2,3-dimethyl-6,6′-diethyl-2,3-dihydropyran was studied by Luk'yanets et al. (27). In the Raman spectrum bands attributed to C=C stretches occur at 1691 cm^{-1}, which has a very strong intensity, and at 1632 cm^{-1} with an intensity about one fourth that of the higher frequency band. These authors also examined the IR and Raman spectra of a series of δ-enollactones including alkylidenetetrahydropyran-2-ones and 3,4-dihydropyran-2-ones. Some characteristic frequencies in the Raman for these compounds are listed in Table 17.2. The δ-enollactones with a double bond exo to the ring have strong bands in both the IR and Raman due to ν(C=C) in the range 1665–1650 cm^{-1}; those with an endocyclic double bond exhibit a band due to ν(C=C) in the 1710–1695 cm^{-1} region which is quite weak in the IR but very strong in the Raman spectra.

The variation in the Raman spectra of coumarin (benzo[*b*] pyr-2-one) (V) with change of state has been investigated by Venkateswaran (28) and by Girijavallabhan and Venkateswarlu (29). In the molten state, the frequencies of the C=C bond stretchings occur at 1626, 1611, and 1568 cm^{-1} whereas in the solid these bands are located at 1619, 1601, and 1563 cm^{-1}. The carbonyl stretching frequency occurs at 1732 cm^{-1} in the Raman spectrum of

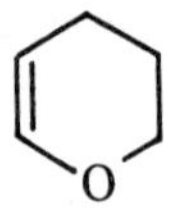

O

IV

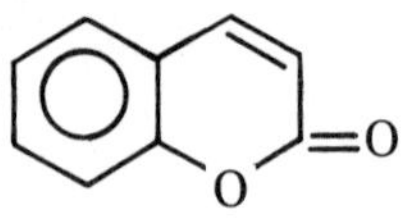

O =O

V

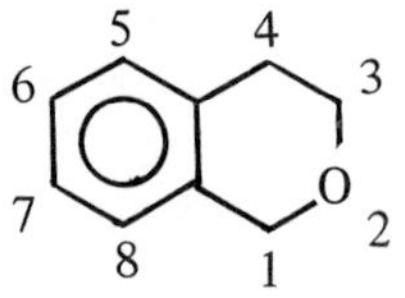

5 4 6 3 O 7 2 8 1

VI

Table 17.2 Characteristic Raman Frequencies for Some δ-Enollactones[a]

Name of Compound	Structure	Frequency (cm^{-1})		
		ν(C=O)	ν(C=C)	Other
1. 5,5-Dimethyl-6-methylenetetrahydropyran-2-one	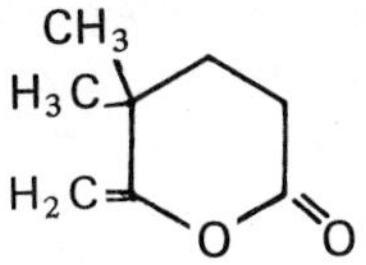 CH_3, H_3C, H_2C=, O, O	1760	1655	862,670,647, 480
2. 5,5-Dimethyl-6-isopropylidenetetrahydropyran-2-one	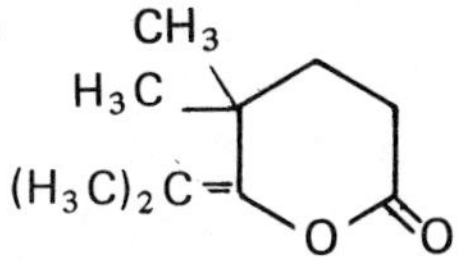 CH_3, H_3C, $(H_3C)_2C$=, O, O	1760–1750	1664	862,670,636, 474
3. 6-Methyl-3,4-dihydropyran-2-one	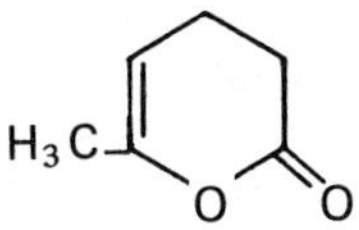 H_3C, O, O	1757	1700	1240,674,547, 444
4. 5,6-Dimethyl-3,4-dihydropyran-2-one	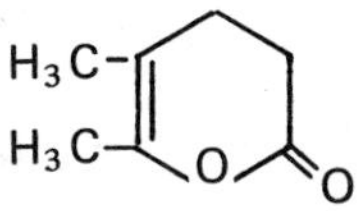 H_3C, H_3C, O, O	1764	1708	1258,674,545, 440
5. 5,6-Cyclohexano-3,4-dihydropyran-2-one	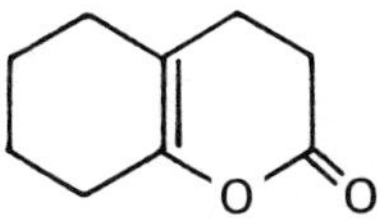 O, O	1764	1711	1258,682,434

[a]Vibrational assignments from ref. 27.

coumarin in the molten state. In the solid state, this band is reduced in intensity and splits into two bands at 1726 and 1706 cm^{-1}. Other pyranones studied by Raman spectroscopy include xanthone (9-xanthenone) (30) and 2,6-dimethyl-4H-pyran-4-one (2,6-dimethyl-γ-pyrone) (30,31). For both these compounds ν(C=O) occurs at 1673 cm^{-1} in dioxane solution.

Freeman (32) has applied Raman spectroscopy to a study of the conformational isomers of some isochromans. Isochroman (VI) and its 1-methyl and 3-methyl derivatives in the liquid state exhibit only a single band at 732,

716, and 740 cm^{-1}, respectively. This highly polarized band is assigned to a ring "breathing" mode of the dihydropyran moiety. In the liquid state, the Raman spectrum of 4-methylisochroman, on the other hand, exhibits two bands at 732 and 710 cm^{-1} indicating the presence of two conformations in which the oxygen atom is free to assume either the cis or trans position relative to the 4-methyl group. Substitution of a methyl group in the 1- or 2-position adjacent to the oxygen atom sterically prevents adoption of one of the two possible conformations due to 1,3-interaction.

The vibrational assignments in the IR and Raman spectra of tetrahydrothiopyran (pentamethylene sulfide) have been discussed by Sheppard (33), Voetter and Tschamler (6), and Hitch and Ross (34). The frequencies of the fundamental vibrations found in the Raman spectrum are listed in Table 17.1. The most intense band in the Raman at 659 cm^{-1} is assigned to the symmetric C–S–C stretch and the antisymmetric C–S–C stretch is assigned a frequency of 692 cm^{-1}. The ring stretches involving mainly the C–C–C bonds that are found in the Raman spectrum are located at 1014, 965, and 906 cm^{-1}. The Raman spectra of Δ^2-dihydrothiopyran has been observed by Akishin et al. (35) who found that ν(C=C) occurred as a very strong band at 1609 cm^{-1}. The frequency of the band assigned to the symmetric C–S–C stretch was located at 664 cm^{-1}.

The Raman spectra of silacyclohexane and its 1,1-dimethyl and 1,1-dichloro derivatives have been investigated by Fogarasi, Torok, and Vdovin (36). The Si–C stretching frequencies found for these compounds are:

Compound	Frequency (cm^{-1}) Antisymmetric CSiC Stretch	Symmetric CSiC Stretch
Silacyclohexane	734	656
1,1-Dimethylsilacyclohexane	695	586
1,1-Dichlorosilacyclohexane	799	694

17.2 RINGS CONTAINING TWO HETEROATOMS

Raman studies of 1,4-dioxane (*p*-dioxane) (Spectrum 104, Appendix 2) have been extensive (20,22,24,30,31,37–50). A complete assignment of the normal vibrations of 1,4-dioxane in both the IR and Raman spectra has been discussed by Ramsay (51,52), Burket and Badger (53), Malherbe and Bernstein (48), Kirchner (49), and Snyder and Zerbi (22). Normal coordinate analyses have been carried out for 1,4-dioxane by the latter authors and for 1,4-dioxane-d_8 by Pickett and Strauss (24). Kimura and Aoki (54) have established that 1,4-dioxane is in the chair conformation (VII), which possesses C_{2h} sym-

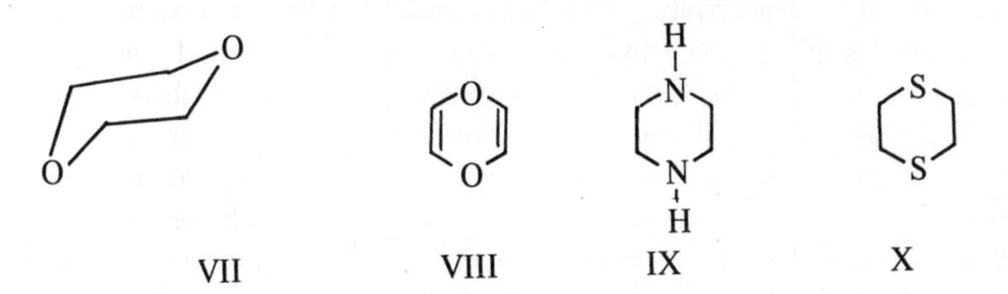

metry. Since this molecule possesses a center of symmetry, the rule of mutual exclusion will apply and transitions that are allowed in the IR will be forbidden in the Raman and those allowed in the Raman will be forbidden in the IR. The distribution of the 36 normal modes of vibration of 1,4-dioxane under C_{2h} symmetry is: $10a_g$ (Raman active, polarized) + $8b_g$ (Raman active, depolarized) + $9a_u$ (IR active) + $9b_u$ (IR active). The assignment of these modes in the vibrational spectra is given in Table 17.3. In the Raman, the symmetric ring "breathing" vibration is assigned to the intense highly polarized band at 834 cm^{-1}. Very strong C–H stretching, characteristic of a methylene group bonded to an oxygen, is located at 2966 and 2854 cm^{-1}. The C–CH_2–O symmetric deformation gives a very strong band characteristic of this moiety at 1443 cm^{-1}.

Snyder and Zerbi (22) have investigated the vibrational spectra and normal vibrations of 2,5(*ee*)-dimethyldioxane(1,4). They found that in the Raman the frequencies of the ring stretches of the unsubstituted compound located at 1109, 1015, and 834 cm^{-1} were shifted to 1154, 1052, and 790 cm^{-1}, respectively, in the disubstituted compound. In a series of *trans*-2,3-dihalogeno-1,4-dioxanes it was found by Altona, Hageman, and Havinga (26) that the two carbon-halogen stretching frequencies occurred characteristically as a strong coupled pair, as shown in the following table:

Compound	ν_{asym}(C–X) (cm^{-1})	ν_{sym}(C–X) (cm^{-1})
2,3-Dichloro-1,4-dioxane	657	757
2,3-Dibromo-1,4-dioxane	568	693
2-Chloro-3-bromo-1,4-dioxane	614	728

The vibrational spectra of 1,4-dioxadiene (VIII) has been investigated by Connett et al. (55) who concluded that the molecule was planar (point group D_{2h}) and not highly conjugated. In the Raman, very strong bands were located at 3121 cm^{-1} (C-H stretch), 1686 cm^{-1} (C=C stretch), 1211 cm^{-1} (C-H in-plane deformation), 928 cm^{-1} (ring "breathing" vibration), and 542 cm^{-1} (ring deformation).

The Raman spectrum of piperazine (IX) has been reported by Kahovec and Kohlrausch (43). The vibrational assignment of the fundamental skeletal modes has been reported by Hendra and Powell (56). An intense Raman band

Table 17.3 Assignments in the Vibrational Spectra of 1,4-Dioxane and 1,4-Dithiane

			1,4-Dioxane[a]			1,4-Dithiane[b]		
C_{2h} Symmetry Species	Vibration Number	Approximate Description of Vibration	Frequency (cm^{-1})	Relative Raman Intensity	Relative IR Intensity	Frequency (cm^{-1})	Relative Raman Intensity	Relative IR Intensity
a_g	1	Antisymmetric C–H stretch	2966	Very strong	—	2906	Very strong	—
	2	Symmetric C–H stretch	2854	Very strong	—	2805	Weak	—
	3	CH_2 deformation	1443	Very strong	—	1407	Strong	—
	4	CH_2 wag	1334	Medium	—	1110	Strong	—
	5	CH_2 twist	1303	Very strong	—	1298	Medium	—
	6	CH_2 rock	1127	Medium	—	821	Strong	—
	7	Ring stretch	1015	Strong	—	999	Medium	—
	8	Ring stretch	834	Very strong	—	628	Very strong	—
	9	Ring deformation	(503)[c]	—	—	374	Medium	—
	10	Ring deformation	427	Medium	—	333	Medium	—
b_g	11	Antisymmetric C–H stretch	(2965)[c]	—	—	2942	Strong	—
	12	Symmetric C–H stretch	(2861)[c]	—	—	2906	Very strong	—
	13	CH_2 deformation	1461	Medium	—	1451	Weak	—
	14	CH_2 wag	1396	Weak	—	1209	Medium	—
	15	CH_2 twist	1216	Strong	—	1408	Strong	—
	16	Ring stretch	1109	Medium	—	694	Strong	—
	17	CH_2 rock	852	Weak	—	945	Medium	—

	18	Ring deformation	486	Medium	—	277	Medium	—
a_u	19	Antisymmetric C–H stretch	2974	—	Very strong	2919	—	Very strong
	20	Symmetric C–H stretch	2867	—	Very strong	2855	—	Very weak
	21	CH_2 deformation	1457	—	Weak	1415	—	Very strong
	22	CH_2 wag	1368	—	Weak	1156	—	—
	23	CH_2 twist	(1285)[c]	—	—	1283	—	Very strong
	24	Ring stretch	1124	—	Very strong	945	—	Weak
	25	CH_2 rock	1083	—	Medium	994	—	Medium
	26	Ring stretch	889	—	Medium	669	—	Medium
	27	Ring deformation	224	—	Very weak	253	—	Strong
b_u	28	Antisymmetric C–H stretch	2974	—	Very strong	2955	—	Very strong
	29	Symmetric C–H stretch	2867	—	Very strong	2816	—	Strong
	30	CH_2 deformation	1457	—	Weak	1418	—	Medium
	31	CH_2 wag	(1395)[c]	—	—	1275	—	Very strong
	32	CH_2 twist	1292	—	Weak	1306	—	Medium
	33	CH_2 rock	1049	—	Medium	904	—	Very strong
	34	Ring stretch	875	—	Strong	626	—	Very weak
	35	Ring deformation	607	—	Medium	472	—	Strong
	36	Ring deformation	276	—	Medium	100	—	—

[a]Assignments from refs. 22,48 for liquid state.
[b]Assignments from ref. 58 for solution in Raman and vapor in IR.
[c]Calculated frequency.

at 836 cm^{-1} is assigned to the ring "breathing" vibration. The observed frequencies of other skeletal stretches are 1109 (strong) and 1184 cm^{-1} (weak). Weak Raman bands in the skeletal deformation region include those at 581, 515, 448, and 404 cm^{-1}.

The vibrational assignments of 1,4-dithiane (X) are given in Table 17.3. Raman spectra of this cyclic sulfide have been measured in solution (41,57, 58) and in the solid state (58–60). Normal coordinate analyses of the ring vibrations have been performed by Hayasaki (57) and by Hitch and Ross (34). IR and Raman data for the vapor, solution, and solid show that the structure of this molecule is the centrosymmetric chair conformation (point group C_{2h}), since there is essentially mutual exclusion of IR and Raman bands in all three states of aggregation. The frequencies of ring vibrations that can be described as predominantly due to C–S stretching occur in the Raman spectra at 694 and 628 cm^{-1} and that due to mainly C–C stretching at 999 cm^{-1}. Those modes occurring at 333 and 277 cm^{-1} can be attributed to S-C-C deformations while that at 374 cm^{-1} is ascribed to a C-S-C deformation.

Raman spectra that have been reported for nonaromatic six-membered heterocyclic rings containing two different heteroatoms in the 1 and 4 positions include morpholine (XI) (31,41,61) and 1,4-thioxane (XII) (31,35,41,60). In morpholine, the ring "breathing" vibration occurs as an intense band at 832 cm^{-1} in the Raman spectrum. In 1,4-dioxane and piperazine, this vibration occurs at 834 and 836 cm^{-1}, respectively. In the Raman spectrum of liquid thioxane the prominent bands due to the carbon-sulfur stretchings occur at 692 and 666 cm^{-1}.

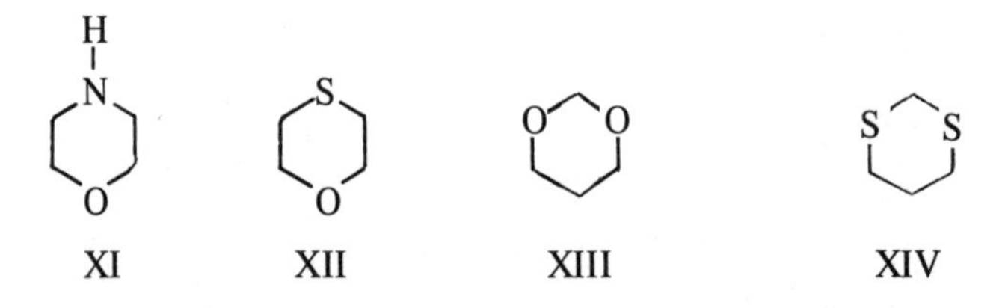

1,3-Dioxane (*m*-dioxane) (XIII) has been studied in the IR and Raman by Pickett and Strauss (24) who also carried out a normal coordinate analysis to aid in the vibrational assignments. Raman bands at 1045, 980, 900, and 825 cm^{-1} were assigned to ring stretching modes and those due to ring deformations were located at 645, 490, 460, 435, and 270 cm^{-1}. 1,3-Dithiane (XIV) has been investigated by Hitch and Ross (34). Bands attributed to C–S stretching occur at 747, 722, 670, and 660 cm^{-1}, whereas ring deformation modes are assigned frequencies at 465, 335, 312, and 217 cm^{-1}.

Heterocyclic six-membered rings containing nitrogens in the 1 and 3 positions include several of the bases that are constituents of deoxyribonucleic acid (DNA) and ribonucleic acid (RNA). These include the pyrimidinones, uracil (XV) and cytosine (XVI), and the purinone, guanine (XVII). The Raman spectra of aqueous solutions of guanine, cytosine, cytidine, uridine, and thymidine (5-methyluracil) have been examined by Malt (62). A comprehensive Raman study of uracil, cytosine, and guanine as well as the corresponding nucleosides, 5′-mononucleotides, and related aklyl derivatives was carried out by Lord and Thomas (63). Spectra were obtained in aqueous solutions over a wide pH range and also in the solid state. Characteristic frequencies in the Raman spectra for derivatives of uracil, cytosine, and guanine are listed in Tables 17.4, 17.5, and 17.6, respectively. In aqueous media, and in the absence of vibrational coupling, the frequency of the C=O stretching of the conjugated carbonyl group in uracil, cytosine, and guanine derivatives falls in the narrow range 1670–1655 cm^{-1}. Under identical conditions, the nonconjugated carbonyl group found in uracil derivatives has a stretching frequency near 1695 cm^{-1}. Stretching frequencies involving C=C and C=N bonds are located in the range 1650–1550 cm^{-1}. The ribose moiety in the nucleoside and nucleotide derivatives generally gave rise to very weak Raman bands. A band at 980 cm^{-1} was characteristic of the mononucleotide bivalent phosphate ion under all conditions of observation.

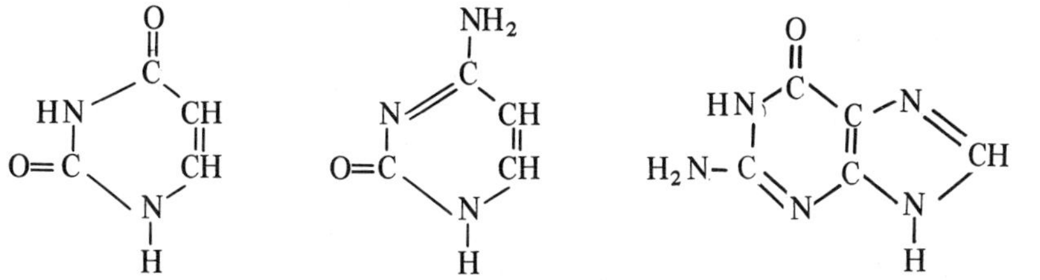

XV XVI XVII

17.3 RINGS CONTAINING THREE OR MORE HETEROATOMS

The Raman spectrum of 1,3,5-trioxane (*s*-trioxane, α-trioxymethylene) has been examined by Kahovec and Kohlrausch (43), Stair and Nielsen (64), and Ward (65). Complete vibrational assignments of this molecule have been carried out (64,65) and a normal coordinate analysis was performed by Pickett and Strauss (24). 1,3,5-Trioxane exists in the chair conformation and possesses C_{3v} symmetry. Its 20 normal modes of vibration are classified as: $7a_1 + 3a_2 + 10e$. The a_1 and e modes are active in both the IR and Raman,

Table 17.4 Characteristic Frequencies in the Raman Spectra[a] of Uracil Derivatives,

$$\begin{array}{c} O \\ \| \\ R_2{-}N{-}C{-}CH \\ O{=}C{-}N(R_1){-}CH \end{array}$$

Name	R_1	R_2	Frequency (cm^{-1})						
			Solid, Neutral, Acidic and Alkaline Solution			Solid, Neutral, and Acidic Solution	Alkaline Solution Only		
1. Uracil	H	H	790 ± 8	—	1390 ± 10	1236 ± 1	1034 ± 1	1212 ± 2	1284 ± 11
2. Uridine	Ribose	H	786 ± 4	—	1392 ± 10	1233 ± 1	1020 ± 2	1209 ± 1	1296 ± 2
3. Uridine-5′-monophosphate	Ribose-5′-phosphate	H	786 ± 4	810 ± 10	1392 ± 11	1233 ± 2	1018 ± 2	1209 ± 1	1296 ± 2
4. 1-Methyluracil	CH_3	H	768 ± 12	802 ± 10	1388 ± 2	1237 ± 5	1011 ± 1	1208 ± 3	1275 ± 1
5. 1,3-Dimethyluracil	CH_3	CH_3	691 ± 4	804 ± 4	804 ± 2	1273 ± 5	—	—	—
Relative Raman Intensity			Strong	Medium	Strong	Very strong	Medium	Strong	Very strong
Polarization			Polarized	Polarized	Polarized	Polarized	Depolarized	Depolarized	Depolarized

[a] Assignments from ref. 63.

Table 17.5 Characteristic Frequencies in the Raman Spectra[a] of Cytosine Derivatives,

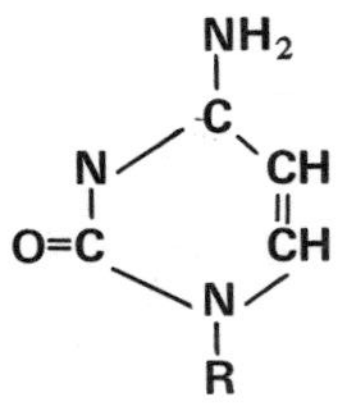

Name	R	Frequency (cm^{-1})							
		Solids, Acidic, Neutral, and Alkaline Solutions							Acidic Solutions
1. Cytosine	H	786 ± 9	970 ± 5	1215 ± 5	—	1282 ± 13	—	1375 ± 10	1420 ± 10
2. Cytidine	Ribose	781 ± 9	995 ± 6	1218 ± 7	1254 ± 11	—	1302 ± 10	1383 ± 8	1430 ± 20
3. Cytidine-5′-monophosphate (disodium)	Ribose-5′-phosphate	780 ± 6	1000 ± 5	1218 ± 8	1254 ± 11	—	1306 ± 6	1383 ± 8	1430 ± 20
4. 1-Methylcytosine	CH_3	785 ± 10	985 ± 10	1209 ± 9	—	1278 ± 15	—	1392 ± 7	1445 ± 5
Relative Raman Intensity		Strong–very strong	Weak-medium	Strong	Very strong	Very strong	Very strong	Weak–medium	Medium–strong
Polarization		Polarized	Depolarized	Polarized	Polarized	Polarized	Polarized	Depolarized	Polarized

[a] Assignments from ref. 63.

Table 17.6 Characteristic Raman Frequencies in the Raman Spectra[a] of Guanine Derivatives,

Name	R	Frequency (cm^{-1}) Acidic, Neutral, and Alkaline Solutions		Neutral and Alkaline Solutions		Alkaline Solution Only	Acidic Solutions Only	
1. Guanine	H	638 ± 13	1361 ± 11	1326 ± 3	1457 ± 2	1326 ± 3	1260	1410
2. Guanosine	Ribose	668 ± 8	1367 ± 5	1315 ± 10	1480 ± 10	1341 ± 4	1285 ± 10	1405 ± 10
3. Guanosine-5′-monophosphate (disodium)	Ribose-5′-phosphate	672 ± 8	1364 ± 4	1313 ± 12	1481 ± 9	1342 ± 3	1285 ± 10	1405 ± 10
4. 9-Ethylguanine	CH_3CH_2-	622 ± 7	1350 ± 2	–	–	–	1280 ± 10	1414 ± 4
Relative Raman Intensity		Medium	Medium-strong	Medium	Very strong	Very strong	Medium-strong	Strong-very strong
Polarization		Polarized	Polarized	Polarized	Polarized	Polarized	Polarized	Polarized

[a] Assignments from ref. 63.

whereas the a_2 modes are inactive in both spectra. The assignments of these normal modes in the Raman spectrum are listed in Table 17.7. The Raman spectra of paraldehyde-I (2,4,6-trimethyl-1,3,5-trioxane) (43,65,66,67,68) and parapropionaldehyde (65) have also been studied. In paraldehyde-I, the methyl groups occupy equatorial positions in the chair conformation. Characteristic ring frequencies for these compounds are listed below:

Compound	Skeletal Frequency (cm^{-1})					
1,3,5-Trioxane	1070 (wk)	962 (vs)	748 (s)	523 (s)	472 (wk)	307 (wk)
Paraldehyde-I	1099 (m)	948 (m)	750 (vwk)	525 (vs)	472 (s)	274 (m)
Parapropionaldehyde	1080 (m)	953 (m)	747 (vwk)	525 (s)	470 (m)	242 (m)

Raman studies of 1,3,5-trithiane (XVIII) have been conducted by Kohlrausch and Rietz (69) and by Klaeboe (58). Assignments of the vibrational fundamentals (58,70) are listed in Table 17.7. The a_2 modes are observed in the spectra of the solid due to the breaking down of the molecular selection rules. Gerding and Karsten (71) have examined the Raman spectra

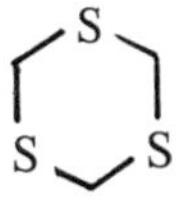

XVIII 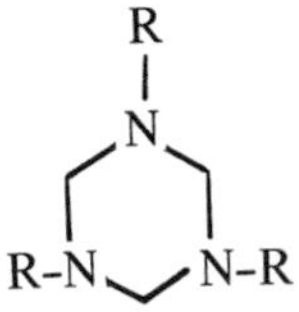 XIX  XX

of the α- and β-isomers of 2,4,6-trimethyl-1,3,5-trithiane (trithioacetaldehyde). The vibrational assignments for these molecules have been discussed by Hitch and Ross (70). In the β-isomer, the three methyl groups occupy equatorial positions whereas in the α-isomer there are two equatorial and one axial methyl group. In the β-isomer, an intense Raman band is found due to S–C–S stretching at 648 cm^{-1} and for the α-isomer two strong bands are located at 694 and 653 cm^{-1}.

Kahovec (72) has reported on the Raman spectra of a series of 1,3,5-trialkylhexahydro-*s*-triazines (XIX) (*N*-alkylated trimethylenetriamines) which were prepared by the reaction of primary amines with formaldehyde under acidic conditions. A strong Raman band was found in the range 920–900 cm^{-1} for all the derivatives studied.

Hexamethylenetetramine (XX) (urotropine, 1,3,5,7-tetrazaadamantane) has point group symmetry T_d with 25 fundamental frequencies. In the IR only the nine fundamental frequencies belonging to the triply degenerate f_2 species are allowed. In the Raman eighteen fundamental frequencies are allowed ($4a_1 + 5e + 9f_2$). The a_2 species vibration and the $6f_1$ frequencies are inactive

Table 17.7 Vibrational Assignments in the Raman Spectra of 1,3,5-Trioxane and 1,3,5-Trithiane

			1,3,5-Trioxane[a]		1,3,5-Trithiane[b]		
Symmetry Species of C_{3v}	Vibration Number	Approximate Description of Vibration	Frequency (cm^{-1})	Relative Raman Intensity	Frequency (cm^{-1})	Physical State	Relative Raman Intensity
a_1	1	CH_2 antisymmetric stretch	3020	Very strong	2904	Melt	Very strong
	2	CH_2 symmetric stretch	2869	Strong	2867	Solid	Weak
	3	CH_2 deformation	1493	Very weak	1462	Solid	Very weak
	4	CH_2 rock	(1228)[c]	—	1375	Melt	Strong
	5	Ring stretch	962	Very strong	654	Melt	Very strong
	6	Ring stretch	748	Strong	405	Melt	Medium
	7	Ring deformation	472	Weak	310	Solid	Very weak
a_2	8	CH_2 wag	(1387)[c]	—	1178	Solid	Strong
	9	CH_2 twist	(1330)[c]	—	930	Solid	Very weak
	10	Ring stretch	(1158)[c]	—	745	Solid	Medium
e	11	CH_2 antisymmetric stretch	3020	Very strong	2976	Melt	Medium
	12	CH_2 symmetric stretch	2869	Strong	2888	Melt	Very strong
	13	CH_2 deformation	1474	Strong	1389	Melt	Strong
	14	CH_2 wag	1411	Medium	1225	Solid	Medium
	15	CH_2 rock	1308	Strong	909	Solid	Very weak
	16	CH_2 twist	1163	Weak	1168	Melt	Medium
	17	Ring stretch	1070	Weak	793	Solid	Weak
	18	Ring stretch	931	Medium	735	Melt	Strong
	19	Ring deformation	523	Strong	291	Melt	Medium
	20	Ring deformation	307	Weak	80	Solid	Strong

[a] Assignments from refs. 64,24 for the liquid state.
[b] Assignments from ref. 58 for the molten or solid state.
[c] Calculated.

in both spectra. The Raman spectra of this molecule has been extensively investigated both in solution and in the crystalline state (73–80). Mecke and Spiesecke (80) have also investigated the Raman spectrum of the analogous compound 1,3,5,7-tetramethyl-2,4,6,8,9,10-hexathiaadamantane. Prominent bands in the Raman spectra of these molecules are:

	Frequency (cm^{-1})	
Compound	Ring Stretches	Ring Deformations
Hexamethylenetetramine	1350,1238,1009, 1048 (polarized)	782 (polarized), 672,511, 460
1,3,5,7-Tetramethyl-2,4,6,8,9,10-hexathiaadamantane	1090 (polarized), 1086,735,726 (polarized),714, 623	491,473,437 (polarized), 419,325,284

REFERENCES

1. L. Kahovec and K. W. F. Kohlrausch, *Z. Phys. Chem., 35B,* 29 (1937).
2. G. B. Bonino and L. Brull, *Gazz. Chim. Ital., 59,* 675 (1929).
3. S. C. Sirkar, *Indian J. Phys., 7,* 61 (1962).
4. S. M. Mitra, *Z. Phys., 102,* 697 (1936).
5. K. W. F. Kohlrausch and W. Stockmair, *Z. Phys. Chem., 31B,* 382 (1936).
6. H. Voetter and H. Tschamler, *Monatsh. Chem., 84,* 134 (1953).
7. O. S. Anisimova, A. N. Kost, Yu. A. Pentin, and L. G. Yudin, *Vestn. Mosk. Univ., Ser. II, Khim. 21,* 19 (1966); *Chem. Abstr. 66,* 23941j (1967).
8. O. S. Anisimova, Yu. A. Pentin, and L. G. Yudin, *Vestn. Mosk. Univ., Ser. II, 22,* 8 (1967); *Chem. Abstr. 68,* 73640x (1968).
9. O. S. Anisimova, A. N. Kost, Yu. A. Pentin, and L. G. Yudin, *Vestn. Mosk. Univ., Ser. II, 22,* 121 (1967); *Chem. Abstr. 68,* 100128v (1968).
10. T. V. Titova, O. S. Anisimova, and Yu. A. Pentin, *Opt. Spectrosc., 23,* 495 (1967).
11. P. J. Krueger and J. Jan, *Can. J. Chem., 48,* 3236 (1970).
12. R. Dulou, E. Elkik, and A. Veillard, *Bull. Soc. Chim. Fr., 1960,* 967.
13. M. Rey-Lafon and M. Forel, *J. Chim. Phys., 67,* 757 (1970).
14. M. Rey-Lafon and M. Forel, *J. Chim. Phys., 67,* 767 (1970).
15. P. P. Shorygin, T. N. Shkurina, M. F. Shostakovskii, F. P. Sidel'kovskaya, and M. G. Zelenskaya, *Bull. Acad. Sci. USSR Div. Chem. Sci., 1959* 2103.

16. G. B. Bonino and R. Manzoni-Ansidei, *Mem. R. Accad. Sci. Inst. Bologna,* (9), 3 (1934).
17. J. Wagner, *Z. Phys. Chem., 48B,* 316 (1941).
18. B. D. Batts and E. Spinner, *Aust. J. Chem., 22,* 2581 (1969).
19. B. D. Batts and E. Spinner, *Aust. J. Chem., 22,* 2595 (1969).
20. P. A. Akishin, N. G. Rambidi, I. K. Korobitsyna, G. Ya. Kondrat'eva, and Yu. K. Yur'ev, *Vestn. Mosk. Univ. 10,* No. 12, *Ser. Fiz. Mat. i Estestv. Nauk No. 8,* 103 (1955); *Chem. Abstr. 50,* 8329i (1956).
21. N. Baggett, S. A. Barker, A. B. Foster, R. H. Moore, and D. H. Whiffen, *J. Chem. Soc.,* 4565 (1960).
22. R. G. Snyder and G. Zerbi, *Spectrochim. Acta, 23A,* 391 (1967).
23. S. C. Burket and R. M. Badger, *J. Am. Chem. Soc., 72,* 4397 (1950).
24. H. M. Pickett and H. L. Strauss, *J. Chem. Phys., 53,* 376 (1970).
25. I. F. Bel'skii, N. I. Shuikin, G. K. Vasilevskaya, and G. K. Gaivoronskaya, *Bull. Acad. Sci. USSR Div. Chem. Sci., 1963,* 1560.
26. C. Altona, H. J. Hageman, and E. Havinga, *Spectrochim. Acta, 24A,* 633 (1968).
27. E. A. Luk'yanets, N. P. Shusherina, E. G. Treshchova, L. A. Kazitsina, and R. Ya. Levina, *J. Org. Chem. USSR, 1,* 1204 (1965).
28. C. S. Venkateswaran, *Curr. Sci., 6,* 328 (1938).
29. C. P. Girijavallabhan and K. Venkateswarlu, *Curr. Sci., 37,* 10 (1968).
30. M. V. Vol'kenshtein and Ya. K. Syrkin, *Acta Physicochim. USSR, 10,* 677 (1939).
31. L. Kahovec and K. W. F. Kohlrausch, *Chem. Ber., 75,* 627 (1942).
32. S. K. Freeman, *Appl. Spectrosc., 24,* 42 (1970).
33. N. Sheppard, *Trans. Faraday Soc., 46,* 429 (1950).
34. M. J. Hitch and S. D. Ross, *Spectrochim. Acta, 25A,* 1041 (1969).
35. P. A. Akishin, N. G. Rambidi, K. Yu. Novitskii, and Yu. K. Yur'ev, *Vestn. Mosk. Univ. 9,* No. 3, *Ser. Fiz. -Mat. i Estestv. Nauk,* No. 2, 77 (1954).
36. G. Fogarasi, F. Torok, and V. M. Vdovin, *Acta Chim. Acad. Sci. Hung., 54,* 277 (1967).
37. C. S. Venkateswaran, *Proc. Indian Acad. Sci., 2,* 279 (1935).
38. A. Simon and F. Feher, *Chem. Ber., 69,* 214 (1936).
39. A. Simon and F. Feher, *Z. Elektrochem., 42,* 688 (1936).
40. K. W. F. Kohlrausch and W. Stockmair, *Z. Phys. Chem., 31B,* 382 (1936).
41. L. Medard, *J. Chim. Phys., 33,* 626 (1936).
42. R. C. Williamson, *J. Chem. Phys., 5,* 666 (1937).
43. L. Kahovec and K. W. F. Kohlrausch, *Z. Phys. Chem., 35B,* 29 (1937).
44. H. W. Hunter, *J. Chem. Phys., 6,* 544 (1938).

45. B. D. Saksena, *Proc. Indian Acad. Sci., 8,* 73 (1938).
46. B. D. Saksena, *Proc. Indian Acad. Sci., 10,* 449 (1939).
47. B. D. Saksena, *Proc. Indian Acad. Sci., 12,* 321 (1940).
48. F. E. Malherbe and H. J. Bernstein, *J. Am. Chem. Soc., 74,* 4408 (1952).
49. H. H. Kirchner, *Z. Phys. Chem. (Frankf.), 29,* 166 (1961).
50. L. Bernard, *J. Chim. Phys., 63,* 641 (1966).
51. D. A. Ramsay, *Proc. Roy. Soc. (Lond.), 190A,* 562 (1947).
52. D. A. Ramsay, *Trans. Faraday Soc., 44,* 289 (1948).
53. S. C. Burket and R. M. Badger, *J. Am. Chem. Soc., 72,* 4397 (1950).
54. M. Kimura and K. Aoki, *J. Chem. Soc. (Jap.), 72,* 169 (1951).
55. J. E. Connett, J. A. Creighton, J. H. S. Green, and W. Kynaston, *Spectrochim. Acta, 22,* 1859 (1966).
56. P. J. Hendra and D. B. Powell, *Spectrochim. Acta, 18,* 299 (1962).
57. K. Hayasaki, *J. Sci. Hiroshima Univ., 34A,* 679 (1960).
58. P. Klaeboe, *Spectrochim. Acta, 25A,* 1437 (1969).
59. G. C. Hayward and P. J. Hendra, *Spectrochim. Acta, 23A,* 1937 (1967).
60. D. A. Rice and R. A. Walton, *Spectrochim. Acta, 27A,* 279 (1971).
61. R. A. Friedel and D. S. McKinney, *J. Am. Chem. Soc., 69,* 604 (1947).
62. R. A. Malt, *Biochim. Biophys. Acta, 120,* 461 (1966).
63. R. C. Lord and G. J. Thomas, Jr., *Spectrochim. Acta, 23A,* 2551 (1967).
64. A. T. Stair, Jr. and J. R. Nielsen, *J. Chem. Phys., 27,* 402 (1957).
65. W. R. Ward, *Spectrochim. Acta, 21,* 1311 (1965).
66. H. Gerding and G. W. A. Rijnders, *Rec. Trav. Chim., 58,* 603 (1939).
67. H. Gerding, W. J. Nijveld, and G. W. A. Rijnders, *Rec. Trav. Chim., 60,* 25 (1941).
68. B. D. Saksena, *Proc. Indian Acad. Sci., 12,* 321 (1940).
69. K. W. F. Kohlrausch and A. W. Reitz, *Z. Phys. Chem., 45B,* 249 (1939).
70. M. J. Hitch and S. D. Ross, *Spectrochim. Acta, 25A,* 1047 (1969).
71. H. Gerding and J. G. A. Karsten, *Rec. Trav. Chim., 58,* 609 (1939).
72. L. Kahovec, *Z. Phys. Chem., 43B,* 364 (1939).
73. P. Krishnamurti, *Indian J. Phys., 6,* 309 (1931).
74. L. Kahovec, K. W. F. Kohlrausch, A. W. Reitz, and J. Wagner, *Z. Phys. Chem., 39B,* 431 (1938).
75. E. Canals and P. Peyrot, *Compt. Rend., 207,* 224 (1938).
76. R. J. W. LeFevre and G. J. Rayner, *J. Chem. Soc., 1938,* 1921.
77. K. S. Bai, *Proc. Indian Acad. Sci., 20A,* 71 (1944).
78. L. Conture-Mathieu, J. P. Mathieu, J. Cremer, and H. Poulet, *J. Chim. Phys., 48,* 1 (1951).

79. R. Mecke and H. Spiesecke, *Chem. Ber., 88,* 1997 (1955).
80. R. Mecke and H. Spiesecke, *Spectrochim. Acta, 7,* 387 (1956).

CHAPTER EIGHTEEN

HETEROCYCLIC COMPOUNDS — AROMATIC SIX-MEMBERED RINGS

18.1 PYRIDINE AND ITS DERIVATIVES

The vibrational assignment for pyridine (Spectrum 105, Appendix 2) as proposed by Wilmhurst and Bernstein (1) and modified slightly by Green, Kynaston, and Paisley (2) is given in Table 18.1. The numerous investigations of the Raman spectra of pyridine prior to 1944 have been reviewed by Herz, Kahovec, and Kohlrausch (3) and by Kline and Turkevich (4). Vibrational studies of pyridine and pyridine-d_8 (5–8), as well as those involving other deuterated derivatives (7,8–11) have been carried out. Normal coordinate calculations for pyridine and some of its deuterated derivatives have also been reported (6–8,10,12,13).

Pyridine belongs to the molecular point group C_{2v} and its 27 fundamental modes of vibration have the following distribution and activity: $10a_1$ (IR; Raman, polarized) + $9b_2$ (IR; Raman, depolarized) + $3a_2$ (Raman, depolarized) + $5b_1$ (IR; Raman, depolarized). The choice of axes for this molecule follows the recommendation of the Joint Commission for Spectroscopy (14) for planar C_{2v} molecules. The alternative notation found in the literature in which the x- and y-axes are interchanged leads only to an interchange of the subscripts for the b_1 and b_2 species.

The normal modes of vibration of pyridine are very similar to those of benzene since the nitrogen atom is isoelectronic with the -CH group and there

Table 18.1 Assignments in the Vibrational Spectra of Pyridine, Pyridinium Chloride, and Pyridine-1-Oxide

			Frequency (cm^{-1})		
Symmetry Species C_{2v}[a]	Wilson Vibration Number	Approximate Description of Vibration	Pyridine[b]	Pyridinium Chloride[c]	Pyridine-1-Oxide[d]
a_1	2	C–H stretch	3054	3081	3102
	13	C–H stretch	3035	3075	3000
	20a	C–H stretch	3054	3025	3070
	8a	Ring stretch	1583	1630	1604
	19a	Ring stretch	1482	1485	1466
	9a	In-plane C–H deformation	1218	1196	1252
	18a	In-plane C–H deformation	1068	1058	1170
	12	Trigonal ring "breathing"	1030	1027	1043
	1	Totally symmetric ring "breathing"	992	1008	1017
	6a	In-plane ring deformation	605	607	837
		N–H stretch	—	2450	—
		Ring–O stretch	—	—	544
b_2	20b	C–H stretch	3080	3090	3102
	7b	C–H stretch	3036	3044	3070
	8b	Ring stretch	1572	1605	1556
	19b	Ring stretch	1439	1532	1482
	14	Ring stretch	1375	1379	1364
	3	In-plane C–H deformation	1288	1321	1329
	15	In-plane C–H deformation	1148	1165	1149
	18b	In-plane C–H deformation	1085	1080	1070

b_2	6b	In-plane ring deformation	652	637	636
		In-plane N–H deformation	—	1249	—
		In-plane N–O deformation	—	—	410
a_2	17a	Out-of-plane C–H deformation	986	—	977
	10a	Out-of-plane C–H deformation	891	883	829
	16a	Out-of-plane ring deformation	375	—	415
b_1	5	Out-of-plane C–H deformation	942	998	977
	10b	Out-of-plane C–H deformation	886	756	909
	4	Out-of-plane ring deformation	749	685	769
	11	Out-of-plane C–H deformation	700	685	675
	16b	Out-of-plane ring deformation	405	400	512
		Out-of-plane N–H deformation	—	945	—
		Out-of-plane N–O and ring deformation	—	—	226

[a]Choice of axes:

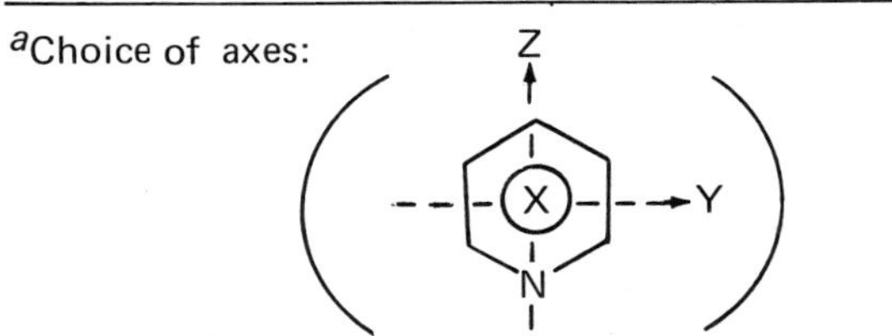

[b]Assignments for liquid state from ref. 2.
[c]Assignments for solid state from ref. 43.
[d]Assignments for liquid state from ref. 52.

are no drastic changes in force constants or large mass effects in going from one molecule to the other, except that three carbon-hydrogen vibrations (a_1 C–H stretch, b_2 C–H in-plane deformation, and b_1 C–H out-of-plane deformation) will be absent in pyridine. Therefore, the numbering of the vibrations of pyridine is generally made by analogy with those proposed by Wilson (15) for benzene and extended to substituted benzenes by Langseth and Lord (16) (see Table 18.1). In the Raman spectra of pyridine and most of its derivatives, bands due to C–H stretching occur in the range typical of aromatic carbon-hydrogen stretches (3100–3000 cm^{-1}) and a doublet arising from the aromatic ring stretches 8a, 8b, involving mainly C–C stretching, occurs in the range 1640–1560 cm^{-1}. The most intense bands in the Raman spectrum of pyridine occur at 1030 and 992 cm^{-1} and are assigned to the trigonal (No. 12) and the whole ring (No. 1) "breathing" vibrations, respectively.

The 2-, 3-, and 4-methylpyridines (α-, β-, γ-picolines) (Spectra 106–108) have been extensively studied using Raman spectroscopy (2,3,8,17–24). Other monosubstituted pyridine compounds investigated include the fluoro- (2,24, 25), chloro- (2,21,24,26), bromo- (2,21,24), ethyl- (27), aldehydic (28), methoxy- (29), cyano- (21), amino- (30), and hydroxy- (31) derivatives. The Raman spectrum of nicotine has also been reported (32). Characteristic frequencies in the Raman spectra of these compounds are listed in Tables 18.2 and 18.3. A doublet (8a, 8b) involving mainly the C–C stretchings in the aromatic ring is located in the Raman spectra in the frequency range 1620–1565 cm^{-1}. For 2- and 3-alkyl substituted pyridines it has been found (33) that the separation between the two bands is close to 20 cm^{-1}; whereas for 4-alkyl derivatives the separation is 40 cm^{-1}. The most intense Raman band in the spectra of monosubstituted pyridines generally arises from the trigonal ring "breathing" vibration (No. 12). In the spectrum of pyridine, this vibration occurs at 1030 cm^{-1}. In 3-substituted derivatives the frequency of this vibration shifts slightly to the range 1030–1010 cm^{-1}. In 2- and 4-substituted derivatives it occurs as an intense Raman band at 1000–985 cm^{-1}. The frequency of the in-plane ring deformation 6b is also useful in further distinguishing the substituent position in monosubstituted pyridines. For 2- and 3-substitution of the ring, this band is found about 615 cm^{-1} whereas for substitution at the 4-position it occurs near 660 cm^{-1}.

Assignments for all the fundamental frequencies in the vibrational spectra of the six dimethylpyridines (lutidines) have been proposed by Green et al. (34). Some of these assignments are given in Table 18.2. Other Raman studies of these compounds have been made (3,18,35,36). In addition to the lutidines, Raman spectra of 2-chloro-5-nitro- (18), 2,6-difluoro- (37), and 2,6-diaminopyridine (38) have been reported. Spinner (30) has investigated the IR and Raman spectra of the 3-, 4-, 5-, and 6-methyl derivatives of 2-aminopyridine. The 2,4- and 2,6-dimethylpyridine derivatives are characterized by

Table 18.2 Characteristic Ring Vibration Frequencies in the Raman Spectra of Substituted Pyridines

Wilson Vibration Number	Frequency (cm^{-1})							
	8a	8b	19a	19b	Ring "Breathing"	4	6b	16a
A. Monosubstituted[a]								
2-Substituted	1620–1570	1580–1560	1480–1450	1440–1415	1000–900	740–720	630–615	410–395
3-Substituted	1595–1570	1585–1560	1480–1465	1430–1410	1030–1010	720–705	630–615	410–395
4-Substituted	1605–1565	1570–1555	1500–1480	1420–1410	1000–985	730–720	670–660	400–385
Relative Raman Intensity	Medium–strong	Medium	Medium	Weak–medium	Very strong	Weak–very weak	Medium	Weak–very weak
B. Disubstituted[b]								
2,3-Dimethyl	1587	1578	1450	1433	(731)[c]	728	—	—
2,4-Dimethyl	1607	1565	1476	1397	999	732	—	—
2,5-Dimethyl	1602	1572	1490	1450	(844)[c]	—	—	—
2,6-Dimethyl	1594	1580	1470	1405	996	728	—	—
3,4-Dimethyl	1594	1563	1493	1403	(758)[c]	733	—	—
3,5-Dimethyl	1599	1579	1452	1432	1032	710	—	—
Relative Raman Intensity	Strong	Variable	Weak–medium	Weak–medium	Very strong	Very weak	—	—
C. Trisubstituted								
2,4,6-Trimethyl	1608	1569	1439	1408	993	—	—	—
Relative Raman Intensity	Medium	Strong	Weak	Weak	Very strong	—	—	—

[a] Assignments suggested in ref. 2.
[b] Assignments given in ref. 34.
[c] These frequencies are highly substituent-dependent.

Table 18.3 Characteristic C–H Vibration Frequencies in the Raman Spectra of Substituted Pyridines

	Frequency (cm^{-1})				
Wilson Vibration Number	3	9a	18b	18a	10a
A. Monosubstituted[a]					
2-Substituted	1305–1285	1155–1145	1110–1085	1050–1040	890–880
3-Substituted	1320–1230	1195–1185	1100–1095	1045–1035	935–915
4-Substituted	1320–1280	1220–1210	1100–1075	1075–1060	915–865
Relative Raman Intensity	Weak-medium	Medium-strong	Weak-medium	Strong-very strong	Weak-very weak
B. Disubstituted[b]					
2,3-Dimethyl	1276	1180	1070	–	927
2,4-Dimethyl	1295	1162	1112	–	–
2,5-Dimethyl	1296	1140	1030	–	923
2,6-Dimethyl	1317	1158	1096	–	905
3,4-Dimethyl	1304	1191	1068	–	920
3,5-Dimethyl	1322	1166	1139	–	935
Relative Raman Intensity	Weak-Very weak	Weak-medium	Weak	–	Weak
C. Trisubstituted					
2,4,6-Trimethyl	1316	1159	1102	–	–
Relative Raman Intensity	Strong	Medium	Weak	–	–

[a] Assignments suggested in ref. 2.
[b] Assignments given in ref. 34.

an intense Raman band near 995 cm^{-1} arising from the trigonal ring "breathing" vibration, as are the 2- and 4-monosubstituted compounds. For 3,5-dimethylpyridine this vibration appear at 1032 cm^{-1} in the Raman spectrum, which is similar to that found for 3-monosubstituted compounds. The frequency of the trigonal ring "breathing" vibration for the disubstituted derivatives (2,3-, 2,5-, 2,4-) is quite substituent-dependent, falling in the range 900–700 cm^{-1}. For the dimethylpyridines, no characteristic bands are found in the range 670–615 cm^{-1} as were found for the monosubstituted derivatives.

Polysubstituted pyridines that have been investigated by Raman spectroscopy include 2,4,6-trimethyl- (*sym*-collidine) (3,18), pentafluoro- (39,40), and pentachloro- (41) pyridine. Assignments for these molecules were made based on the character of the observed bands and on a comparison with the corresponding penta- and hexa-substituted benzenes.

The vibrational spectra of the pyridinium ion, $C_5H_5NH^+$, and several of its deuterated analogues have been studied in both the solid state (42,43) and in aqueous solution (43-46). Raman spectra that have been reported for monosubstituted pyridinium chlorides include those derived from 1-methyl- (46,47), 2-, 3-, 4-methyl- (3,21,28), aldehydic- (28), hydroxy- (45), methoxy- (29), amino- (30), and halogeno- (21) substituted pyridines. The hydrochlorides of some dimethyl- and trimethylpyridines (3,48), as well as those of the methyl derivatives of 2-aminopyridine (30), have also been the subject of Raman investigations. The assignment of the fundamental frequencies of pyridinium chloride are given in Table 18.1. Characteristic frequencies in the Raman spectra of pyridinium and monosubstituted pyridinium ions are listed in Tables 18.4 and 18.5. In most cases, cation formation produces only small variations in the frequencies and relative intensities observed in the Raman spectra of the neutral compounds except that the ring stretching vibrations (8a, 8b, 19a, 19b) for the cations generally occur 30–50 cm^{-1} higher in frequency. The N–H stretching vibration exhibits a large variation in frequency depending upon the nature of the anion; for example, in the solid pyridinium halides $C_5H_5NH^+Cl^-$, $C_5H_5NH^+Br^-$, and $C_5H_5NH^+I^-$, it is located at 2380, 3800, and 3000 cm^{-1}, respectively. The frequency of the N–H in-plane deformation varies only slightly and is found generally as a medium intensity Raman band at 1250–1240 cm^{-1}. The out-of-plane N–H deformation is variable in both frequency (940–880 cm^{-1}) and intensity.

The Raman spectrum of pyridine-1-oxide has been reported by Ito and Hata (49), Ramaiah and Srinivasan (50,51) and Mirone (52). The latter was an extensive study of the IR spectra of the solid, liquid, and gaseous states together with the Raman spectrum of liquid pyridine-1-oxide (see Table 18.1). A normal coordinate analysis of this compound was carried out by Berezin (53). It was found that the vibrational spectra of pyridine-1-oxide closely resembled those for the isoelectronic and nearly isobaric molecules fluorobenzene and phenol. In the Raman spectrum of the liquid, bands at 1252 (m), 835 (s), and 541 cm^{-1} (w) were attributed to ring vibrations involving considerable N–O stretching. The in-plane deformation mode of the N–O bond was assigned a frequency at 469 cm^{-1} (weak Raman intensity). Weak to very weak Raman bands at 511 and 226 cm^{-1} were found to involve some N–O out-of-plane deformation. The strongest Raman bands occurred at 1602 (ring stretch) and 1020 cm^{-1} (ring "breathing" mode). In the Raman spectra of 2-methyl- and 3-methylpyridine-1-oxides (54) a band near 1252 cm^{-1} in both instances was assigned to the N–O stretching vibration. For 2-methylpyridine-1-oxide the strongest Raman bands occurred at 1612 (ring

Table 18.4 Characteristic Ring Vibration Frequencies in the Raman Spectra of Pyridinium and Substituted Pyridinium Ions

Wilson Vibration Number		Frequency (cm^{-1})						
		8a	8b	19b	19a	Ring "Breathing" Modes	6b	6a
1. $C_5H_5NH^+X^-$ ($X = Cl, Br, I$)	(Solid)	1640–1630	1615–1600	1540–1530	1490–1480	1030–1028, 1010–1005	640–630	610–605
2. 1-Methylpyridine hydrochloride	(Aqueous solution)	1637	1586	1499	1484	1030	649	531
3. 2-Substituted-pyridine hydrochlorides	(Aqueous solution)	1635–1600	1620–1585	1540–1520	1475–1455	1015–1010	630–620	—
4. 3-Substituted-pyridine hydrochlorides	(Aqueous solution)	1635–1620	1615–1605	1550–1520	1475–1460	1050–1035	630–620	—
5. 4-Substituted-pyridine hydrochlorides	(Aqueous solution)	1640–1620	1610–1590	1520–1505	1500–1485	1011–1007	655–640	—
Relative Raman Intensity		Strong–very strong	Medium	Medium	Medium	Very strong	Medium	Medium

Table 18.5 Characteristic C-H and N-H Vibrations in the Raman Spectra of Pyridinium and Substituted Pyridinium Ions

		Frequency (cm^{-1})					
Wilson Vibration Number		3	9a	18b	18a	In-plane N-H Deformation	Out-of-plane N-H Deformation
1. $C_5H_5NH^+X^-$ (X = Cl, Br, I)	(Solid)	1330–1320	1210–1185	1085–1075	1055–1050	1255–1240	940–860
2. 1-Methylpyridine-hydrochloride	(Aqueous solution)	1287	1218	—	1191	—	—
3. 2-Substituted-pyridine hydrochlorides	(Aqueous solution)	1310–1280	1166–1164	1105–1085	1075–1040	1250–1220	—
4. 3-Substituted-pyridine hydrochlorides	(Aqueous solution)	1330–1315	1190–1185	1120–1100	1050–1035	1270–1260	—
5. 4-Substituted-pyridine hydrochlorides	(Aqueous solution)	1335–1325	1210–1205	1110–1090	1065–1060	1260–1250	—
Relative Raman Intensity		Weak-medium	Medium	Medium	Medium-strong	Medium	Weak

stretch) and 1049 cm^{-1} (ring "breathing" vibration). For the 3-methyl derivative, the corresponding intense Raman bands were located at 1603 and 1017 cm^{-1}, respectively.

The Raman spectra of quinoline (55-62) and isoquinoline (57-59, 62,63) have been extensively examined. The assignments of the most prominent bands observed in the Raman spectra of these two molecules, as reported by Wait and McNerney (62), are:

Quinoline: 1372 cm^{-1}–ν(ring), 1034 cm^{-1}–ring "breathing", 760 and 521 cm^{-1}–β(ring).
Isoquinoline: 1383 cm^{-1}–ν(ring), 1034 cm^{-1}–ring "breathing", 781, 523, and 505 cm^{-1}–β(ring).

Some of the methyl-substituted quinolines have been the subject of Raman investigations by Bonino and Manzoni-Ansidei (58) and by Jatkar (57). Raman spectra of 1,10-phenanthroline (*p*-phenanthroline) and its complexes with Zn (II) and Hg(II) have been reported by Krishnan and Plane (64).

18.2 DIAZINES

Partial assignments of the fundamental frequencies in the IR and Raman spectra of the diazines, pyrazine (I), pyrimidine (II), and pyridazine (III) were first reported by Ito et al. (65). A more detailed investigation of these three diazines by Lord, Marston, and Miller (66) led to a more complete vibrational assignment. Vibrational assignments in the IR and Raman spectra of the liquids pyridazine, pyridazine-d_4, pyridazine-3,6-d_2, and pyridazine-4,5-d_2 were discussed by Stidham and Tucci (67). A normal coordinate analysis for pyridazine has been reported by Berezin (68). Further spectroscopic investigations of pyrazine include that of Simmons, Innes, and Begun (69) on the IR of the liquid and vapor and Raman spectra of liquid and solid pyrazine and pyrazine-d_4 and that of Califano, Adembri, and Sbrana (70) on the IR spectra of the vapor and solid states of pyrazine, *cis*-pyrazine-d_2 and pyrazine-d_4. A

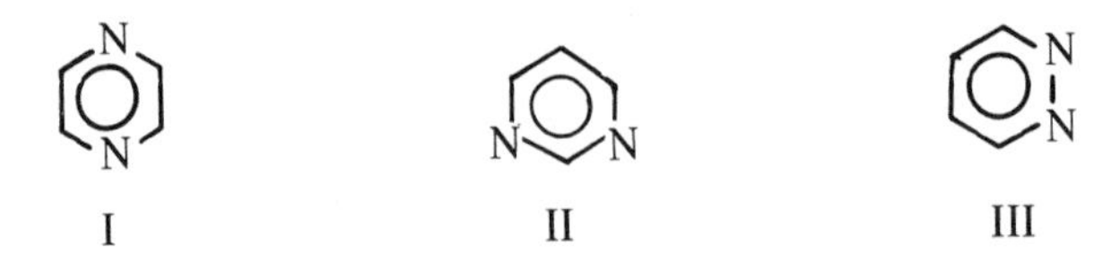

valence force field calculation of the frequencies of pyrazine has been carried out by Scully (71) for both in-plane and out-of-plane vibrations and by Berezin (72) for the planar vibrations only. The lattice vibrations of crystalline pyrazine have been investigated by Ito and Shigoeka (73). More recent work on the vibrational assignments in the spectra of pyrimidine includes that of Simmons and Innes (74) and that of Sbrana, Adembri, and Califano (75). Berezin and Potapov (76) have performed a normal coordinate analysis for pyrimidine. The IR and Raman spectra of crystalline pyrimidine at liquid nitrogen temperature have been reported by Foglizzo and Novak (77,78).

Assignments of the fundamental frequencies of pyrazine, pyrimidine, and pyridazine are given in Table 18.6. The pyrazine molecule is planar and belongs to the point group D_{2h}. The molecule possesses a center of symmetry and, therefore, satisfies the rule of mutual exclusion (no necessary coincidences occur in the Raman and IR spectra). The distribution of the 24 normal vibrations under D_{2h} symmetry is: $5a_g + b_{1g} + 2b_{2g} + 4b_{3g} + 2a_u + 4b_{1u} + 4b_{2u} + 2b_{3u}$. All g species are Raman-active only. All u species except a_u are IR-active only. Pyrimidine possesses C_{2v} symmetry and its 24 normal vibrations are distributed: $9a_1 + 2a_2 + 5b_1 + 8b_2$. All vibrations are IR- and Raman-active except for a_2 which is Raman-active only. Pyridazine belongs to the point group C^*_{2v} (also termed C^x_{2v}). Its normal vibrations are distributed: $9a_1 + 4a_2 + 3b_1 + 8b_2$. Characteristic Raman bands for these three diazines and some of their derivatives are listed in Table 18.7.

The IR and Raman spectra of pyrazine-di-*N*-oxide have been investigated by Szoke, Varsanyi, and Baitz (79) who found that the normal frequencies of this molecule were similar to those of *p*-difluorobenzene, which has an isoelectronic structure. A very strong band, in the IR only, at 1265 cm^{-1} was assigned to the stretching of the N-O bond. The ring "breathing" vibration was observed as an intense Raman band at 888 cm^{-1}. Foglizzo and Novak (80, 81) have studied the IR and Raman spectra of the pyrazinium halides, $C_4H_4N_2H^+X^-$ (X = Cl, Br, I) in both the solid state and in solution. For these ions, the N-H stretching frequency is observed only in the IR. In the solid state Raman spectra, the in-plane N-H deformation occurs as a strong band near 1240 cm^{-1} and the out-of-plane N-H deformation as a medium-to-strong band at 975-965 cm^{-1}. An increase in the planar skeletal frequencies was found for the pyrazinium ion as compared with those for the pyrazine molecule (see Table 18.7).

Derivatives of pyrimidine that have been investigated by Raman spectroscopy include 2,4,6-trifluoropyrimidine (82), 2,4,6-trichloropyrimidine (83), and fifteen various substituted pyrimidines, most of which contain an amino group (84). Spinner et al. (31,45,85) have employed IR and Raman spectra in a study of the lactim-lactam tautomerism in the hydroxypyrimidines and their hydrochloride and sodium salts. The IR and Raman spectra of the pyrimidinium halides in the range 300-30 cm^{-1} have been reported (81). Characteristic Raman frequencies for substituted pyrimidines are listed in Table 18.7. Most of the substituted pyrimidines exhibit one or more Raman bands in the aromatic C-H stretching region 3060-3000 cm^{-1} and four bands in the range

Table 18.6 Fundamental Frequencies in the Vibrational Spectra of the Diazines[a]

			Pyrazine[b]		Pyrimidine[c]		Pyridazine[d]	
Type of Vibra-Tion	Wilson Vibration Number	Approximate Description of Vibration	D_{2h} Symmetry Species	Frequency (cm^{-1})	C_{2v} Symmetry Species	Frequency (cm^{-1})	C_{2v}^* Symmetry Species	Frequency (cm^{-1})
Planar	2	C–H stretch	a_g	3060	a_1	3048	a_1	3043
	8a	Ring stretch		1584		1570		1572
	9a	In-plane C–H deformation		1232		1141		1160
	1	Ring stretch		1015		991		964
	6a	In-plane ring deformation		609		677		619
	7b	C–H stretch	b_{3g}	3045	b_2	3095	b_2	3043
	8b	Ring stretch		1523		1570		1566
	3	In-plane C–H deformation		1118		1227		1239
	6b	In-plane ring deformation		641		621		664
	13	C–H stretch	b_{1u}	3066	a_1	3001	b_2	3075
	19a	Ring stretch		1409		1402		1444
	18a	In-plane C–H deformation		1067		—		1052
	12	Ring stretch		1022		1066		1032
	20a	C–H stretch		—		3083		—
	20b	C–H stretch	b_{2u}	3066	b_2	—	a_1	3063
	19b	Ring stretch		1418		1467		1414
	14	Ring stretch		1342		1371		1347
	15	In-plane C–H deformation		1148		1161		1063
	18b	In-plane C–H deformation		—		1075		—

Non-planar	10a	Out-of-plane C–H deformation	b_{1g}	925	a_2	—	b_1	842
	5	Out-of-plane C–H deformation	b_{2g}	753	b_1	987	a_2	861
	4	Out-of-plane ring deformation		703		705		751
	10b	Out-of-plane C–H deformation		—		718		—
	17a	Out-of-plane C–H deformation	a_u	950	a_2	870	a_2	938
	16a	Out-of-plane ring deformation		340		394		421
	11	Out-of-plane C–H deformation	b_{3u}	804	b_1	806	b_1	760
	16b	Out-of-plane ring deformation		417		344		370

[a]Choice of axes for pyrimidine (1) and for pyridazine (2).

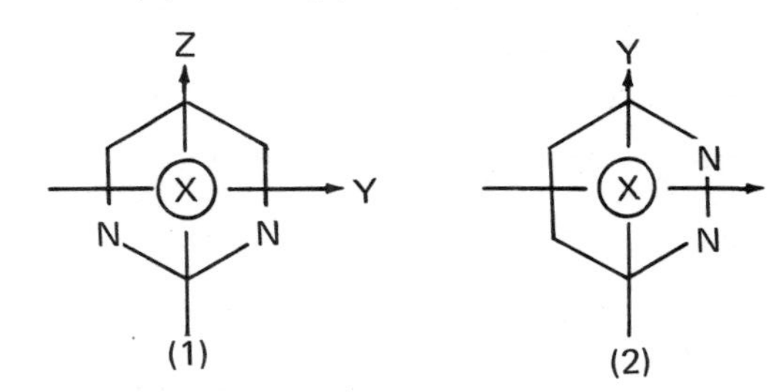

[b]Assignment for liquid state from refs. 66,69,70.
[c]Assignment for liquid state from refs. 66,75.
[d]Assignment for liquid state from refs. 66,67.

Table 18.7 Characteristic Frequencies in the Raman Spectra of Substituted Diazines

	Frequency (cm^{-1})								
	Ring Stretching Modes				Ring "Breathing" Modes	C–H In-plane Deformation			Ring Deformation Mode
Pyrazine	1584	1523	1490	1418	1022, 1015	1232	1148	1067	641,609
Pyrazinum halides	1620-1605	1595–1575	1490–1480	1460	1060–1030 1030–1020	1225–1220	1175–1170	1120–1105	690–680 603
Pyrazine-di-*N*-oxide	1644	1620	1492	1450	888	1300	1181	1125	630
Pyridazine'	1572	1566	1444	1414	1032,964	1329	1160	1063	664,619
Pyrimidine	1570	1570	1467	1402	1066,991	1227	1161	1075	677,621
2-Substituted	1590–1555	1565–1520	1480–1400	1410–1375	1005–990	1225–1205	1190–1170	1090–1085	645–625
4-Substituted	1590–1555	1565–1520	1480–1400	1410–1375	ca. 1000	ca. 1260	ca.1190	ca. 1070	685–660
5-Substituted	1590–1555	1565–1520	1480–1400	1410–1375	ca. 1050	ca. 1225	ca. 1200	ca. 1120	—
2,4-Disubstituted	1590–1555	1565–1520	1480–1400	1410–1375	1005–980	1205–1180	1140–1090	—	—
2,4,6-Trisubstituted	1590–1555	1565–1520	1480–1400	1410–1375	1005–980	—	1150–1110	—	—
2,4,5,6-Tetrasubstituted	1590–1555	1565–1520	1480–1400	1410–1375	ca. 1050	—	—	—	—

1600–1375 cm^{-1} due to aromatic ring stretches. Those derivatives substituted in the 2-, 4-, 2,4-, and 2,4,6-positions show a strong Raman band due to the trigonal ring "breathing" mode at 1005-980 cm^{-1}. Those substituted at the 5-position and the tetrasubstituted compounds display this vibration in the Raman spectra about 1050 cm^{-1}. 2-Substituted pyrimidines are distinguished by the ring deformation at 645–625 cm^{-1} whereas in 4-substituted derivatives, the frequency of this vibration is located at 685–660 cm^{-1}. Spectra of pyrimidine derivatives containing an amino-group have a band at 1680–1620 cm^{-1} arising from the NH_2 scissoring deformation.

Vibrational assignments in the IR and Raman spectra of purine (IV) and purine-*N-d* in the solid state and aqueous solution have been reported by

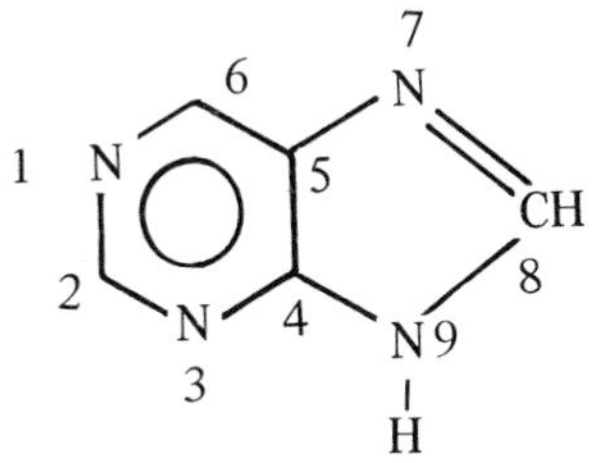

IV 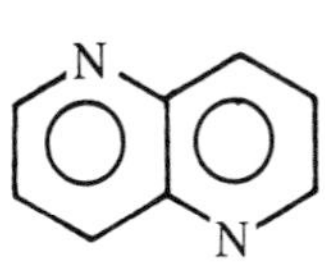 V 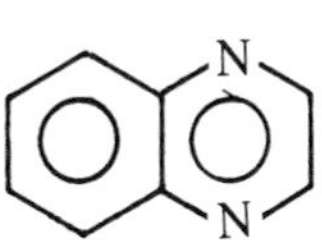 VI

Lautie and Novak (86,87). Purine derivatives that have been investigated by Raman spectroscopy include adenine (6-aminopurine) (88,89), 9-methyladenine (89), adenosine (88,89), adenosine-5′-monophosphate (88,89), and adenosine triphosphate (ATP) (88). Characteristic Raman frequencies for these molecules are given in Table 18.8.

Table 18.8 Characteristic Frequencies in the Raman Spectra of Purine and Its Derivatives [a]

	Frequency (cm^{-1})						
Physical State	Solids, Neutral, Acidic and Alkaline Solutions		Acidic Solution	Solids, Neutral, and Alkaline Solutions		Neutral and Alkaline Solutions	
Purine	1330	–	1405	1488	1300	1420	–
Adenine	1338 ± 8	714 ± 11	1412 ± 2	1456 ± 6	1313 ± 2	–	1388 ± 3
9-Methyladenine	1335 ± 5	714 ± 11	1412 ± 9	1487 ± 3	1313 ± 2	1427 ± 1	1388 ± 3
Adenosine	1336 ± 12	726 ± 7	1415 ± 2	1485 ± 5	1308 ± 2	1425 ± 2	1383 ± 5
Adenosine-5′-monophosphate	1337 ± 7	723 ± 5	1414 ± 1	1478 ± 8	1309 ± 1	1426 ± 4	1383 ± 3

[a]Assignments from refs. 86,89.

The fundamental frequencies in the Raman spectrum of 1,5-diazanaphthalene (1,5-naphthyridine) (V) have been assigned by Armarego, Barlin, and Spinner (90) and by Merritt and Pirkle (91). Vibrational assignments in the IR and Raman spectra of 1,4-diazanaphthalene (quinoxaline) (VI), 2,3-diazanaphthalene (phthalazine) (VII), 1,3-diazanaphthalene (quinazoline) (VIII), and 1,2-diazanaphthalene (cinnoline) (IX) have been made by Mitchell, Glass, and Merritt (92). Prominent Raman bands for these diazanaphthalenes in aqueous solution are listed in Table 18.9.

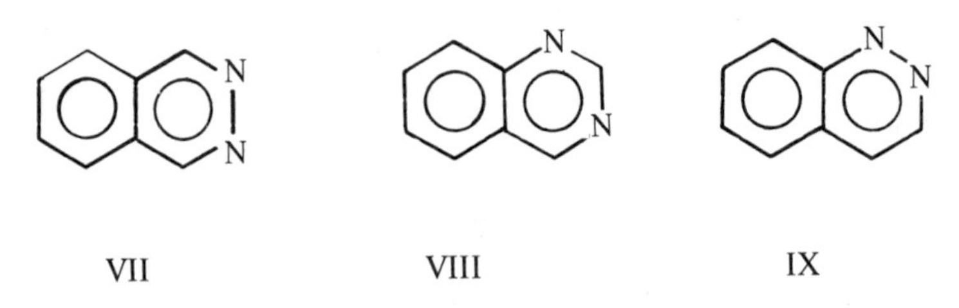

18.3 TRIAZINES

The vibrational spectra of *s*-triazine (1,3,5-triazine) (X) has been investigated by

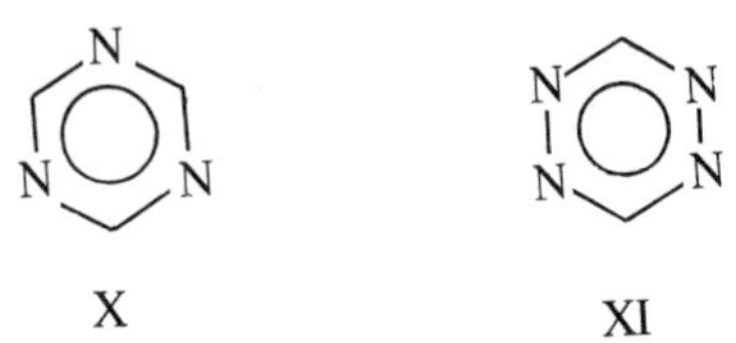

by several authors (72,93–95). Lancaster, Stamm, and Colthup (96) examined the IR of the vapor, liquid, and solid and the Raman spectrum of the liquid together with that of *s*-triazine-d_3. Their vibrational assignment was supported by a valence force field normal coordinate calculation. *s*-Triazine has the symmetry D_{3h} and of the ten fundamental frequencies allowed in the Raman, eight were observed and assigned as follows:

Type of Vibration	Frequency (cm^{-1})
C–H stretch	3042
Ring stretches	1555,1410
Ring “breathing”	1132,992
In-plane C–H deformation	1176
In-plane ring deformation	676
Out-of-plane ring deformation	340

Table 18.9 Prominent Bands in the Raman Spectra of Aqueous Solutions of Some Diazanaphthalenes[a]

	Frequency (cm^{-1})									
	Ring Stretching Modes			Ring "Breathing" Mode	In-plane C–H Deformations			In-plane Ring Deformations	
1,5-Diazanaphthalene	1582	1483	1375	1052	1408	1272	1153	771	533
Quinoxaline	1580	1504	1367	1033	1423	1211	1140	765	535
Phthalazine	1583	1494	1380	1020	1448	1218	1160	802	521
Quinazoline	1564	1496	1380	1082	1412	1215	1158	784	528
Cinnoline	1586	1496	1400	1048	1420	1184	1162	778	530

[a] Assignments from refs. 91,92.

Raman spectroscopy has been employed to characterize the following derivatives of *s*-triazine: 2,4,6-trifluoro-*s*-triazine (97,98), 2,4,6-trichloro-*s*-triazine (cyanuric chloride) (98,99), tris (trifluoromethyl)-*s*-triazine (100), 2-fluoro-4,6-dichloro-*s*-triazine (101), 2,4-difluoro-6-chloro-*s*-triazine (101), and 2,4-dichloro-6-isocyanodichloro-*s*-triazine (tetrameric cyanogen chloride) (102). Characteristic Raman bands in the spectra of these trisubstituted derivatives of *s*-triazine appear at 1600–1510 and 1500–1460 cm^{-1} (ring stretches) and at 1000–980 cm^{-1} (ring "breathing" vibration).

18.4 TETRAZINES

s-Tetrazine (XI) has the point group symmetry D_{2h} and of the 18 normal vibrations, 8 are IR-active only, 9 are Raman-active only, and 1 is inactive in both spectra. IR studies of *s*-tetrazine, and *s*-tetrazine-d_1 and -d_2 were conducted by Spencer, Cross, and Wiberg (103) and a normal coordinate analysis of the planar vibrations of these molecules was carried out by Berezin (72). However, because of the intense red color of *s*-tetrazine, the Raman spectrum could not be obtained until the advent of laser sources. Using a helium-neon laser (λ 6328 Å) source, Franks, Merer, and Innes (104) obtained the Raman spectra of solid *s*-tetrazine and *s*-tetrazine-d_2 and observed eight of the nine Raman-active vibrations of *s*-tetrazine:

Type of Vibration	Frequency (cm^{-1})
C–H stretch	3090
Ring stretches	1521,1418
In-plane C–H deformation	1303
Ring "breathing"	1017
Out-of-plane ring deformation	800
In-plane ring deformations	737,651

REFERENCES

1. J. K. Wilmhurst and H. J. Bernstein, *Can. J. Chem., 35,* 1183 (1957).
2. J. H. S. Green, W. Kynaston, and H. M. Paisley, *Spectrochim. Acta, 19,* 549 (1963).
3. E. Herz, L. Kahovec and K. W. F. Kohlrausch, *Z. Phys. Chem., 53B,* 124 (1943).
4. C. H. Kline, Jr. and J. Turkevich, *J. Chem. Phys., 12,* 300 (1944).
5. L. Corrsin, B. J. Fax, and R. C. Lord, *J. Chem. Phys., 21,* 1170 (1953).
6. E. Wachsmann and E. W. Schmid, *Z. Phys. Chem. (N. F.), 27,* 145 (1961).
7. M. A. Kovner, Yu. S. Korostelev, and V. I. Berezin, *Opt. Spectrosc., 10,* 233 (1961).
8. D. A. Long and W. O. George, *Spectrochim. Acta, 19,* 1777 (1963).
9. F. A. Andersen, B. Bak, S. Brodersen, and J. Rastrup-Andersen, *J. Chem. Phys., 23,* 1047 (1955).
10. V. I. Berezin, *Opt. Spectrosc., 15,* 167 (1963).
11. M. A. Kovner, S. K. Potapov, G. A. Rekhen, N. A. Sakhnov, and I. V. Shevchenko, *Opt. Spectrosc., 29,* 281 (1970).
12. D. A. Long, F. S. Murfin, and E. L. Thomas, *Trans. Faraday Soc., 59,* 12 (1963).
13. D. A. Long and E. L. Thomas, *Trans. Faraday Soc., 59,* 783 (1963).
14. R. S. Mulliken, *J. Chem. Phys., 23,* 1997 (1955).
15. E. B. Wilson, Jr., *Phys. Rev., 45,* 706 (1934).
16. A. Langseth and R. C. Lord, *Kgl. Dan. Vidensk. Selsk., Mat.-fys. Medd., 16,* No. 6, 1938.
17. S. K. K. Jatkar, *Indian. J. Phys., 10,* 23 (1936).
18. R. Manzoni-Ansidei, *Bull. Sci. Fac. Chim. Ind. Bologna, 1,* 137 (1940).
19. D. A. Long, F. S. Murfin, J. L. Hales, and W. Kynaston, *Trans. Faraday Soc., 53,* 1171 (1957).
20. V. Lorenzelli and G. Randi, *Lincei-Rend. Sc. Fis. Mat. Nat., 34,* 527 (1963).

21. E. Spinner, *J. Chem. Soc., 1963,* 3860.
22. D. W. Scott, W. N. Hubbard, J. F. Messerly, S. S. Todd, I. A. Hossenlopp, W. D. Good, D. R. Douslin, and J. P. McCullough, *J. Phys. Chem., 67,* 680 (1963).
23. D. W. Scott, W. D. Good, G. B. Guthrie, S. S. Todd, I. A. Hossenlopp, A. G. Osborn, and J. P. McCullough, *J. Phys. Chem., 67,* 685 (1963).
24. G. Varsanyi, T. Farago, and S. Holly, *Acta Chem. Acad. Sci. Hung., 43,* 205 (1965).
25. H. P. Stephenson and F. L. Voelz, *J. Chem. Phys., 22,* 1945 (1954).
26. J. W. Murray and D. H. Andrews, *J. Chem. Phys., 1,* 406 (1933).
27. J. H. S. Green and P. W. B. Barnard, *J. Chem. Soc., 1963* , 640.
28. P. Chiorboli, P. Mirone, and V. Lorenzelli, *Ann. Chim. (Rome), 48,* 355 (1958).
29. E. Spinner and J. C. B. White, *J. Chem. Soc., 1962,* 3115.
30. E. Spinner, *J. Chem. Soc., 1962,* 3119.
31. E. Spinner and A. Albert, *J. Chem. Soc., 1960,* 1221.
32. V. Brustier and P. Blanc, *Bull. Soc. Chim. Fr., (5) 10,* 58 (1943).
33. G. L. Cook and F. M. Church, *J. Phys. Chem., 61,* 458 (1957).
34. J. H. S. Green, D. J. Harrison, W. Kynaston, and H. M. Paisley, *Spectrochim. Acta, 26A,* 2139 (1970).
35. K. C. Medhi, *Opt. Spectrosc., 19,* 24 (1965).
36. K. C. Medhi and D. K. Mukherjee, *Spectrochim. Acta, 21,* 895 (1965).
37. R. T. Bailey and D. Steele, *Spectrochim. Acta, 23A,* 2997 (1967).
38. G. D. Baruah, R. A. Amma, P. S. Dube, and S. N. Rai, *Indian J. Pure Appl. Phys., 8,* 761 (1970).
39. D. A. Long and R. T. Bailey, *Trans. Faraday Soc., 59,* 599 (1963).
40. D. A. Long and D. Steele, *Spectrochim. Acta, 19,* 1791 (1963).
41. R. T. Bailey and G. P. Strachan, *Spectrochim. Acta, 26A,* 1129 (1970).
42. D. Cook, *Can. J. Chem., 39,* 3009 (1961).
43. R. Foglizzo and A. Novak, *J. Chim. Phys., 66,* 1539 (1969).
44. H. J. Bernstein and V. H. Martin, *Trans. Roy. Soc. Can., 31,* 95 (1937).
45. E. Spinner, *J. Chem. Soc., 1960,* 1226.
46. E. Spinner, *J. Chem. Soc., 1963,* 3870.
47. E. Spinner, *Aust. J. Chem., 20,* 1805 (1967).
48. R. Manzoni-Ansidei, *Bull. Sci. Fac. Chem. Ind. Bologna, 1,* 184 (1940).
49. M. Ito and N. Hata, *Bull. Chem. Soc. Jap., 28,* 353 (1955).
50. K. Ramaiah and V. R. Srinivasan, *Curr. Sci., 27,* 340 (1958).
51. K. Ramaiah and V. R. Srinivasan, *Proc. Ind. Acad. Sci., 50A,* 213 (1959).
52. P. Mirone, *Atti. Accad. Nazl. Lincei Rend., Cl. Sci. Fis., Mat. Nat., 35,* 530 (1963).

53. V. I. Berezin, *Opt. Spectrosc., 18,* 119 (1965).
54. K. Ramaiah and V. R. Srinivasan, *Proc. Ind. Acad. Sci., 55A,* 221 (1962).
55. A. S. Ganesan and S. Venkateswaran, *Indian J. Phys., 4,* 195 (1929).
56. G. B. Bonino and P. Cella, *Atti. Accad. Lincei, 15,* 385 (1932).
57. S. K. K. Jatkar, *Indian J. Phys., 10,* 23 (1936).
58. G. B. Bonino and R. Manzoni-Ansidei, *Mem. Accad. Sci. Inst. Bologna, Cl. Sci. Fis. [9], 1,* Sep. 7 (February 18, 1934).
59. H. Luther, D. Mootz, and F. Radwitz, *J. Prakt. Chem., 5,* 242 (1958).
60. K. K. Deb, *Indian J. Phys., 35,* 535 (1961).
61. P. Chiorboli and A. Bertoluzza, *Ann. Chim. (Rome), 49,* 245 (1959).
62. S. C. Wait, Jr. and J. C. McNerney, *J. Mol. Spectrosc., 34,* 56 (1970).
63. K. K. Deb, *Indian J. Phys., 36,* 557 (1962).
64. K. Krishnan and R. A. Plane, *Spectrochim. Acta, 25A,* 831 (1969).
65. M. Ito, R. Shimada, T. Kuraishi, and W. Mizushima, *J. Chem. Phys., 25,* 597 (1956).
66. R. C. Lord, A. L. Marston, and F. A. Miller, *Spectrochim. Acta, 9,* 113 (1957).
67. H. D. Stidham and J. V. Tucci, *Spectrochim. Acta, 23A,* 2233 (1967).
68. V. I. Berezin, *Opt. Spectrosc., 18,* 71 (1965).
69. J. D. Simmons, K. K. Innes, and G. M. Begun, *J. Mol. Spectrosc., 14,* 190 (1964).
70. S. Califano, G. Adembri, and G. Sbrana, *Spectrochim. Acta, 20,* 385 (1964).
71. D. B. Scully, *Spectrochim. Acta, 17,* 233 (1961).
72. V. I. Berezin, *Opt. Spectrosc., 16,* 131 (1964).
73. M. Ito and T. Shigoeka, *J. Chem. Phys., 44,* 1001 (1966).
74. J. D. Simmons and K. K. Innes, *J. Mol. Spectrosc., 13,* 435 (1964).
75. G. Sbrana, G. Adembri, and S. Califano, *Spectrochim. Acta, 22,* 1831 (1966).
76. V. I. Berezin and S. K. Potapov, *Opt. Spectrosc., 18,* 22 (1965).
77. R. Foglizzo and A. Novak, *J. Chim. Phys., 64,* 1484 (1967).
78. R. Foglizzo and A. Novak, *Spectrosc., Lett., 2,* 165 (1969).
79. S. Szoke, G. Varsanyi, and E. Baitz, *Acta Chim. Acad. Sci. Hung., 54,* 145 (1966).
80. R. Foglizzo and A. Novak, *Appl. Spectrosc., 24,* 601 (1970).
81. R. Foglizzo and A. Novak, *J. Mol. Struct., 7,* 205 (1971).
82. R. T. Bailey and D. Steele, *Spectrochim. Acta, 23A,* 2989 (1967).
83. R. T. Bailey and D. Steele, *Spectrochim. Acta, 25A,* 219 (1969).
84. A. J. Lafaix and J. M. Lebas, *Spectrochim. Acta, 26A,* 1243 (1970).
85. E. Spinner, *J. Chem. Soc., 1960,* 1232.

86. A. Lautie and A. Novak, *J. Chim. Phys., 65,* 1359 (1968).
87. A. Lautie and A. Novak, *Compt. Rend., 272B,* 599 (1971).
88. R. A. Malt, *Biochim. Biophys. Acta, 120,* 461 (1966).
89. R. C. Lord and G. J. Thomas, Jr., *Spectrochim. Acta, 23A,* 2551 (1967).
90. W. L. F. Armarego, G. B. Barlin, and E. Spinner, *Spectrochim. Acta, 22,* 117 (1966).
91. J. A. Merritt and R. J. Pirkle, *J. Mol. Spectrosc., 35,* 251 (1970).
92. R. W. Mitchell, R. W. Glass, and J. A. Merritt, *J. Mol. Spectrosc., 36,* 310 (1970),
93. J. Goubeau, E. L. John, A. Kreutzberger, and C. J. Grundmann, *J. Phys. Chem., 58,* 1078 (1954).
94. J. E. Lancaster and N. B. Colthup, *J. Chem. Phys., 22,* 1149 (1954).
95. R. F. Stamm and J. E. Lancaster, *J. Chem. Phys., 22,* 1280 (1954).
96. J. E. Lancaster, R. F. Stamm, and N. B. Colthup, *Spectrochim. Acta, 17,* 155 (1961).
97. D. A. Long, J. Y. H. Chau, and R. B. Gravenor, *Trans. Faraday Soc., 58,* 2316 (1962).
98. E. S. Levin and N. P. Vinogradova, *Zh. Prikl. Spektrosk., 4,* 330 (1966).
99. D. M. Thomas, J. B. Bates, A. Bandy, and E. R. Lippincott, *J. Chem. Phys., 53,* 3698 (1970).
100. J. W. Dawson, J. B. Hynes, K. Niedenzu, and W. Sawodny, *Spectrochim; Acta, 23A,* 1211 (1967).
101. W. Sawodny, K. Niedenzu, J. B. Hynes, and J. W. Dawson, *Spectrochim. Acta, 23A,* 1327 (1967).
102. K. Kawai, Y. Kodama, and F. Mizukami, *Spectrochim. Acta, 24A,* 1013 (1968).
103. G. H. Spencer, P. C. Cross, and K. B. Wiberg, *J. Chem. Phys., 35,* 1939 (1961).
104. L. A. Franks, A. J. Merer, and K. K. Innes, *J. Mol. Spectrosc., 26,* 458 (1968).

APPENDIX ONE

A SUMMARY OF CHARACTERISTIC RAMAN FREQUENCIES

Frequency (cm^{-1})	Vibration	Compound	Refer to Section
3400–3330	Bonded antisymmetric NH_2 stretch	Primary amines	4.1
3380–3340	Bonded OH stretch	Aliphatic alcohols	3.1
3374	CH stretch	Acetylene (gas)	12.1
3355–3325	Bonded antisymmetric NH_2 stretch	Primary amides	9.2
3350–3300	Bonded NH stretch	Secondary amines	4.1
3335–3300	≡CH stretch	Alkyl acetylenes	12.1
3300–3250	Bonded symmetric NH_2 stretch	Primary amines	4.1
3310–3290	Bonded NH stretch	Secondary amides	9.3
3190–3145	Bonded symmetric NH_2 stretch	Primary amides	9.2
3175–3154	Bonded NH stretch	Pyrazoles	16.2
3103	Antisymmetric $=CH_2$ stretch	Ethylene (gas)	6.1
3100–3020	CH_2 stretches	Cyclopropane	1.5

3100–3000	Aromatic CH stretch	Benzene derivatives	13.3
3095–3070	Antisymmetric $=CH_2$ stretch	$C=CH_2$ derivatives	6.2
3062	CH stretch	Benzene	13.1
3057	Aromatic CH stretch	Alkyl benzenes	13.3
3040–3000	CH stretch	C=CHR derivatives	6.2
3026	Symmetric $=CH_2$ stretch	Ethylene (gas)	6.1
2990–2980	Symmetric $=CH_2$ stretch	$C=CH_2$ derivatives	6.2
2986–2974	Symmetric NH_3^+ stretch	Alkyl ammonium chlorides (aqueous solution)	4.2
2969–2965	Antisymmetric CH_3 stretch	*n*-Alkanes	1.2
2929–2912	Antisymmetric CH_2 stretch	*n*-Alkanes	1.2
2884–2883	Symmetric CH_3 stretch	*n*-Alkanes	1.2
2861–2849	Symmetric CH_2 stretch	*n*-Alkanes	1.2
2850–2700	CHO group (2 bands)	Aliphatic aldehydes	7.2
2590–2560	SH stretch	Thiols	5.1
2316–2233	C≡C stretch (2 bands)	$R-C \equiv C-CH_3$	12.1
2301–2231	C≡C stretch (2 bands)	$R-C \equiv C-R'$	12.1
2300–2250	Pseudoantisymmetric N=C=O stretch	Isocyanates	11.5
2264–2251	Symmetric C≡C–C≡C stretch	Alkyl diacetylenes	12.1
2259	C≡N stretch	Cyanamide	12.4
2251–2232	C≡N stretch	Aliphatic nitriles	12.2
2220–2100	Pseudoantisymmetric N=C=S stretch (2 bands)	Alkyl isothiocyanates	11.6
2220–2000	C≡N stretch	Diaklyl cyanamides	12.4
2172	Symmetric C≡C–C≡C stretch	Diacetylene	12.1
2161-2134	$\overset{+}{N} \equiv \overset{-}{C}$ stretch	Aliphatic isonitriles	12.3
2160–2100	C≡C stretch	Alkyl acetylenes	12.1
2156–2140	C≡N stretch	Alkyl thiocyanates	12.5
2104	Antisymmetric N=N=N stretch	CH_3N_3	11.8
2094	C≡N stretch	HCN	12.2
2049	Pseudoantisymmetric C=C=O stretch	Ketene	11.4
1974	C≡C stretch	Acetylene (gas)	12.1
1964–1958	Antisymmetric C=C=C stretch	Allenes	11.1
1870–1840	Symmetric C=O stretch	Saturated 5-membered ring cyclic anhydrides	15.1
1820	Symmetric C=O stretch	Acetic anhydride	8.5
1810–1788	C=O stretch	Acid halides	8.6
1807	C=O stretch	Phosgene	8.6

1805–1799	Symmetric C=O stretch	Noncyclic anhydrides	8.5
1800	C=C stretch	$F_2C{=}CF_2$ (gas)	6.3
1795	C=O stretch	Ethylene carbonate	15.2
1792	C=C stretch	$F_2C{=}CFCH_3$	6.3
1782	C=O stretch	Cyclobutanone	7.3
1770–1730	C=O stretch	Halogenated aldehydes	7.2
1744	C=O stretch	Cyclopentanone	7.3
1743–1729	C=O stretch	Cationic α-amino acids (aqueous solution)	9.6
1741–1734	C=O stretch	*O*-Alkyl acetates	8.4
1740–1720	C=O stretch	Aliphatic aldehydes	7.2
1739–1714	C=C stretch	$C{=}CF_2$ derivatives	6.3
1736	C=C stretch	Methylene cyclopropane	6.9
1734–1727	C=O stretch	*O*-Alkyl propionates	8.4
1725–1700	C=O stretch	Aliphatic ketones	7.3
1720–1715	C=O stretch	*O*-Alkyl formates	8.4
1712–1694	C=C stretch	RCF=CFR	6.3
1695	Nonconjugated C=O stretch	Uracil derivatives (aqueous solution)	17.2
1689–1644	C=C stretch	Monofluoroalkenes	6.3
1687–1651	C=C stretch	Alkylidene cyclopentanes	6.9
1686–1636	Amide I band	Primary amides (solids)	9.2
1680–1665	C=C stretch	Tetraalkyl ethylenes	6.2
1679	C=C stretch	Methylene cyclobutane	6.9
1678–1664	C=C stretch	Trialkyl ethylenes	6.2
1676–1665	C=C stretch	*trans*-Dialkyl ethylenes	6.2
1675	Symmetric C=O stretch (cyclic dimer)	Acetic acid	8.1
1673–1666	C=N stretch	Aldimines	10.1
1672	Symmetric C=O stretch (cyclic dimer)	Formic acid (aqueous solution)	8.1
1670–1655	Conjugated C=O stretch	Uracil, cytosine, and guanine derivatives (aqueous solution)	17.2
1670–1630	Amide I band	Tertiary amides	9.4
1666–1652	C=N stretch	Ketoximes	10.3
1665–1650	C=N stretch	Semicarbazones (solid)	10.4
1663–1636	Symmetric C=N stretch	Aldazines, ketazines	10.5
1660–1654	C=C stretch	*cis*-Dialkyl ethylenes	6.2
1660–1650	Amide I band	Secondary amides	9.3

1660–1649	C=N stretch	Aldoximes	10.3
1660–1610	C=N stretch	Hydrazones (solid)	10.4
1658–1644	C=C stretch	$R_2C{=}CH_2$	6.2
1656	C=C stretch	Cyclohexene, cycloheptene	6.13
1654–1649	Symmetric C=O stretch (cyclic dimer)	Carboxylic acids	8.1
1652–1642	C=N stretch	Thiosemicarbazones (solid)	10.4
1650–1590	NH_2 scissors	Primary amines	4.1
1649–1625	C=C stretch	Allyl derivatives	6.4
1648–1640	N=O stretch	Alkyl nitrites	4.7
1648–1638	C=C stretch	$H_2C{=}CHR$	6.2
1647	C=C stretch	Cyclopropene	6.10
1638	C=O stretch	Ethylene diothiocarbonate	15.4
1637	Symmetric C=C stretch	Isoprene	6.7
1634–1622	Antisymmetric NO_2 stretch	Alkyl nitrates	4.6
1630–1550	Ring stretches (doublet)	Benzene derivatives	13.3
1623	C=C stretch	Ethylene (gas)	6.1
1620–1540	Three or more coupled C=C stretches	Polyenes	6.8
1616–1571	C=C stretch	Chloroalkenes	6.3
1614	C=C stretch	Cyclopentene	6.12
1596–1547	C=C stretch	Bromoalkenes	6.3
1581–1465	C=C stretch	Iodoalkenes	6.3
1575	Symmetric C=C stretch	1,3-Cyclohexadiene	6.13
1573	N=N stretch	Azomethane (in solution)	10.6
1566	C=C stretch	Cyclobutene	6.11
1560–1550	Antisymmetric NO_2 stretch	Primary nitroalkanes	4.5
1555–1550	Antisymmetric NO_2 stretch	Secondary nitroalkanes	4.5
1548	N=N stretch	1-Pyrazoline	15.3
1545–1535	Antisymmetric NO_2 stretch	Tertiary nitroalkanes	4.5
1515–1490	Ring stretch	2-Furfuryl group	16.1
1500	Symmetric C=C stretch	Cyclopentadiene	6.12
1480–1470	OCH_3, OCH_2 deformations	Aliphatic ethers	3.2
1480–1460	Ring stretch	2 Furfurylidene or 2-furoyl group	16.1
1473–1446	CH_3, CH_2 deformations	*n*-Alkanes	1.2
1466–1465	CH_3 deformation	*n*-Alkanes	1.2

1450–1400	Pseudoantisymmetric N=C=O stretch	Isocyanates	11.5
1443–1398	Ring stretch	2-Substituted thiophenes	16.1
1442	N=N stretch	Azobenzene	10.6
1440–1340	Symmetric CO_2^- stretch	Carboxylate ions (aqueous solution)	8.3
1415–1400	Symmetric CO_2^- stretch	Dipolar and anionic α-amino acids (aqueous solution)	9.6
1415–1385	Ring stretch	Anthracenes	13.9
1395–1380	Symmetric NO_2 stretch	Primary nitroalkanes	4.5
1390–1370	Ring stretch	Naphthalenes	13.8
1385–1368	CH_3 symmetric deformation	*n*-Alkanes	1.2
1375–1360	Symmetric NO_2 stretch	Secondary nitroalkanes	4.5
1355–1345	Symmetric NO_2 stretch	Tertiary nitroalkanes	4.5
1350–1330	CH deformation	Isopropyl group	1.3
1320	Ring vibration	1,1-Dialkyl cyclopropanes	1.5
1314–1290	In-plane CH deformation	*trans*-Dialkyl ethylenes	6.2
1310–1250	Amide III band	Secondary amides	9.3
1310–1175	CH_2 twist and rock	*n*-Alkanes	1.2
1305–1295	CH_2 in-phase twist	*n*-Alkanes	1.2
1300–1280	CC bridge bond stretch	Biphenyls	13.7
1282–1275	Symmetric NO_2 stretch	Alkyl nitrates	4.6
1280–1240	Ring stretch	Epoxy derivatives	14.1
1276	Symmetric N=N=N stretch	CH_3N_3	11.8
1270–1251	In-plane CH deformation	*cis*-Dialkyl ethylenes	6.2
1266	Ring "breathing"	Ethylene oxide (oxirane)	14.1
1230–1200	Ring vibration	Para-disubstituted benzenes	13.4
1220–1200	Ring vibration	Mono- and 1,2-dialkyl cyclopropanes	1.5
1212	Ring "breathing"	Ethylene imine (aziridine)	14.1
1205	C_6H_5–C vibration	Alkyl benzenes	13.3
1196–1188	Symmetric SO_2 stretch	Alkyl sulfates	5.9
1188	Ring "breathing"	Cyclopropane	1.5
1172–1165	Symmetric SO_2 stretch	Alkyl sulfonates	5.9
1150–950	CC stretches	*n*-Alkanes	1.2

1145–1125	Symmetric SO_2 stretch	Dialkyl sulfones	5.9
1144	Ring "breathing"	Pyrrole	16.1
1140	Ring "breathing"	Furan	16.1
1130–1100	Symmetric C=C=C stretch (2 bands)	Allenes	11.1
1130	Pseudosymmetric C=C=O stretch	Ketene	11.4
1112	Ring "breathing"	Ethylene sulfide	14.1
1111	NN stretch	Hydrazine	4.4
1070–1040	S=O stretch (1 or 2 bands)	Aliphatic sulfoxides	5.8
1065	C=S stretch	Ethylene trithio-carbonate	15.4
1060–1020	Ring vibration	Ortho-disubstituted benzenes	13.4
1040–990	Ring vibration	Pyrazoles	16.2
1030–1015	In-plane CH deformation	Monosubstituted benzenes	13.3
1030–1010	Trigonal ring "breathing"	3-Substituted pyri-dines	18.1
1030	Trigonal ring "breathing"	Pyridine	18.1
1029	Ring "breathing"	Trimethylene oxide (oxetane)	14.2
1026	Ring "breathing"	Trimethylene imine (azetidine)	14.2
1010–990	Trigonal ring "breathing"	Mono-, meta-, and 1,3,5 substituted benzenes	13.3
1001	Ring "breathing"	Cyclobutane	1.6
1000–985	Trigonal ring "breathing"	2- and 4-Substituted pyridines	18.1
992	Ring "breathing"	Benzene	13.1
992	Ring "breathing"	Pyridine	18.1
939	Ring "breathing"	1,3-Dioxolane	15.2
933	Ring vibration	Alkyl cyclobutanes	1.6
930–830	Symmetric COC stretch	Aliphatic ethers	3.2
914	Ring "breathing"	Tetrahydrofuran	15.1
906	ON stretch	Hydroxylamine	4.3
905–837	CC skeletal stretch	*n*-Alkanes	1.2
900–890	Ring vibration	Alkyl cyclopentanes	1.7
900–850	Symmetric CNC stretch	Secondary amines	4.1
899	Ring "breathing"	Pyrrolidine	15.1
866	Ring "breathing"	Cyclopentane	1.7

877	OO stretch	Hydrogen peroxide	3.4
851-840	Pseudosymmetric CON stretch	*O*-Alkyl hydroxylamines	4.3
836	Ring "breathing"	Piperazine	17.2
835-749	Skeletal stretch	Isopropyl group	1.3
834	Ring "breathing"	1,4-Dioxane	17.2
832	Ring "breathing"	Thiophene	16.1
832	Ring "breathing"	Morpholine	17.2
830-720	Ring vibration	Para-disubstituted benzenes	13.4
825-820	C_3O skeletal stretch	Secondary alcohols	3.1
818	Ring "breathing"	Tetrahydropyran	17.1
815	Ring "breathing"	Piperidine	17.1
802	Ring "breathing"	Cyclohexane (chair form)	1.8
785-700	Ring vibration	Alkyl cyclohexanes	1.8
760-730	C_4O skeletal stretch	Tertiary alcohols	3.1
760-650	Symmetric skeletal stretch	*tert*-Butyl group	1.3
740-585	CS stretch (1 or more bands)	Alkyl sulfides	5.2
735-690	"C=S stretch"	Thioamides, thioureas (solid)	5.5
733	Ring "breathing"	Cycloheptane	1.4
730-720	CCl stretch, P_C conformation	Primary chloroalkanes	2.2
715-620	CS stretch (1 or more bands)	Dialkyl disulfides	5.3
709	CCl stretch	CH_3Cl	2.1
703	Ring "breathing"	Cyclooctane	1.4
703	Symmetric CCl_2 stretch	CH_2Cl_2	2.1
690-650	Pseudosymmetric N=C=S stretch	Alkyl isothiocyanates	11.6
688	Ring "breathing"	Tetrahydrothiophene	15.1
668	Symmetric CCl_3 stretch	$CHCl_3$	2.1
660-650	CCl stretch, P_H conformation	Primary chloroalkanes	2.2
659	Symmetric CSC stretch	Pentamethylene sulfide	17.1
655-640	CBr stretch, P_C conformation	Primary bromoalkanes	2.2
630-615	Ring deformation	Monosubstituted benzenes	13.3
615-605	CCl stretch, S_{HH} conformation	Secondary chloroalkanes	2.2
610-590	CI stretch, P_C conformation	Primary iodoalkanes	2.2
609	CBr stretch	CH_3Br	2.1
577	Symmetric CBr_2 stretch	CH_2Br_2	2.1
570-560	CCl stretch, T_{HHH} conformation	Tertiary chloroalkanes	2.2

565-560	CBr stretch, P_H conformation	Primary bromo-alkanes	2.2
540-535	CBr stretch, S_{HH} conformation	Secondary bromo-alkanes	2.2
539	Symmetric CBr_3 stretch	$CHBr_3$	2.1
525-510	SS stretch	Dialkyl disulfides	5.3
523	CI stretch	CH_3I	2.1
520-510	CBr stretch, T_{HHH} conformation	Tertiary bromo-alkanes	2.2
510-500	CI stretch, P_H conformation	Primary iodoalkanes	2.2
510-480	SS stretch	Dialkyl trisulfides	5.3
495-485	CI stretch, S_{HH} conformation	Secondary iodoalkanes	2.2
495-485	CI stretch, T_{HHH} conformation	Tertiary iodoalkanes	2.2
484-475	Skeletal deformation	Dialkyl diacetylenes	12.1
483	Symmetric CI_2 stretch	CH_2I_2	2.1
459	Symmetric CCl_4 stretch	CCl_4	2.1
437	Symmetric CI_3 stretch	CHI_3 (in solution)	2.1
425-150	"Chain expansion"	*n*-Alkanes	1.2
355-335	Skeletal deformation	Monoalkyl acetylenes	12.1
267	Symmetric CBr_4 stretch	CBr_4 (in solution)	2.1
200-160	Skeletal deformation	Aliphatic nitriles	12.2
178	Symmetric CI_4 stretch	CI_4 (solid)	2.1

APPENDIX TWO

REPRESENTATIVE RAMAN SPECTRA

The following 108 Raman spectra are presented to familiarize the reader with the appearance of some typical Raman spectra and to illustrate the characteristic frequencies reported in the text. The spectra were obtained with a Cary Model 81 Raman Spectrometer modified for laser excitation. A Spectra-Physics Model 165/265 Argon Ion Laser operating on the 4880 Å line was used. The high-purity samples were contained in a Pyrex capillary and were illuminated coaxially to the laser beam. Both a numerical and a molecular formula index are provided.

Numerical Index for Raman Spectra

Spectrum Number	Compound	Physical State	Refer to Section:
1	*n*-Heptane	Liquid	1.2
2	2-Methylhexane	Liquid	1.3
3	3-Methylhexane	Liquid	1.3
4.	3-Ethylpentane	Liquid	1.3
5	2,3-Dimethylpentane	Liquid	1.3
6	2,4-Dimethylpentane	Liquid	1.3
7	2,2-Dimethylpentane	Liquid	1.3

8	3,3-Dimethylpentane	Liquid	1.3
9	2,2,3-Trimethylbutane	Liquid	1.3
10	Cyclohexane	Liquid	1.4
11	1-Chlorobutane	Liquid	2.2
12	1-Bromo-2-methylpropane	Liquid	2.2
13	1-Chloro-3-methylbutane	Liquid	2.2
14	2-Chlorobutane	Liquid	2.2
15	2-Bromo-2-methylpropane	Liquid	2.2
16	1-Propanol	Liquid	3.1
17	Dipropyl ether	Liquid	3.2
18	Acetal	Liquid	3.3
19	*n*-Butylamine	Liquid[a]	4.1
20	Dipropylamine	Liquid	4.1
21	Triethylamine	Liquid[a]	4.1
22	1-Nitropentane	Liquid	4.5
23	*n*-Propyl nitrate	Liquid	4.6
24	1-Hexanethiol	Liquid	5.1
25	Benzyl sulfide	Molten	5.2
26	Dipropyl disulfide	Liquid	5.3
27	Thioacetamide	Powder	5.5
28	Dimethyl sulfoxide	Liquid	5.8
29	1-Octene	Liquid	6.2
30	2,3-Dimethyl-1-butene	Liquid	6.2
31	2-Heptene (cis and trans)	Liquid	6.2
32	*cis*-3-Octene	Liquid	6.2
33	*trans*-3-Octene	Liquid	6.2
34	2-Methyl-2-pentene	Liquid	6.2
35	Allyl bromide	Liquid	6.4
36	1,5-Hexadiene	Liquid	6.7
37	2-Methyl-1,3-butadiene	Liquid	6.7
38	Cyclohexene	Liquid	6.13
39	Octanal	Liquid	7.2
40	2-Pentanone	Liquid	7.3
41	4-Methyl-3-hexanone	Liquid	7.3
42	Fenchone	Liquid	7.3
43	2,4-Pentanedione	Liquid	7.3
44	Propionic acid	Liquid	8.1
45	Methyl formate	Liquid	8.4
46	Methyl acetate	Liquid	8.4
47	*n*-Butyl acrylate	Liquid	8.4
48	Acetic anhydride	Liquid	8.5
49	Formamide	Liquid	9.2

50	Acetamide	Powder[b]	9.2
51	*N*-Methylformamide	Liquid	9.3
52	*N*-Methylacetamide	Liquid	9.3
53	*N,N*-Dimethylformamide	Liquid	9.4
54	*N,N*-Dimethylacetamide	Liquid	9.4
55	Butyraldoxime	Liquid	10.3
56	2-Butanone oxime	Liquid	10.3
57	Phenyl isocyanate	Liquid	11.5
58	1-Adamantyl isothiocyanate	Powder[b]	11.6
59	1-Hexyne	Liquid	12.1
60	3-Methyl-1-butyne	Liquid	12.1
61	2-Hexyne	Liquid	12.1
62	2-Methyl-1-hexen-3-yne	Liquid	12.1
63	Acetonitrile	Liquid	12.2
64	*n*-Butyronitrile	Liquid	12.2
65	Benzonitrile	Liquid	12.2
66	Benzene	Liquid	13.1
67	Methylbenzene	Liquid	13.3
68	*n*-Butylbenzene	Liquid	13.3
69	Isopropylbenzene	Liquid	13.3
70	*tert*-Butylbenzene	Liquid	13.3
71	Fluorobenzene	Liquid	13.3
72	Chlorobenzene	Liquid	13.3
73	Bromobenzene	Liquid	13.3
74	Phenol	Molten	13.3
75	Nitrobenzene	Liquid	13.3
76	α-Methylstyrene	Liquid	13.3
77	Methyl benzoate	Liquid	13.3
78	1,2-Dimethylbenzene	Liquid	13.4
79	1-Chloro-2-methylbenzene	Liquid	13.4
80	1,2-Dichlorobenzene	Liquid	13.4
81	1,3-Dimethylbenzene	Liquid	13.4
82	1-Bromo-3-methylbenzene	Liquid	13.4
83	1,3-Dichlorobenzene	Liquid	13.4
84	1,4-Dimethylbenzene	Liquid	13.4
85	1-Chloro-4-methylbenzene	Liquid	13.4
86	1,4-Dichlorobenzene	Molten	13.4
87	1,2,3-Trimethylbenzene	Liquid	13.5
88	1,2-Dichloro-3-methylbenzene	Liquid	13.5
89	1,2,4-Trimethylbenzene	Liquid	13.5
90	1,2,4-Trichlorobenzene	Liquid	13.5
91	1,3,5-Trimethylbenzene	Liquid	13.5

92	1,3-Dichloro-5-methylbenzene	Liquid	13.5
93	1,2,4,5-Tetramethylbenzene	Powder[b]	13.6
94	3,4,5-Trimethylphenol	Powder[b]	13.6
95	Pentamethylbenzene	Powder[b]	13.6
96	Biphenyl	Powder	13.7
97	1-Methyl-2-phenylbenzene	Liquid	13.7
98	Naphthalene	Powder	13.8
99	1-Ethylnaphthalene	Liquid	13.8
100	Propylene oxide	Liquid	14.1
101	Trimethylene oxide	Liquid	14.2
102	Tetrahydrothiophene	Liquid	15.1
103	Thiophene	Liquid	16.1
104	1,4-Dioxane	Liquid	17.2
105	Pyridine	Liquid[a]	18.1
106	2-Methylpyridine	Liquid[a]	18.1
107	3-Methylpyridine	Liquid[a]	18.1
108	4-Methylpyridine	Liquid[a]	18.1

[a]Sample freshly distilled
[b]Bands appearing in the 3900–3700 cm^{-1} region are instrumental artifacts.

Molecular Formula Index for Raman Spectra

Molecular Formula	Spectra	Molecular Formula	Spectra
CH_3NO	49	C_6H_{12}	10,30,34
C_2H_3N	63	$C_6H_{14}O$	17,18
$C_2H_4O_2$	45	$C_6H_{14}S$	24
C_2H_5NO	50,51	$C_6H_{14}S_2$	26
C_2H_5NS	27	$C_6H_{15}N$	20,21
C_2H_6OS	28	C_7H_5N	65
C_3H_5Br	35	C_7H_5NO	57
C_3H_6O	100,101	$C_7H_6Cl_2$	88,92
$C_3H_6O_2$	44,46	C_7H_7Br	82
C_3H_7NO	52,53	C_7H_7Cl	79,85
$C_3H_7NO_3$	23	C_7H_8	67
C_3H_8O	16	C_7H_{10}	62
C_4H_4S	103	$C_7H_{12}O_2$	47
$C_4H_6O_3$	48	C_7H_{14}	31
C_4H_7N	64	$C_7H_{14}O$	41
$C_4H_8O_2$	104	C_7H_{16}	1-9
C_4H_8S	102	$C_8H_8O_2$	77
C_4H_9Br	12,15	C_8H_{10}	78,81,84
C_4H_9Cl	11,14	C_8H_{16}	29,32,33
C_4H_9NO	54-56	$C_8H_{16}O$	39
$C_4H_{11}N$	19	C_9H_{10}	76
C_5H_5N	105	C_9H_{12}	69,87,89,91
C_5H_8	37,60	$C_9H_{12}O$	94
$C_5H_8O_2$	43	$C_{10}H_8$	98
$C_5H_{10}O$	40	$C_{10}H_{14}$	68,70,93
$C_5H_{11}Cl$	13	$C_{10}H_{16}O$	42
$C_5H_{11}NO_2$	22	$C_{11}H_{15}NS$	58
$C_6H_3Cl_3$	90	$C_{11}H_{16}$	95
$C_6H_4Cl_2$	80,83,86	$C_{12}H_{10}$	96
C_6H_5Br	73	$C_{12}H_{12}$	99
C_6H_5Cl	72	$C_{13}H_{12}$	97
C_6H_5F	71	$C_{14}H_{14}S$	25
$C_6H_5NO_2$	75		
C_6H_6	66		
C_6H_6O	74		
C_6H_7N	106-108		
C_6H_{10}	36,38,59,61		

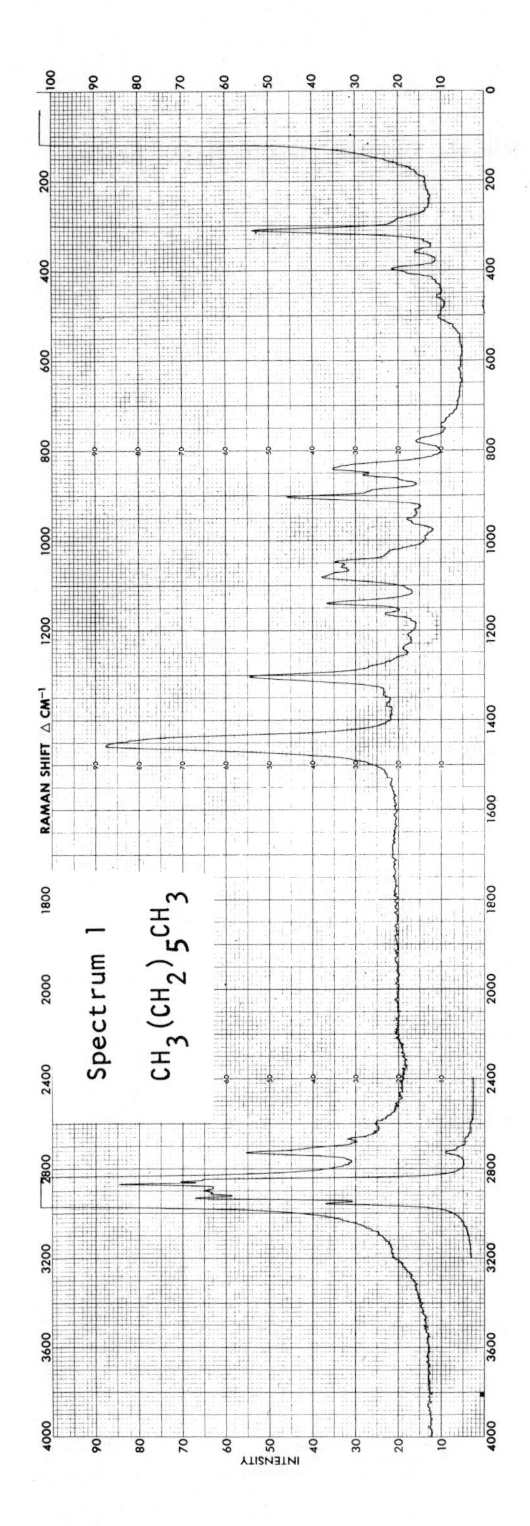
Spectrum 1
$CH_3(CH_2)_5CH_3$
RAMAN SHIFT Δ CM^{-1}
INTENSITY

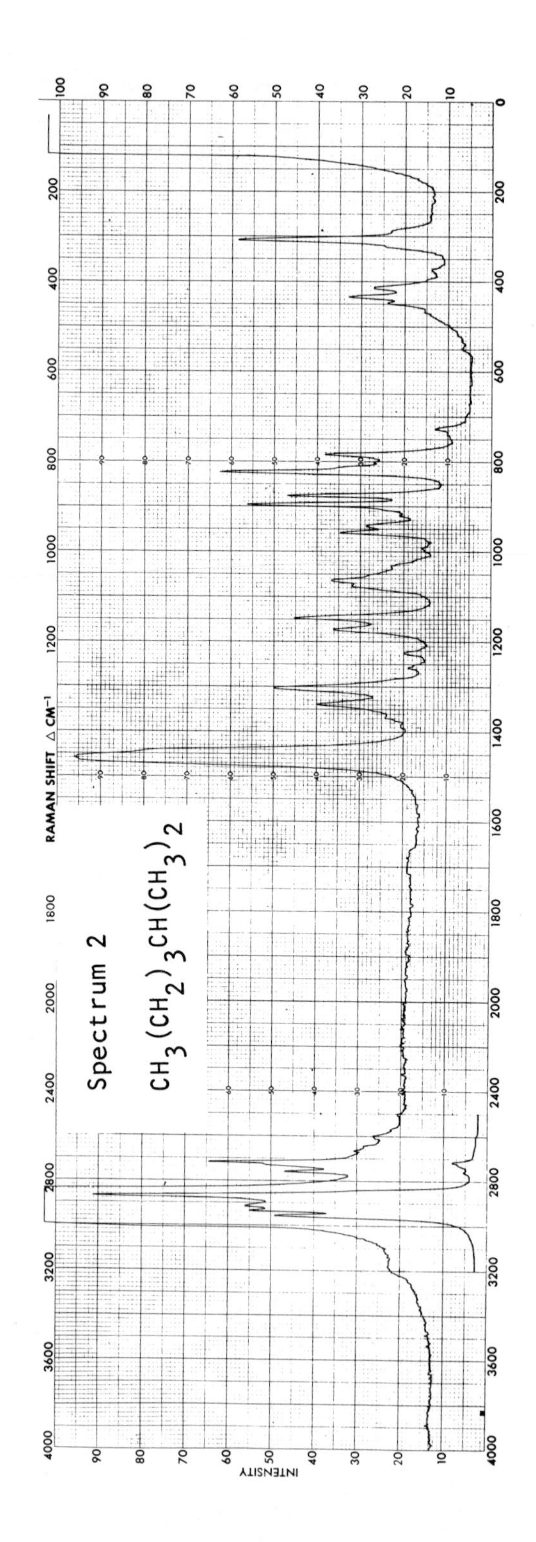

Spectrum 2
$CH_3(CH_2)_3CH(CH_3)_2$
RAMAN SHIFT Δ CM^{-1}
INTENSITY

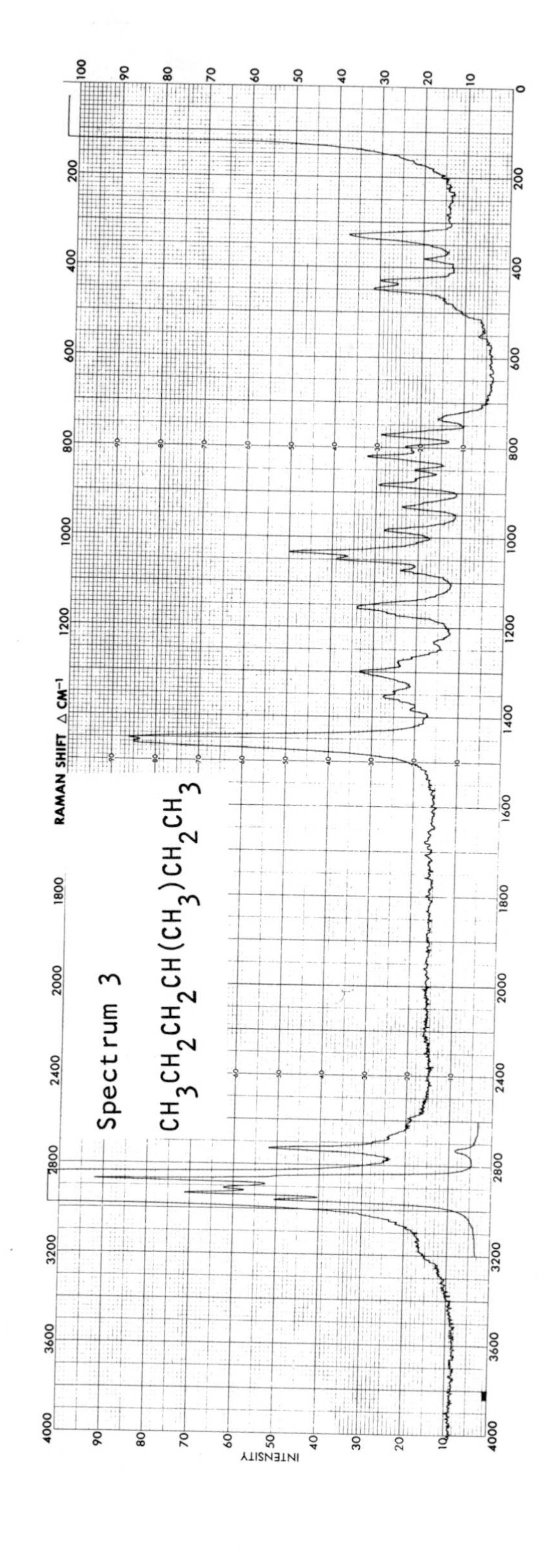
Spectrum 3
CH3CH2CH2CH(CH3)CH2CH3
RAMAN SHIFT Δ CM-1
INTENSITY

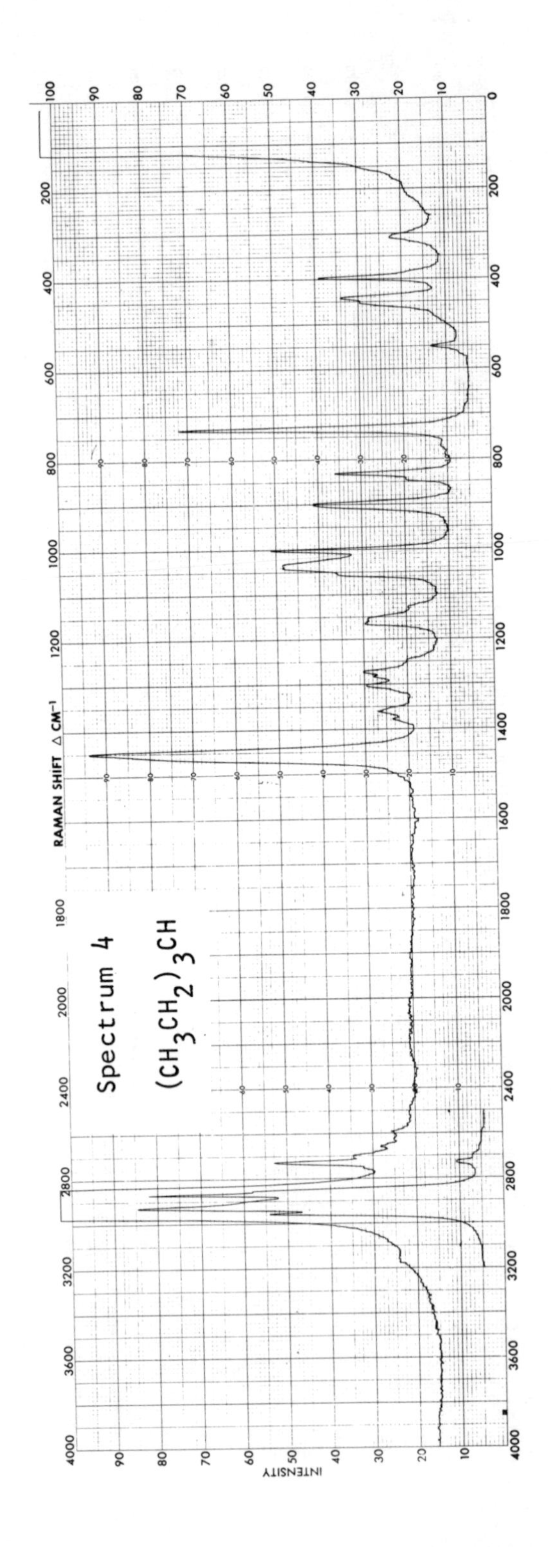

Spectrum 4
$(CH_3CH_2)_3CH$
RAMAN SHIFT Δ CM^{-1}
INTENSITY

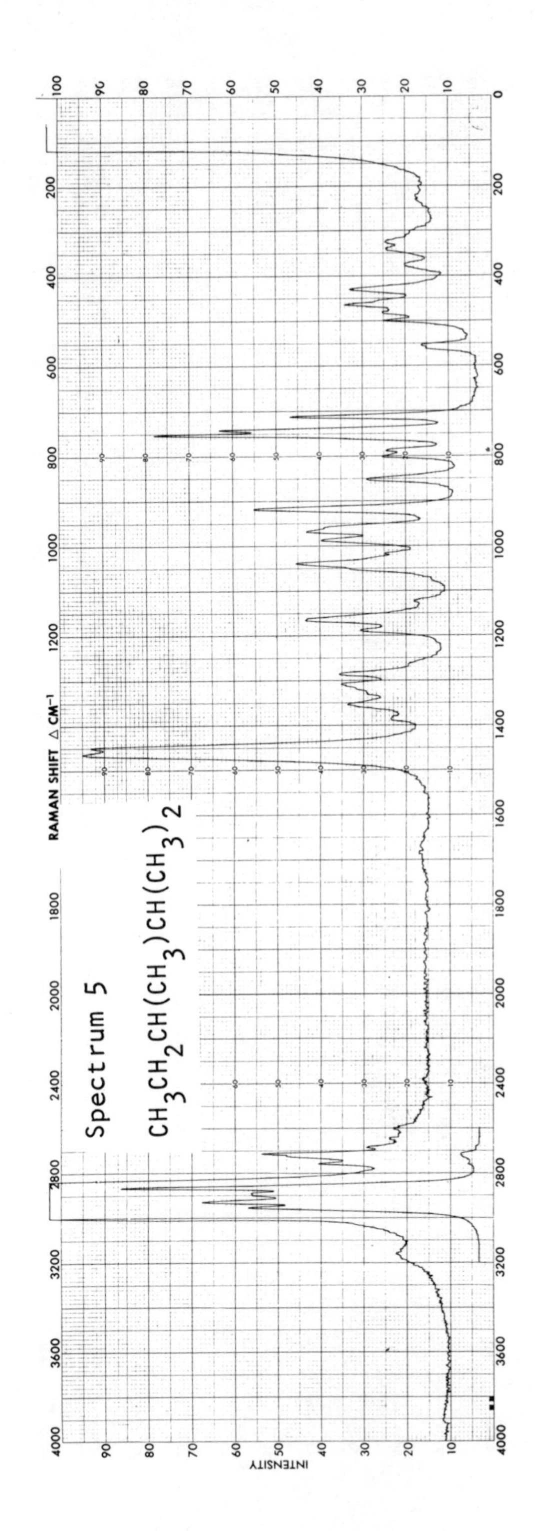
Spectrum 5
$CH_3CH_2CH(CH_3)CH(CH_3)_2$
RAMAN SHIFT Δ CM^{-1}
INTENSITY

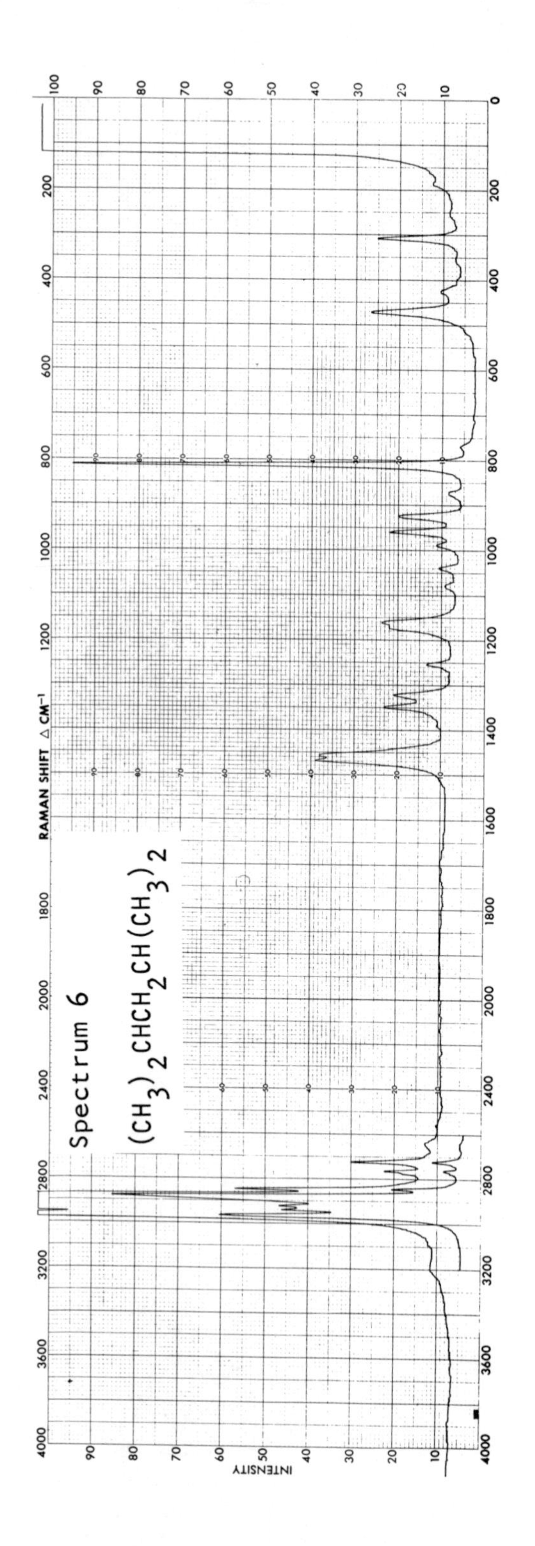

Spectrum 6
(CH3)2CHCH2CH(CH3)2
RAMAN SHIFT Δ CM⁻¹
INTENSITY

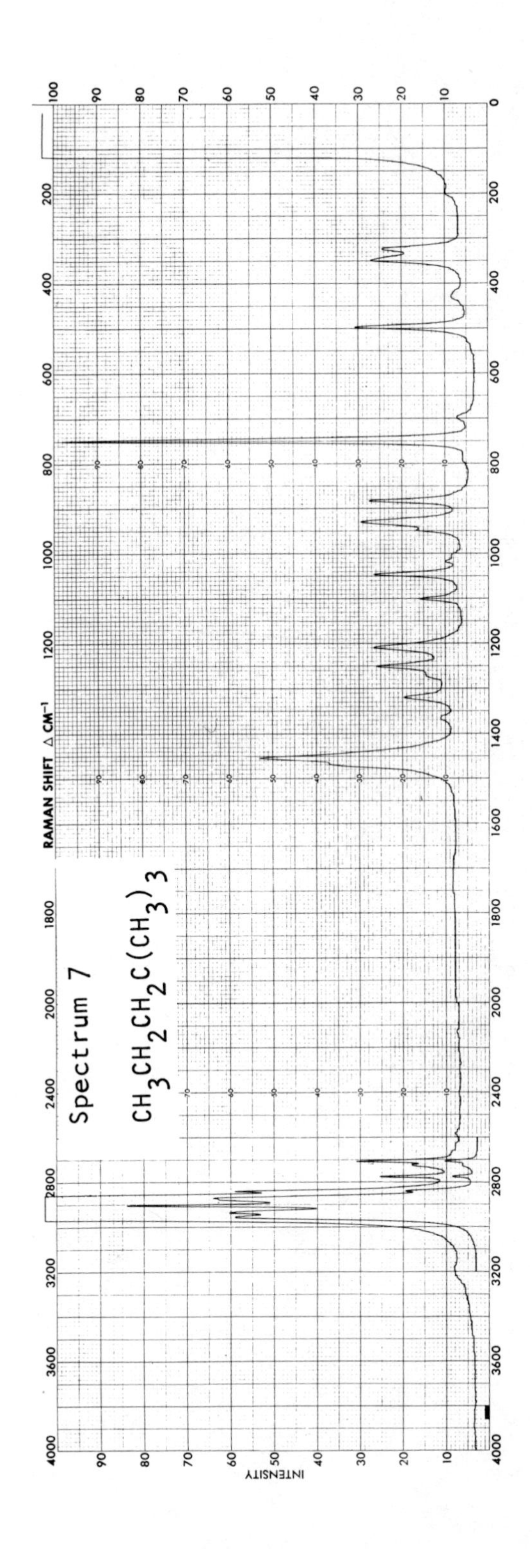
Spectrum 7
$CH_3CH_2CH_2C(CH_3)_3$
RAMAN SHIFT Δ CM⁻¹
INTENSITY
4000
3600
3200
2800
2400
2000
1800
1600
1400
1200
1000
800
600
400
200
0
10
20
30
40
50
60
70
80
90
100

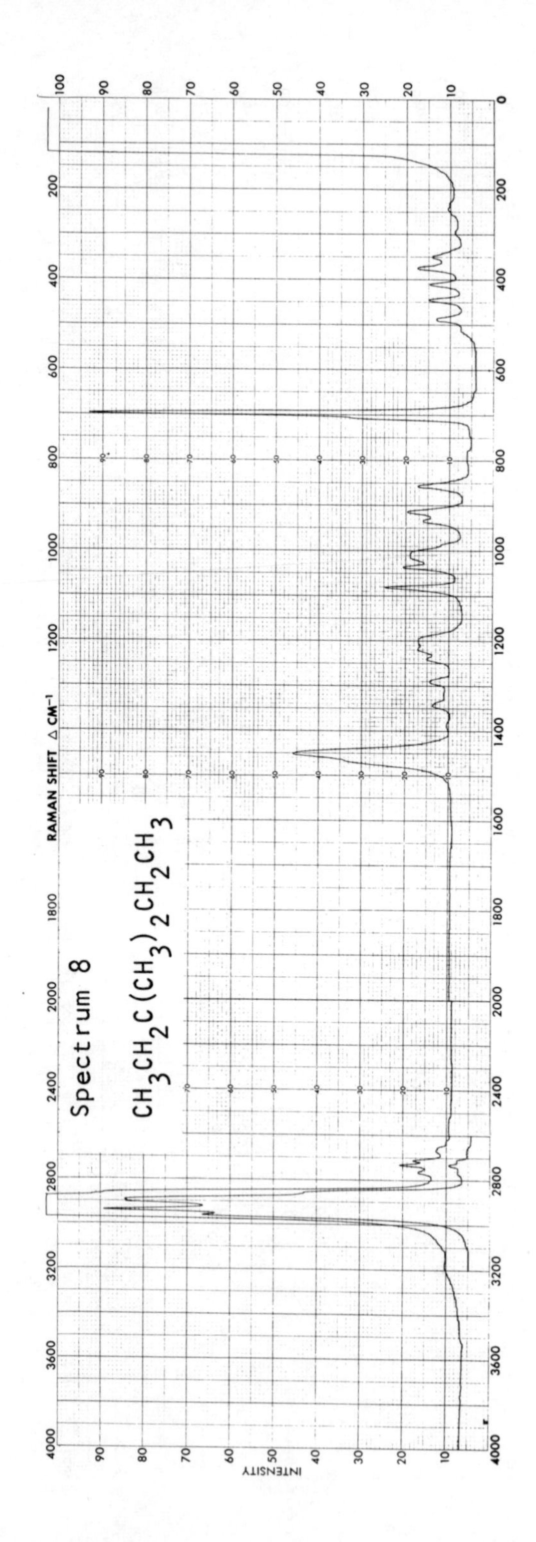
Spectrum 8
$CH_3CH_2C(CH_3)_2CH_2CH_3$
RAMAN SHIFT Δ CM^{-1}
INTENSITY

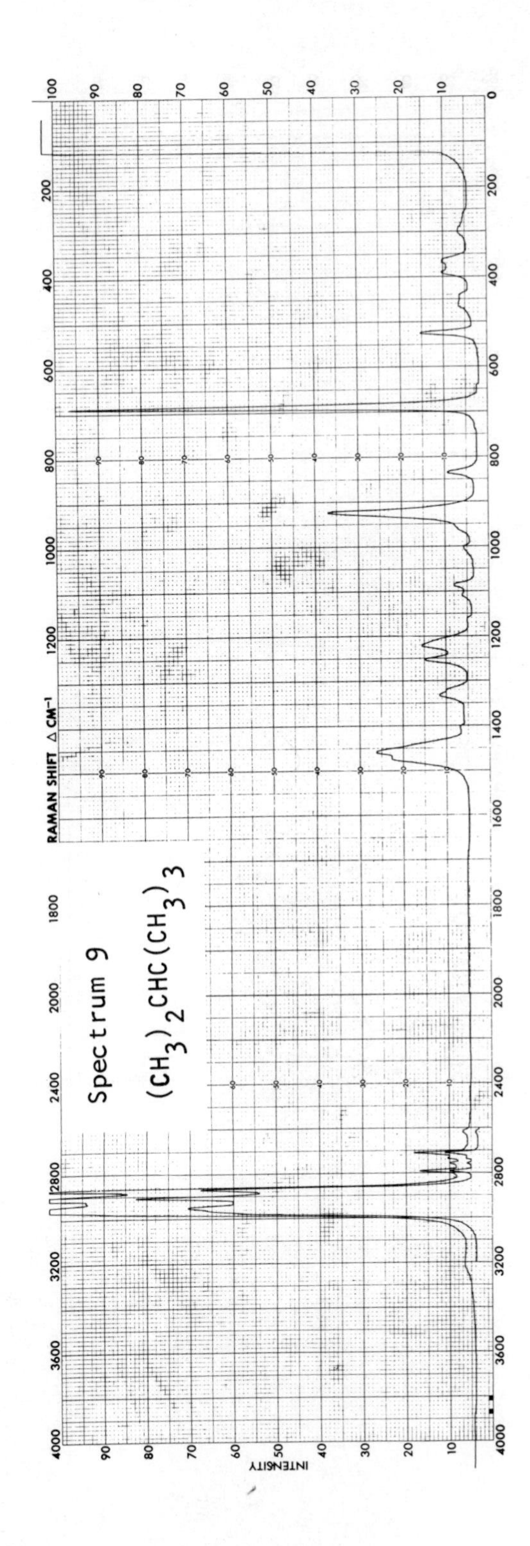

Spectrum 9
$(CH_3)_2CHC(CH_3)_3$
RAMAN SHIFT Δ CM^{-1}
INTENSITY

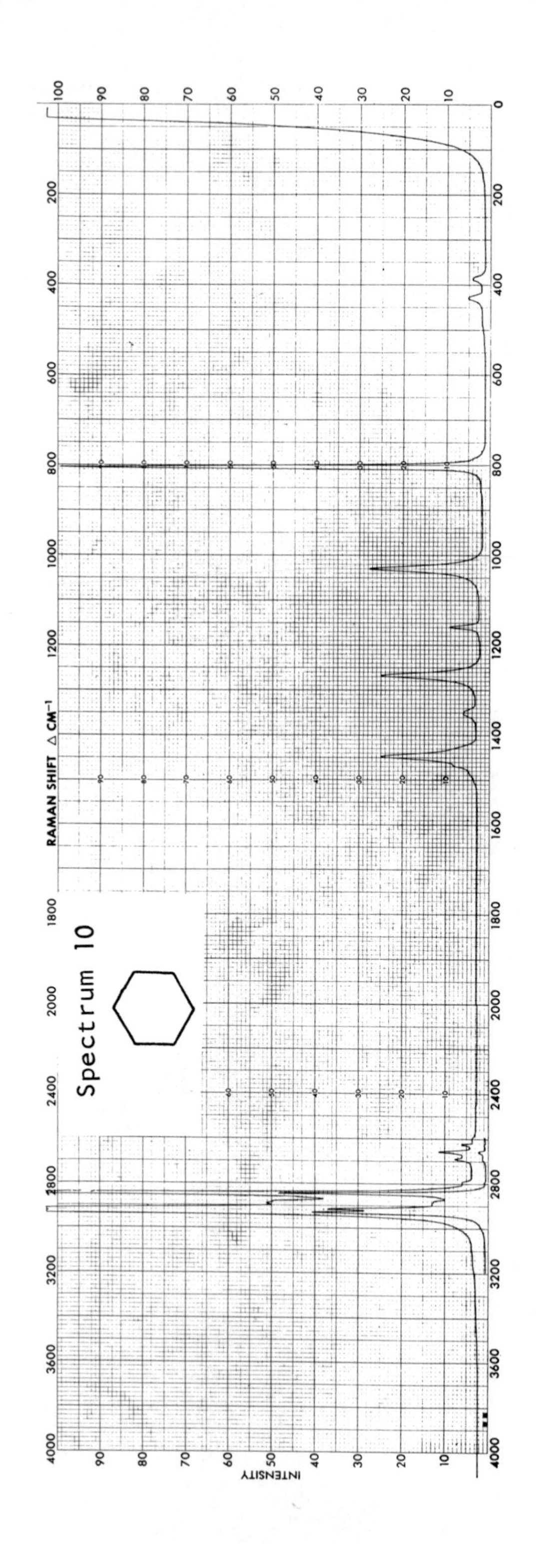

Spectrum 10
RAMAN SHIFT Δ CM^{-1}
INTENSITY
0
200
400
600
800
1000
1200
1400
1600
1800
2000
2400
2800
3200
3600
4000
10
20
30
40
50
60
70
80
90
100

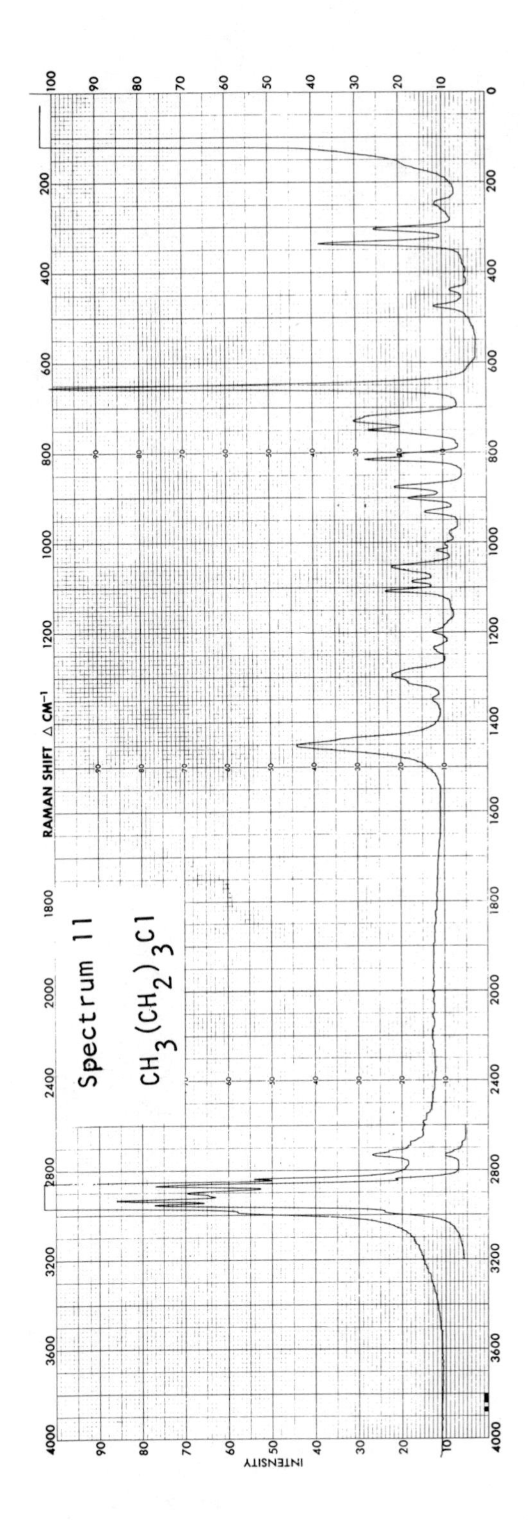

Spectrum 11
$CH_3(CH_2)_3Cl$
RAMAN SHIFT Δ CM^{-1}
INTENSITY

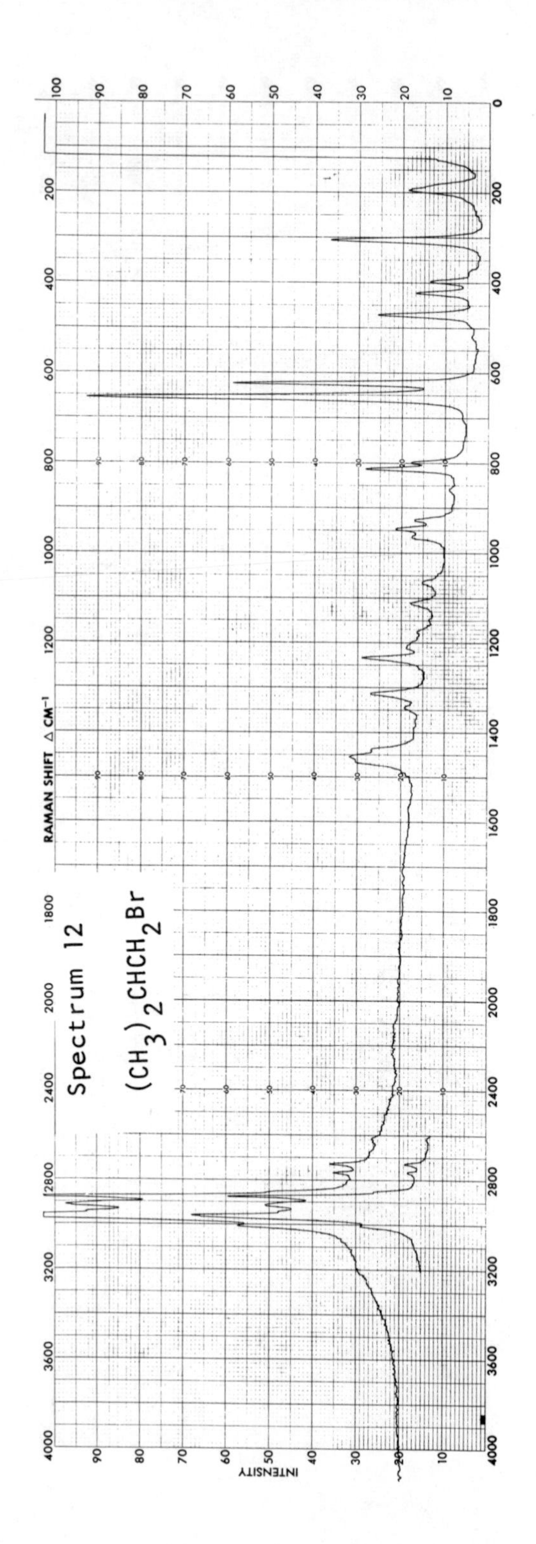

Spectrum 12
$(CH_3)_2CHCH_2Br$
RAMAN SHIFT Δ CM^{-1}
INTENSITY

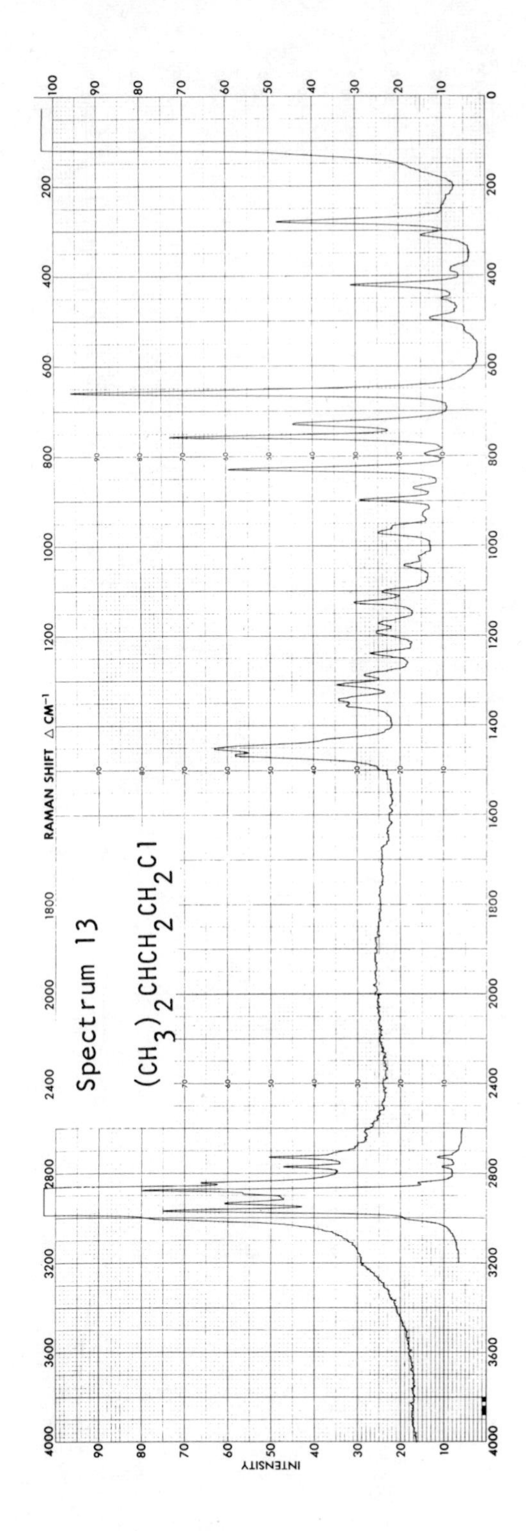
Spectrum 13
(CH3)2CHCH2CH2Cl
RAMAN SHIFT Δ CM⁻¹
INTENSITY

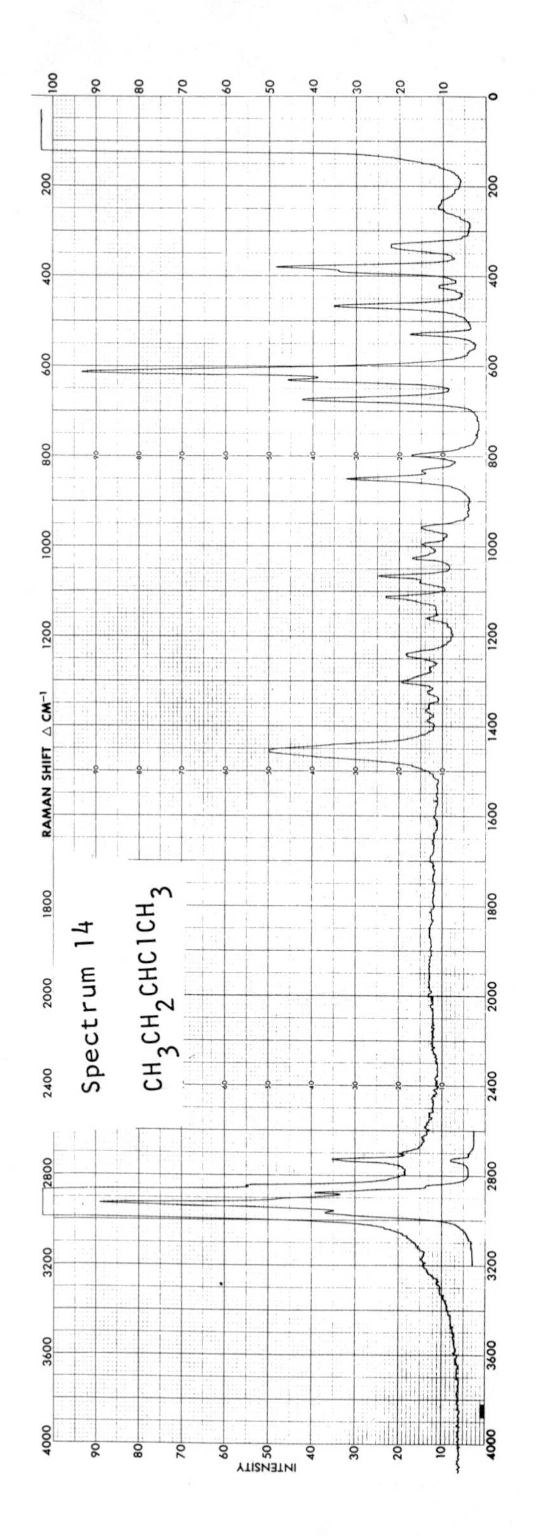

Spectrum 14
CH3CH2CHClCH3
RAMAN SHIFT Δ CM-1
INTENSITY

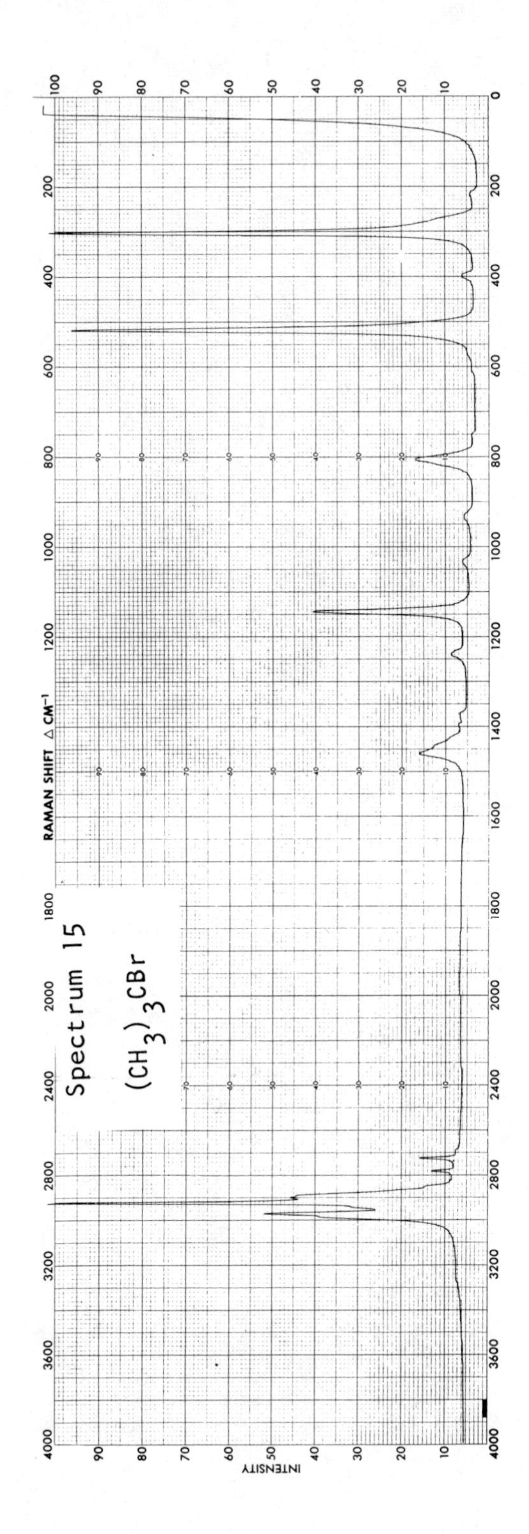

Spectrum 15
$(CH_3)_3CBr$
RAMAN SHIFT Δ CM⁻¹
INTENSITY

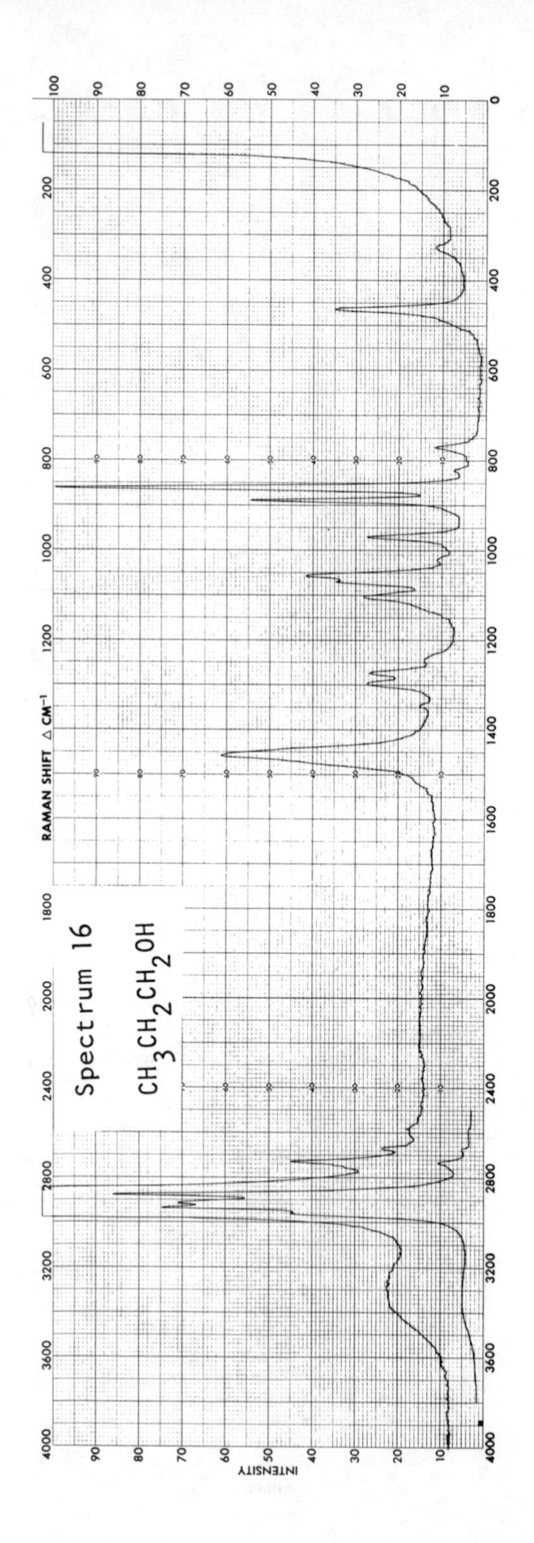
Spectrum 16
$CH_3CH_2CH_2OH$
RAMAN SHIFT Δ CM^{-1}
INTENSITY

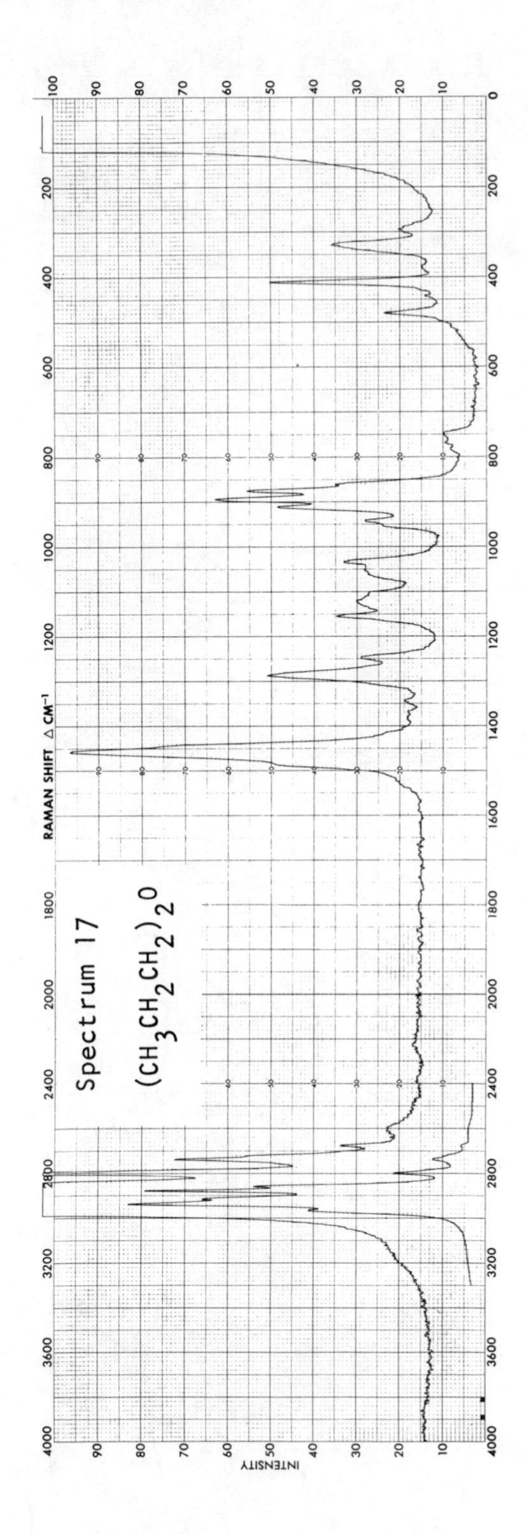
Spectrum 17
$(CH_3CH_2CH_2)_2O$
RAMAN SHIFT Δ CM^{-1}
INTENSITY
4000
3600
3200
2800
2400
2000
1800
1600
1400
1200
1000
800
600
400
200
0
100
90
80
70
60
50
40
30
20
10

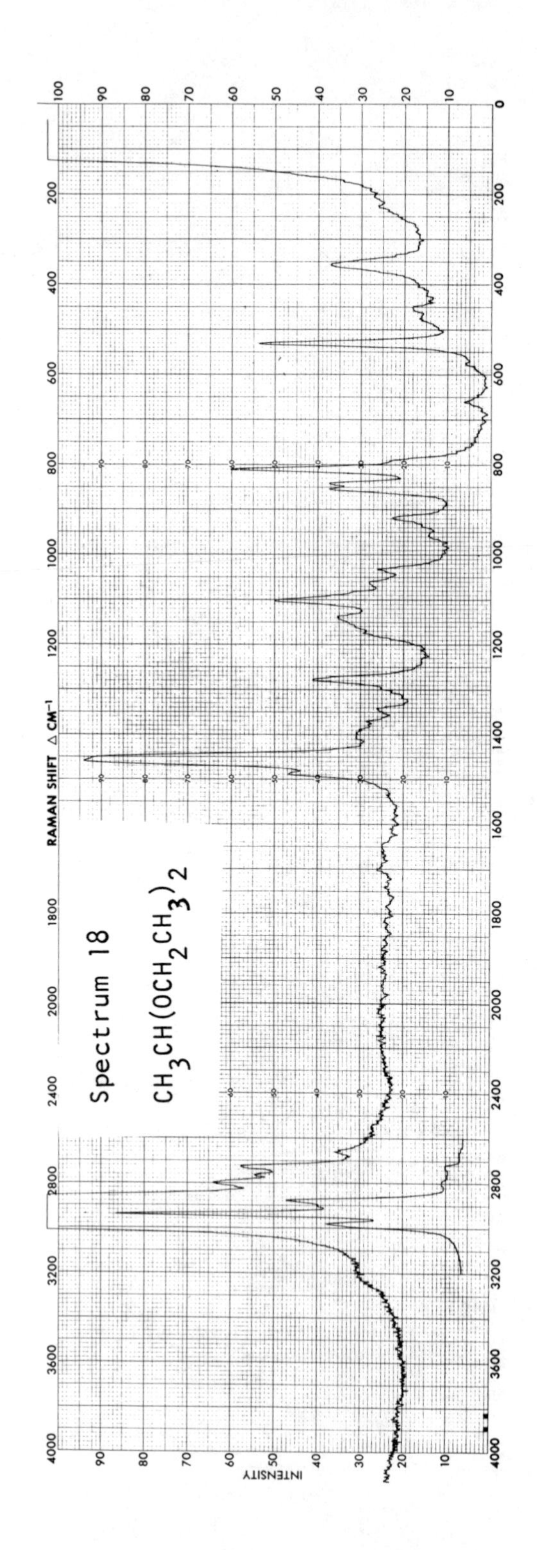
Spectrum 18
CH3CH(OCH2CH3)2
RAMAN SHIFT Δ CM⁻¹
INTENSITY

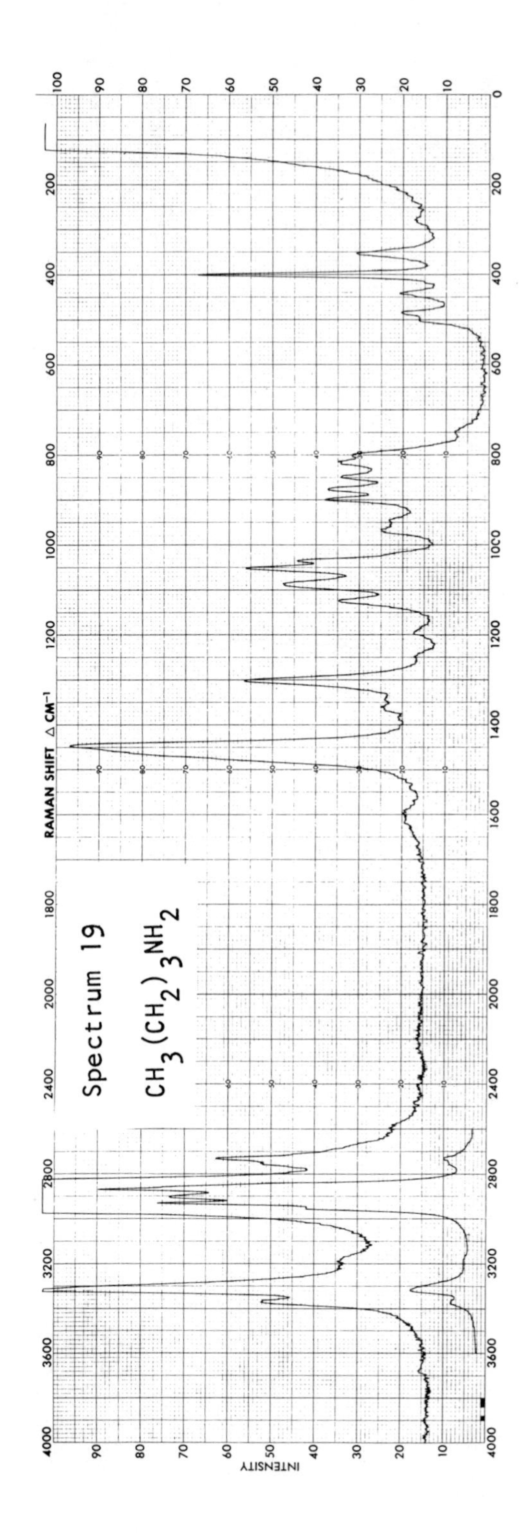
Spectrum 19
$CH_3(CH_2)_3NH_2$
RAMAN SHIFT Δ CM^{-1}
INTENSITY
4000 3600 3200 2800 2400 2000 1800 1600 1400 1200 1000 800 600 400 200
0 10 20 30 40 50 60 70 80 90 100

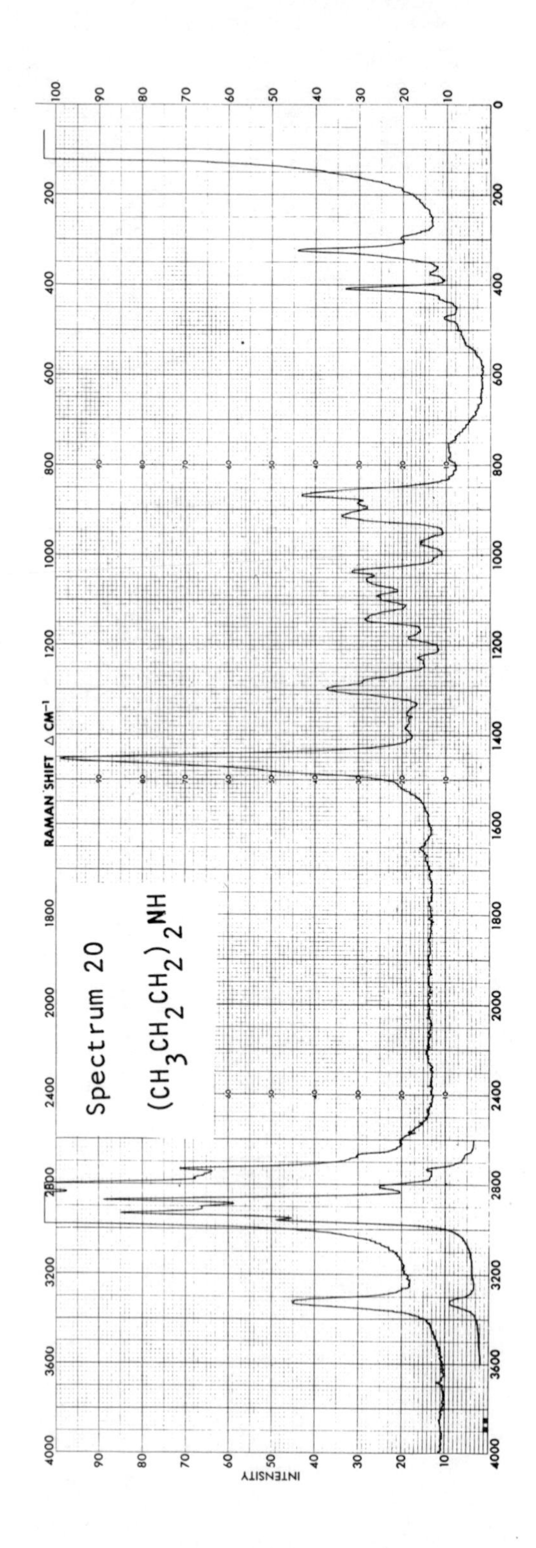
Spectrum 20
$(CH_3CH_2CH_2)_2NH$
RAMAN SHIFT Δ CM^{-1}
INTENSITY

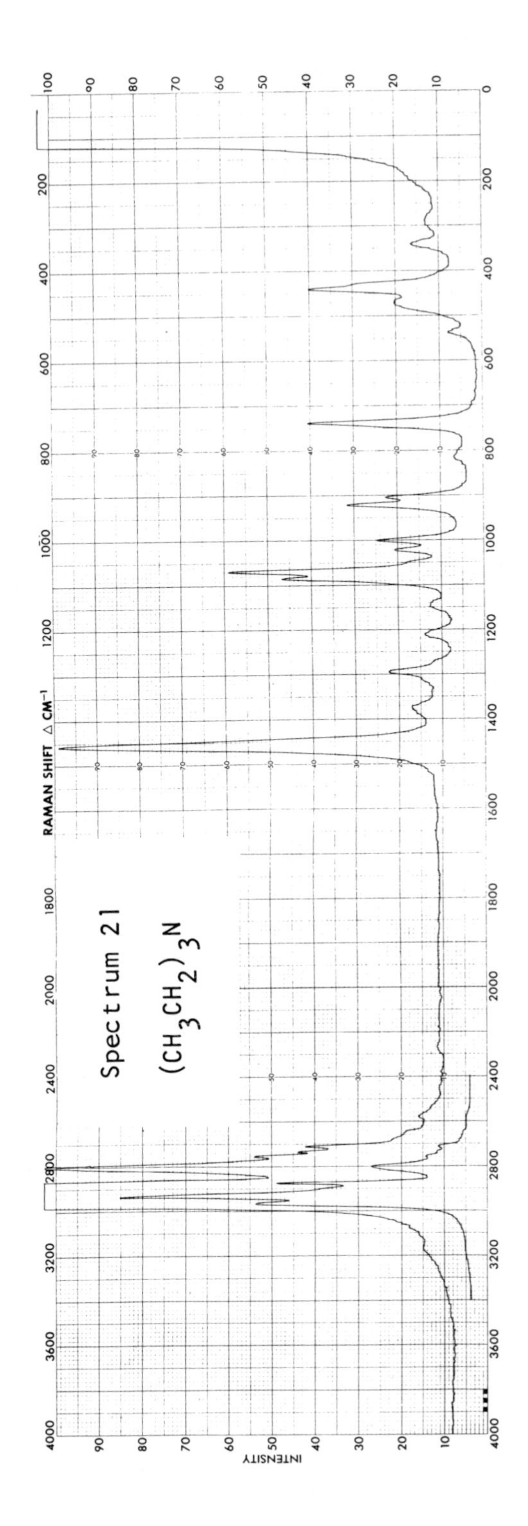
Spectrum 21
$(CH_3CH_2)_3N$
RAMAN SHIFT Δ CM^{-1}
INTENSITY

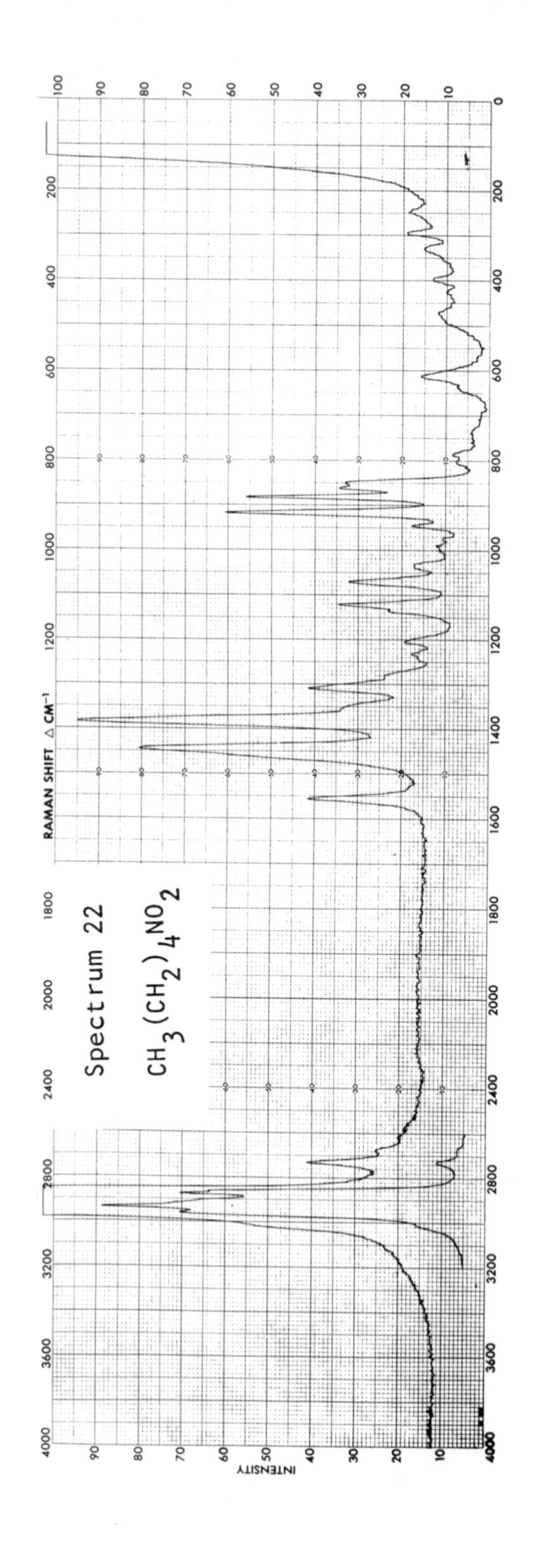

Spectrum 22
$CH_3(CH_2)_4NO_2$
RAMAN SHIFT Δ CM^{-1}
INTENSITY

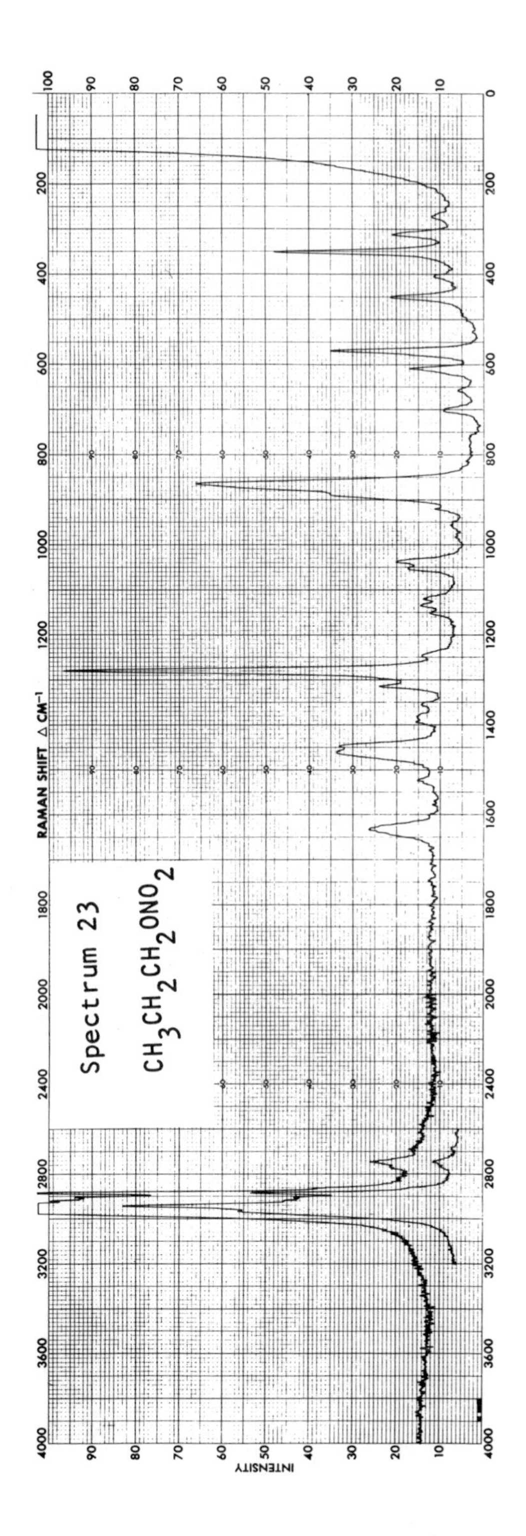
Spectrum 23
$CH_3CH_2CH_2ONO_2$
RAMAN SHIFT Δ CM^{-1}
INTENSITY

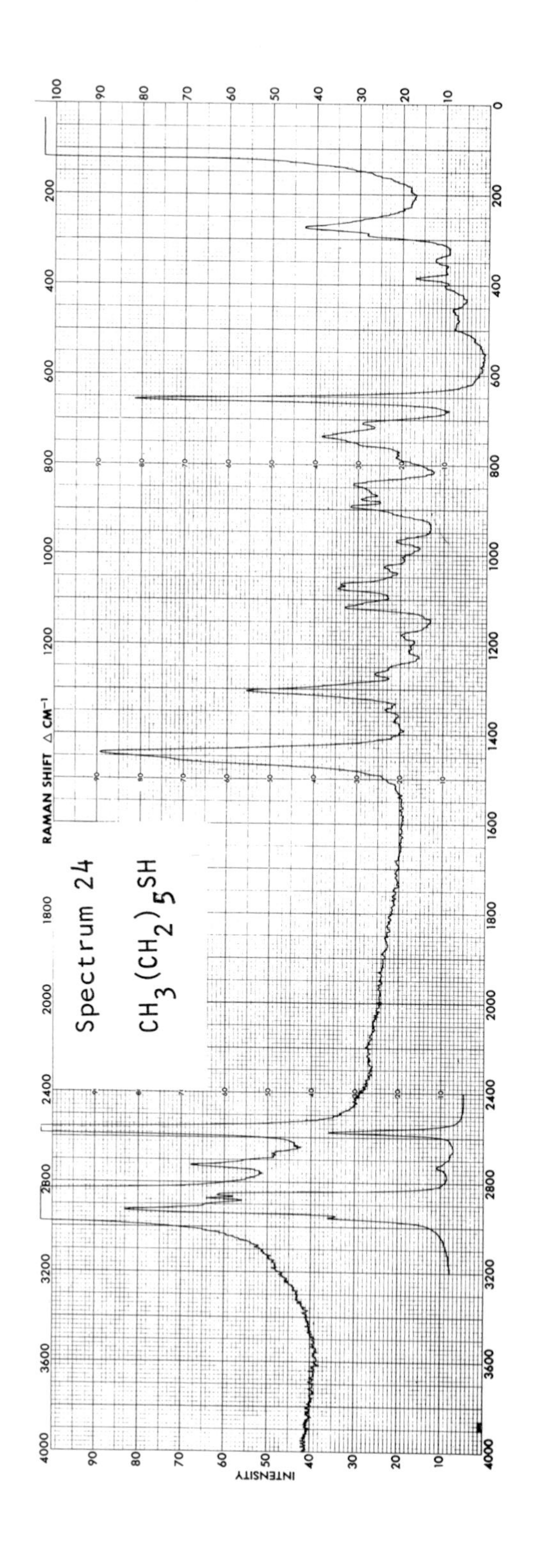
Spectrum 24
$CH_3(CH_2)_5SH$
RAMAN SHIFT Δ CM^{-1}
INTENSITY

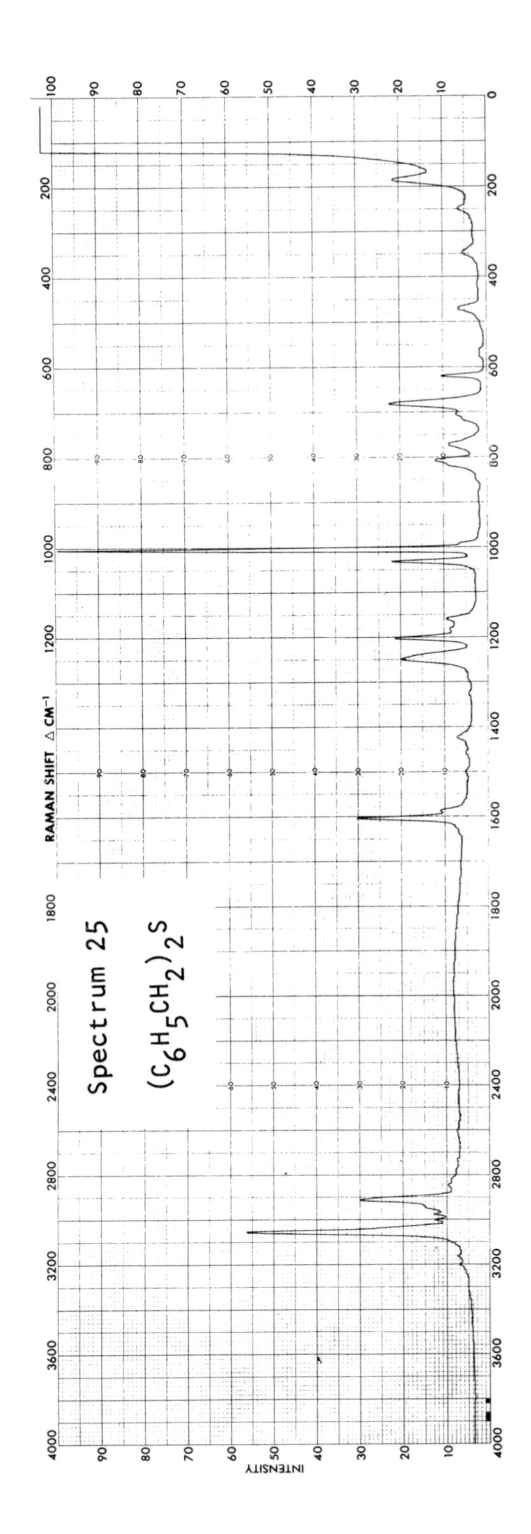
Spectrum 25
$(C_6H_5CH_2)_2S$
RAMAN SHIFT Δ CM^{-1}
INTENSITY

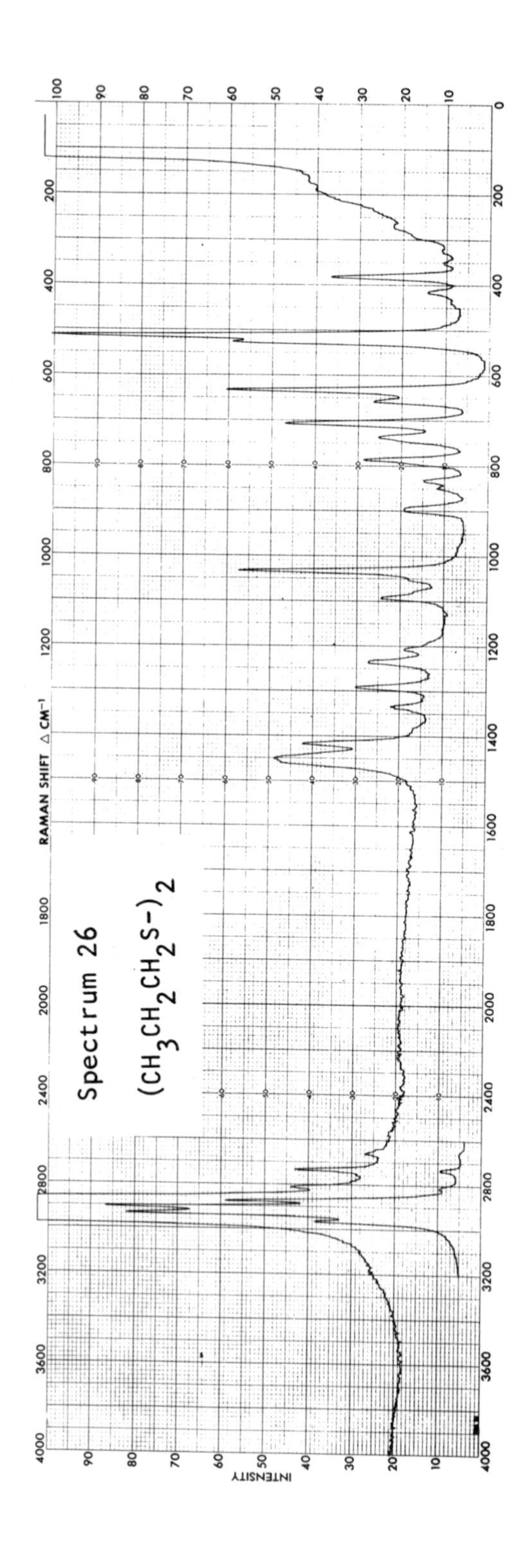

Spectrum 26
$(CH_3CH_2CH_2S-)_2$
RAMAN SHIFT Δ CM^{-1}
INTENSITY

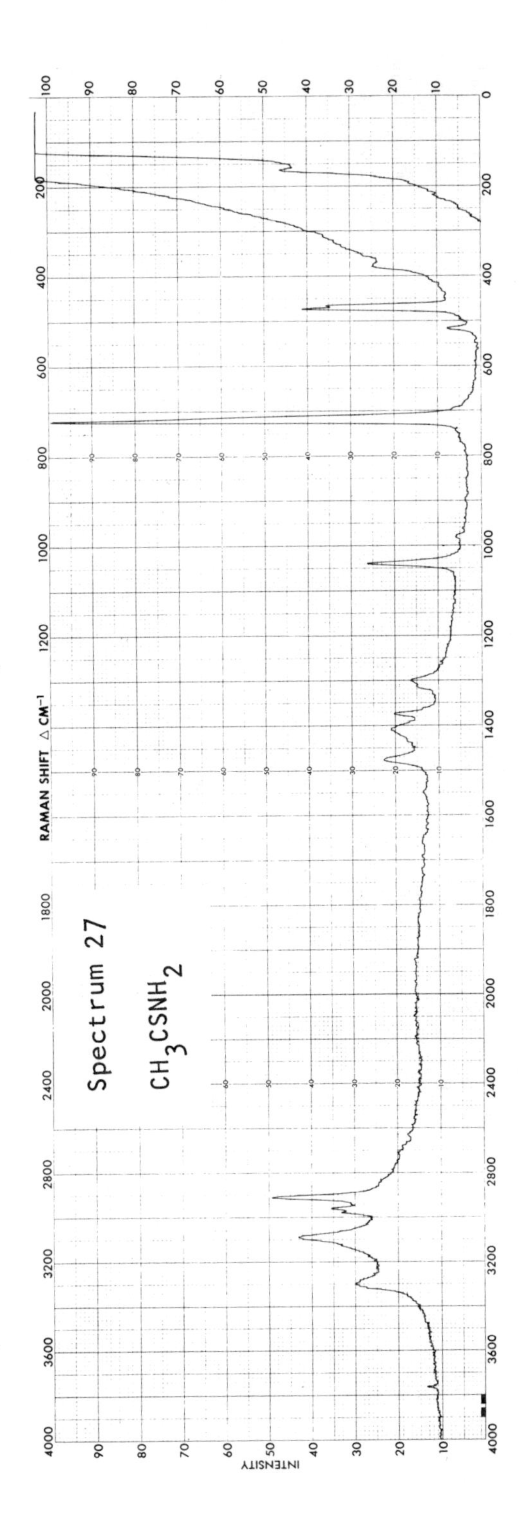

Spectrum 27
CH_3CSNH_2
RAMAN SHIFT Δ CM^{-1}
INTENSITY
0
200
400
600
800
1000
1200
1400
1600
1800
2000
2400
2800
3200
3600
4000
10
20
30
40
50
60
70
80
90
100

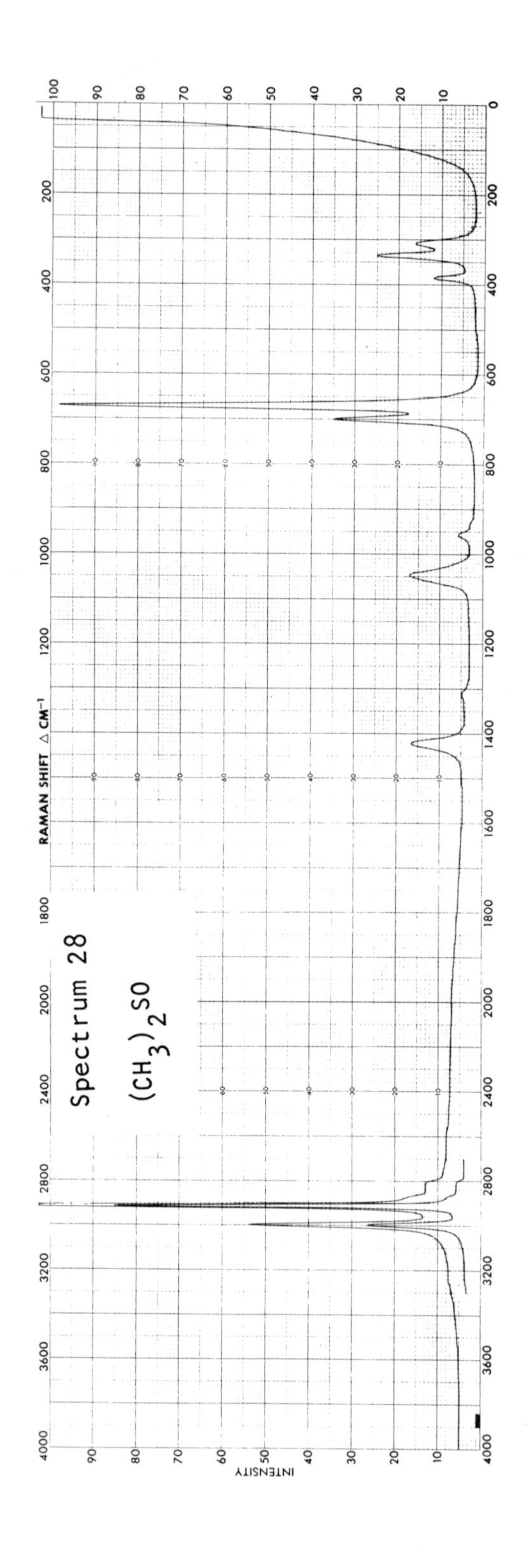

Spectrum 28
$(CH_3)_2SO$
RAMAN SHIFT Δ CM^{-1}
INTENSITY

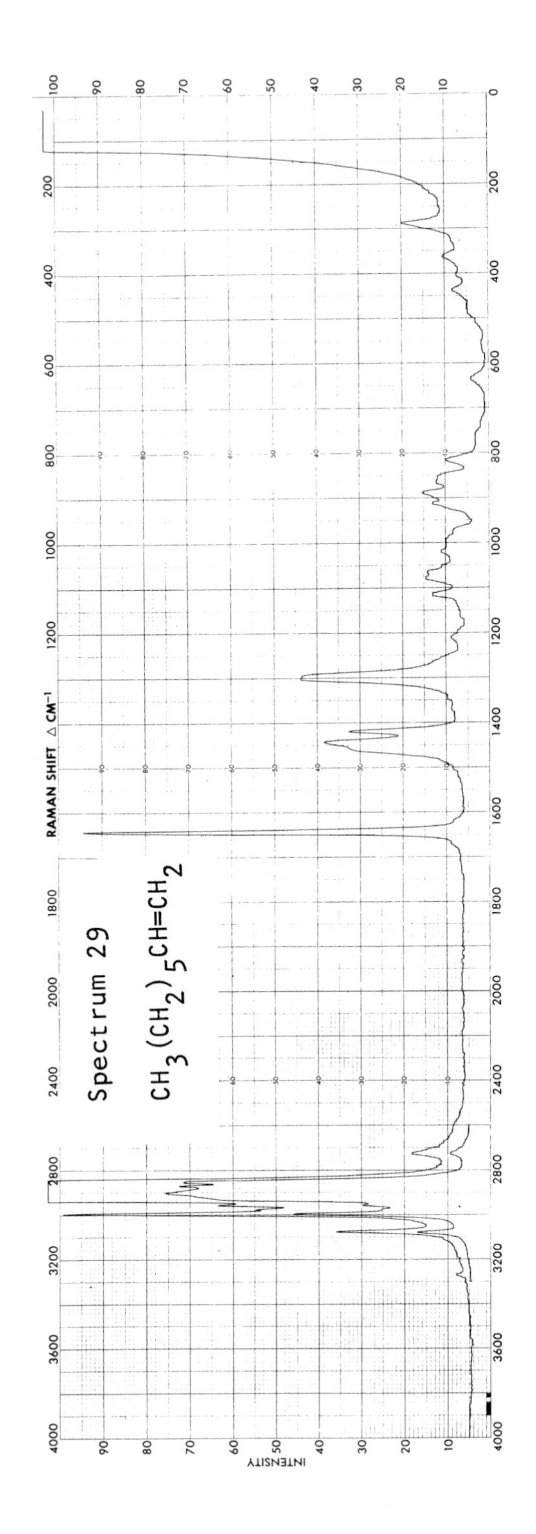
Spectrum 29
$CH_3(CH_2)_5CH=CH_2$
RAMAN SHIFT Δ CM^{-1}
INTENSITY

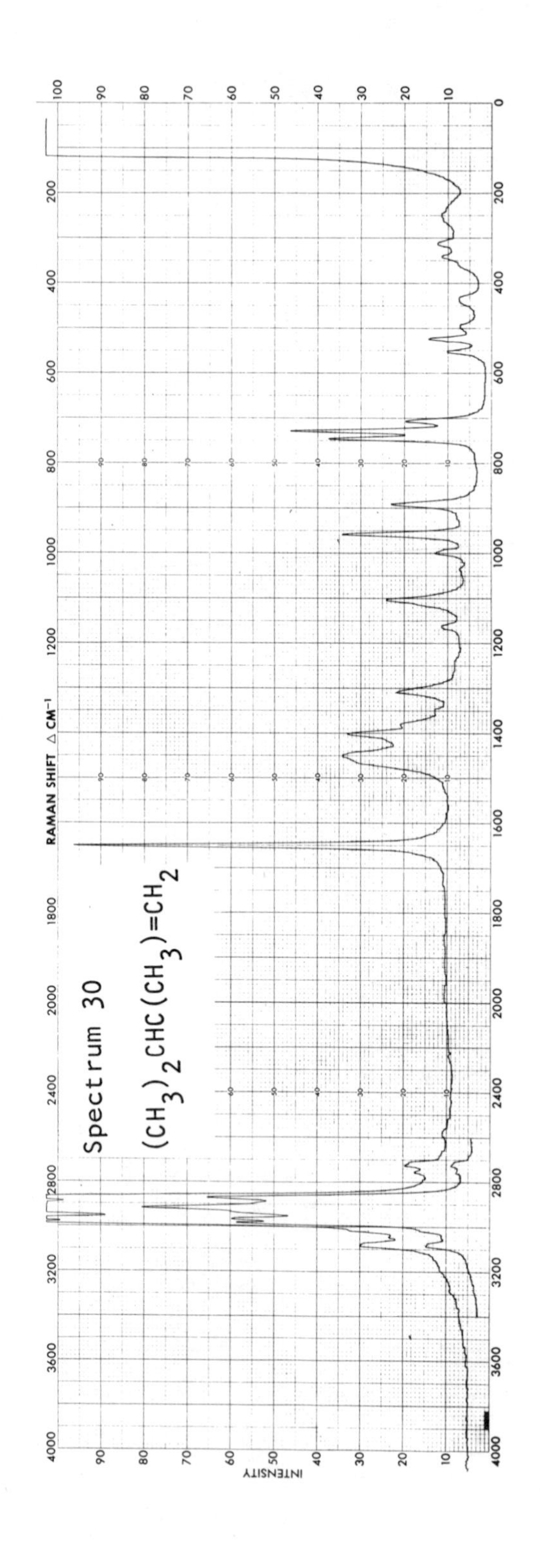
Spectrum 30
$(CH_3)_2CHC(CH_3)=CH_2$
RAMAN SHIFT Δ CM⁻¹
INTENSITY

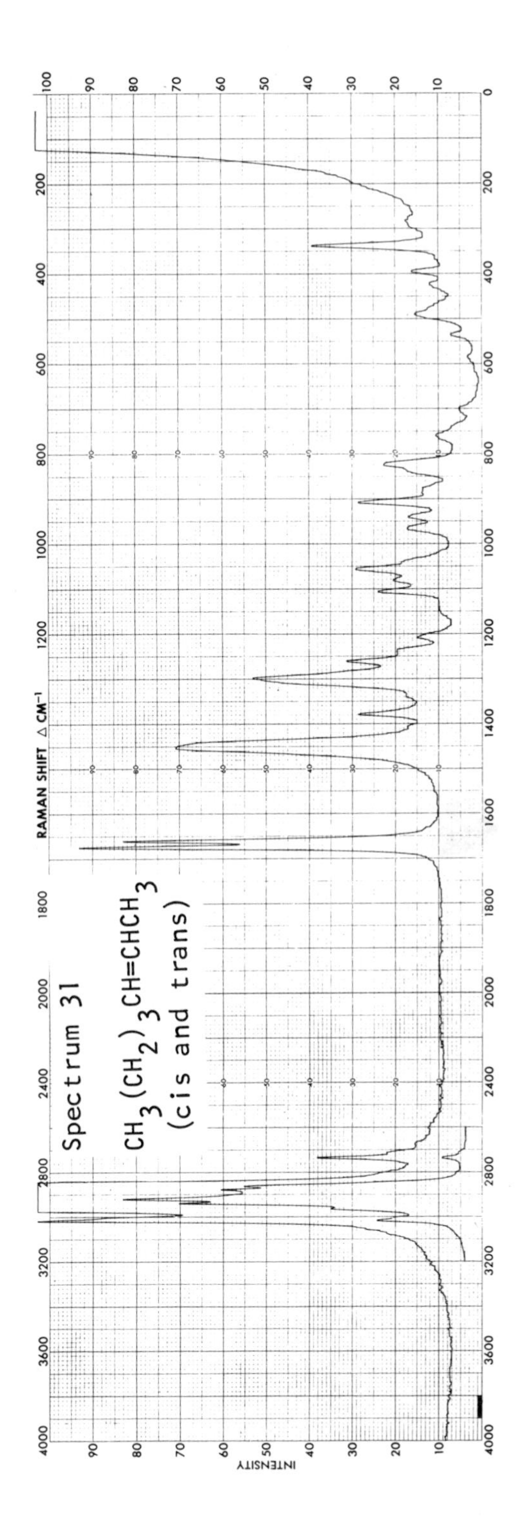
Spectrum 31
CH3(CH2)3CH=CHCH3
(cis and trans)
RAMAN SHIFT Δ CM⁻¹
INTENSITY

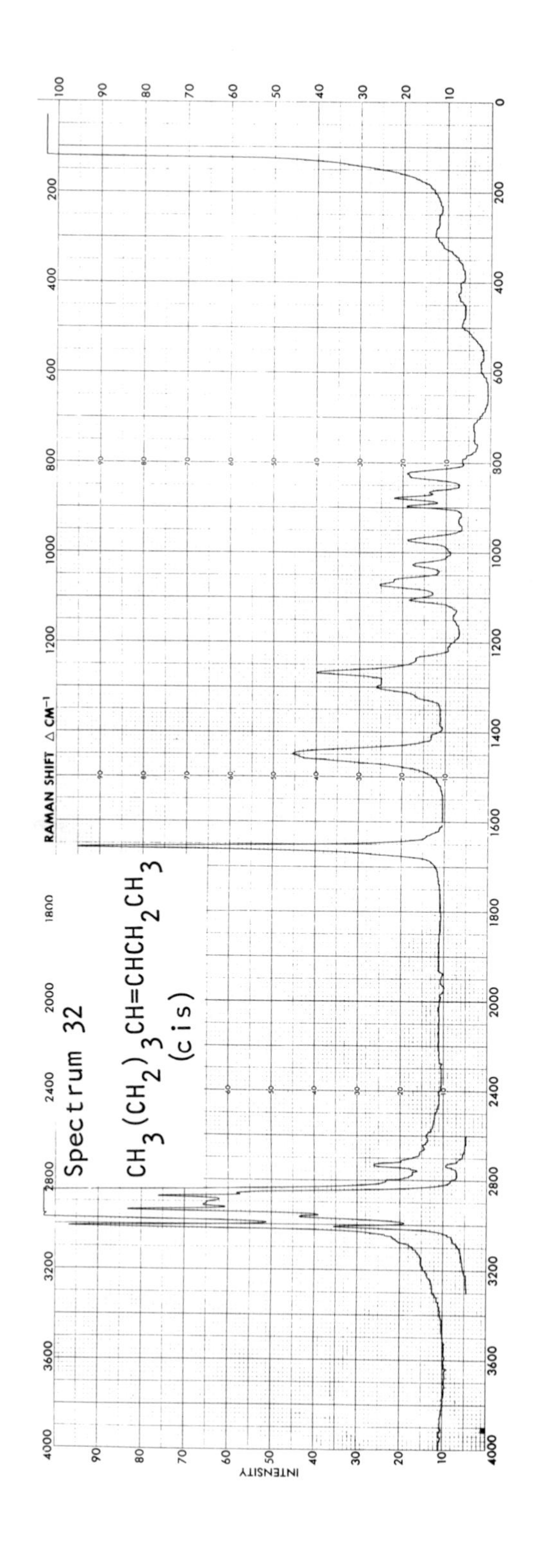
Spectrum 32
CH3(CH2)3CH=CHCH2CH3
(cis)
RAMAN SHIFT Δ CM⁻¹
INTENSITY

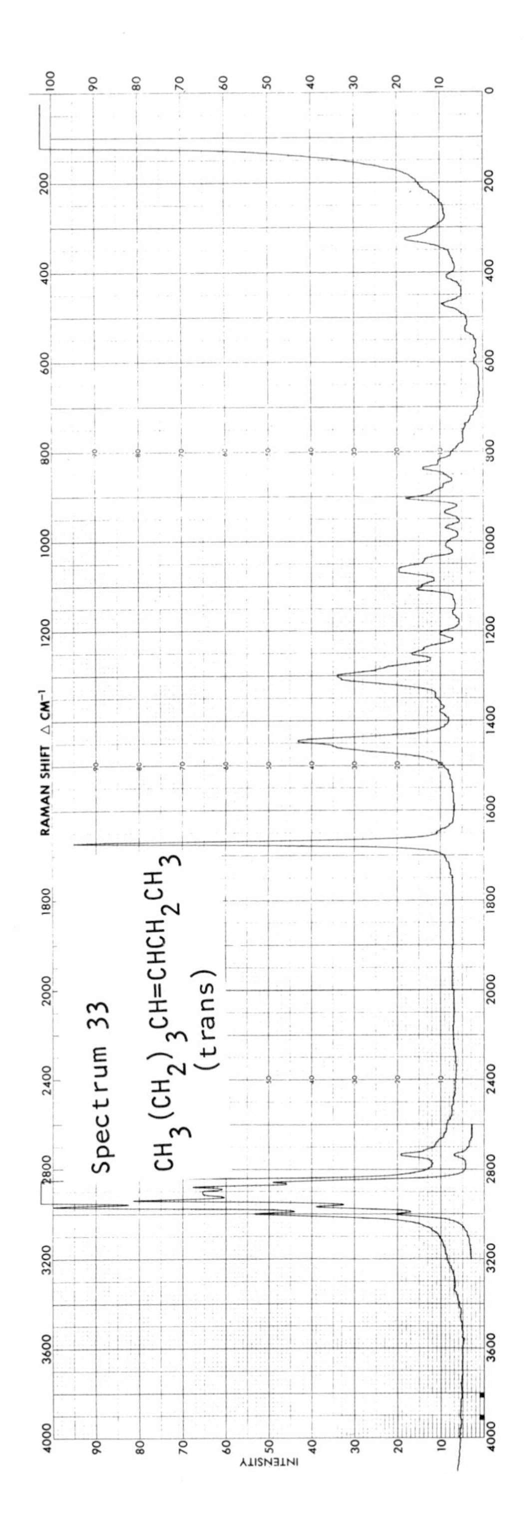
Spectrum 33
$CH_3(CH_2)_3CH=CHCH_2CH_3$
(trans)
RAMAN SHIFT Δ CM^{-1}
INTENSITY
4000 3600 3200 2800 2400 2000 1800 1600 1400 1200 1000 800 600 400 200
0 10 20 30 40 50 60 70 80 90 100

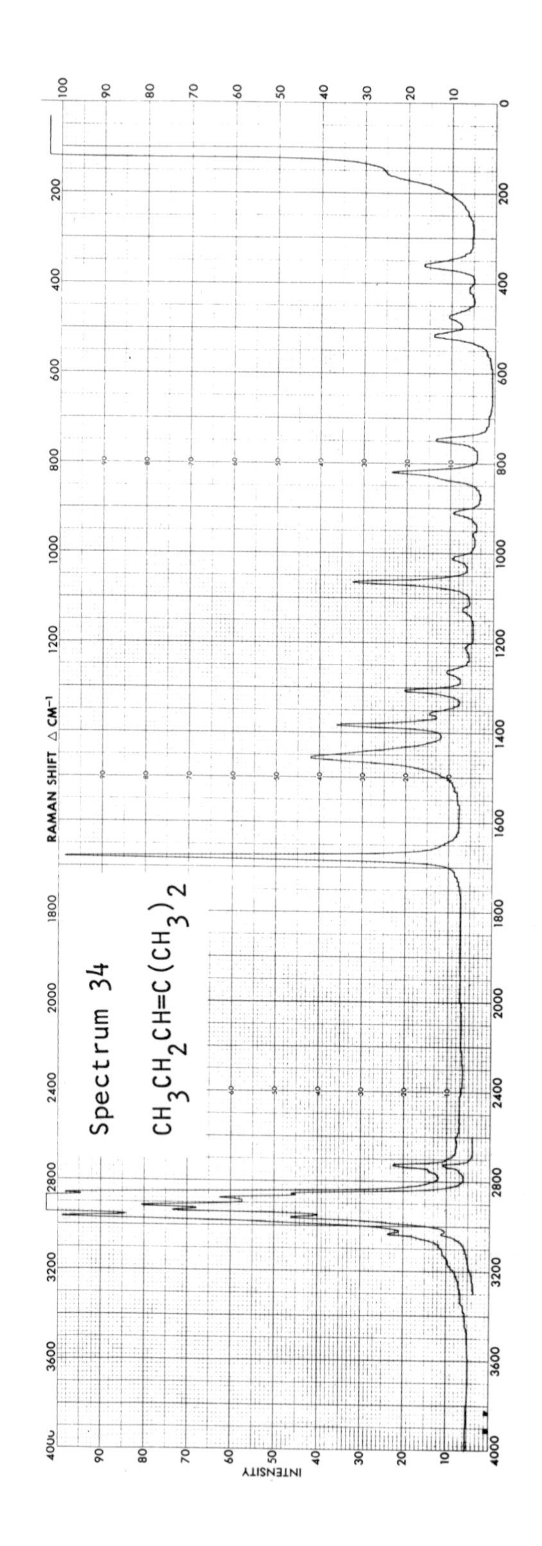
Spectrum 34
CH3CH2CH=C(CH3)2
RAMAN SHIFT Δ CM⁻¹
INTENSITY

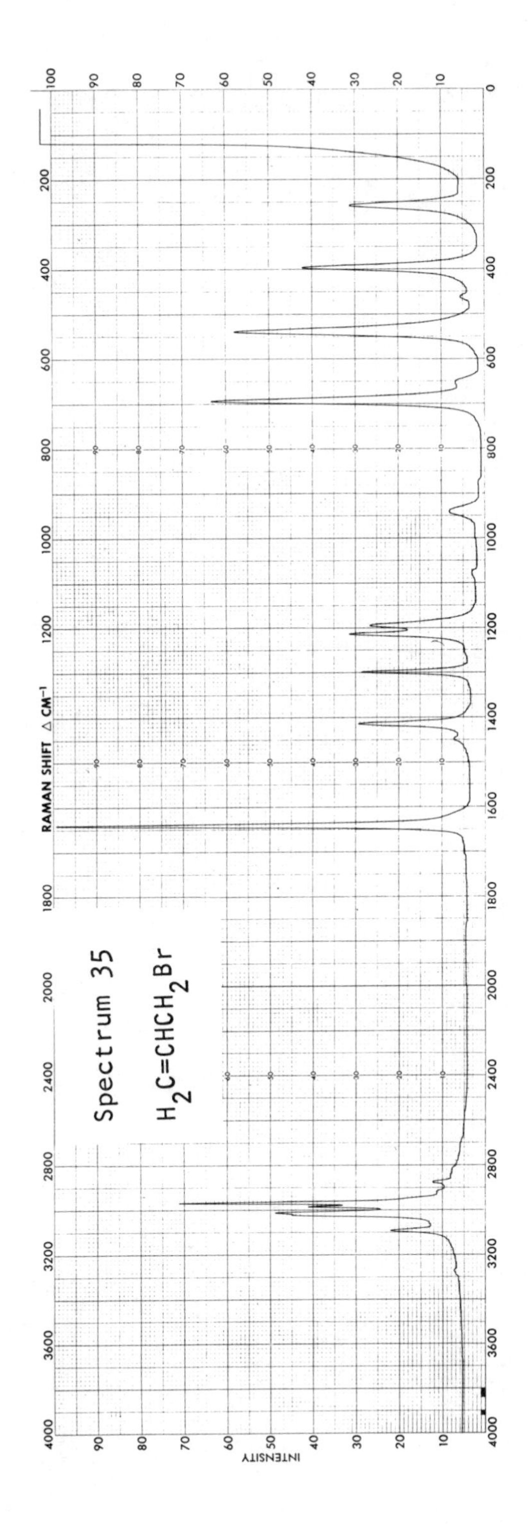
Spectrum 35
H2C=CHCH2Br
RAMAN SHIFT Δ CM⁻¹
INTENSITY

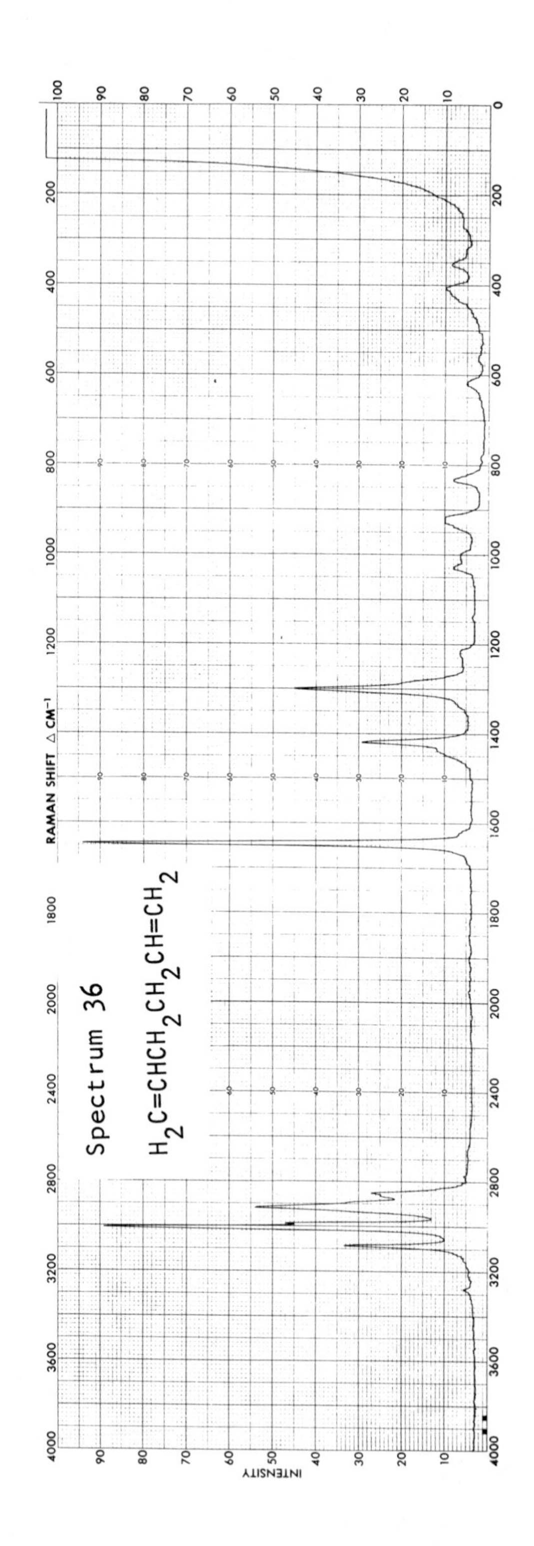
Spectrum 36
$H_2C{=}CHCH_2CH_2CH{=}CH_2$
RAMAN SHIFT Δ CM^{-1}
INTENSITY

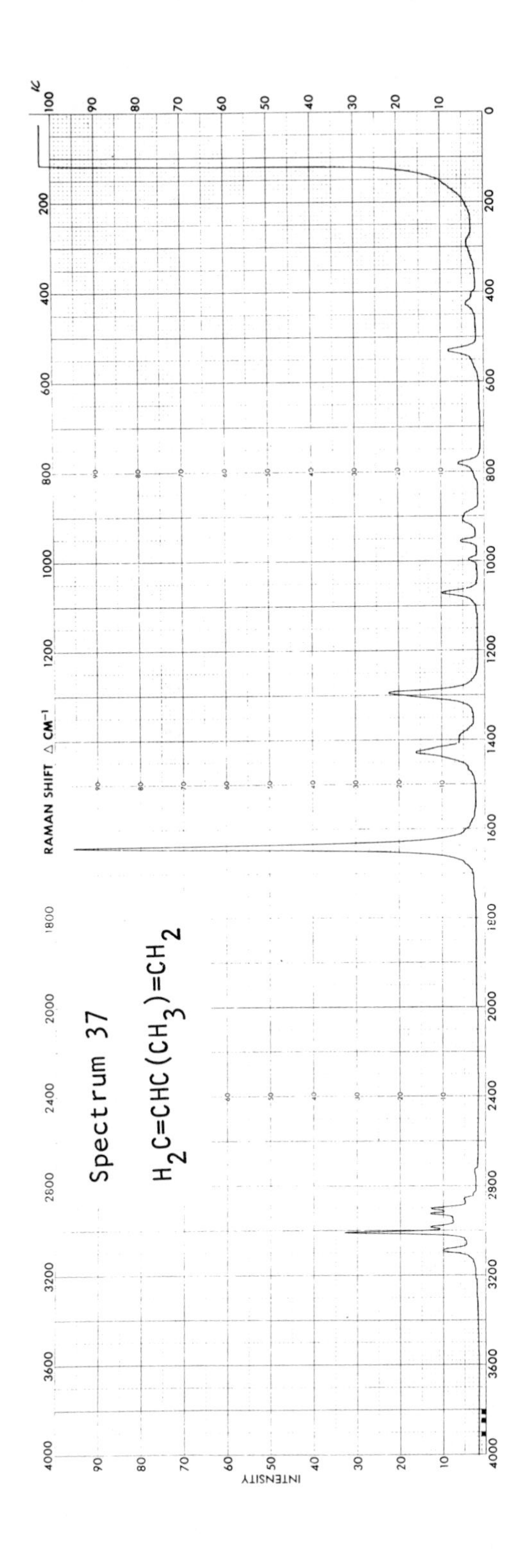

RAMAN SHIFT Δ CM⁻¹
INTENSITY
Spectrum 37
$H_2C=CHC(CH_3)=CH_2$

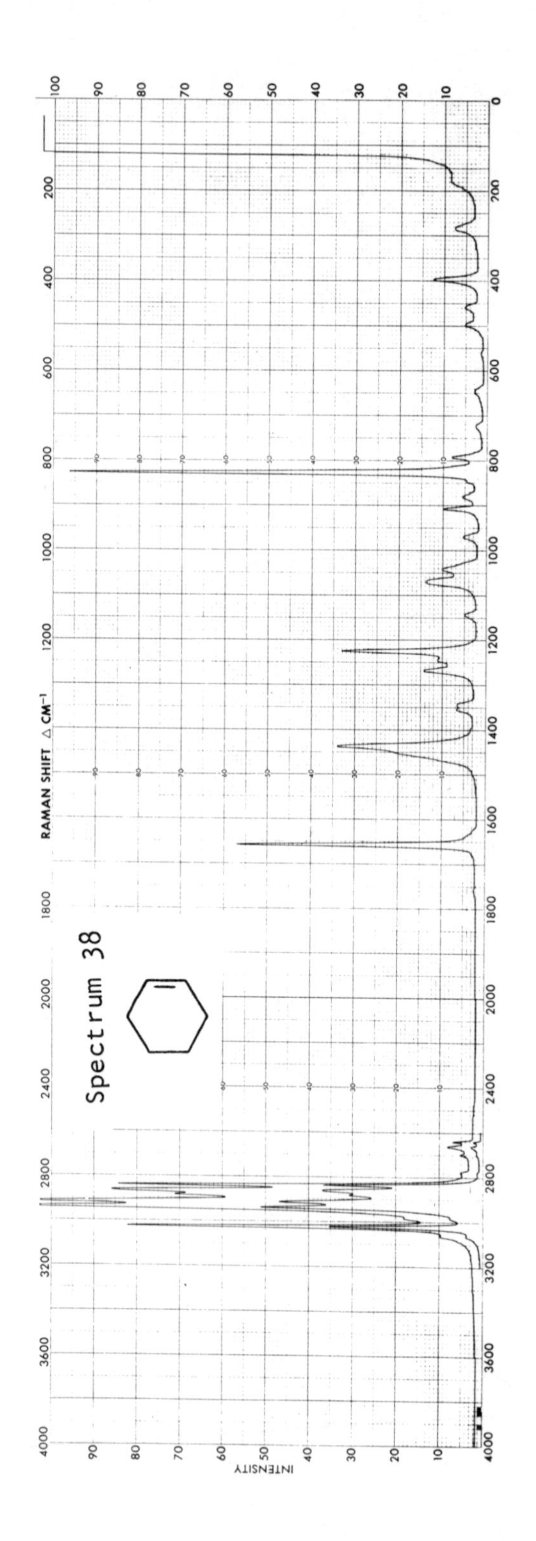
Spectrum 38
RAMAN SHIFT Δ CM⁻¹
INTENSITY
4000 3600 3200 2800 2400 2000 1800 1600 1400 1200 1000 800 600 400 200
100 90 80 70 60 50 40 30 20 10 0

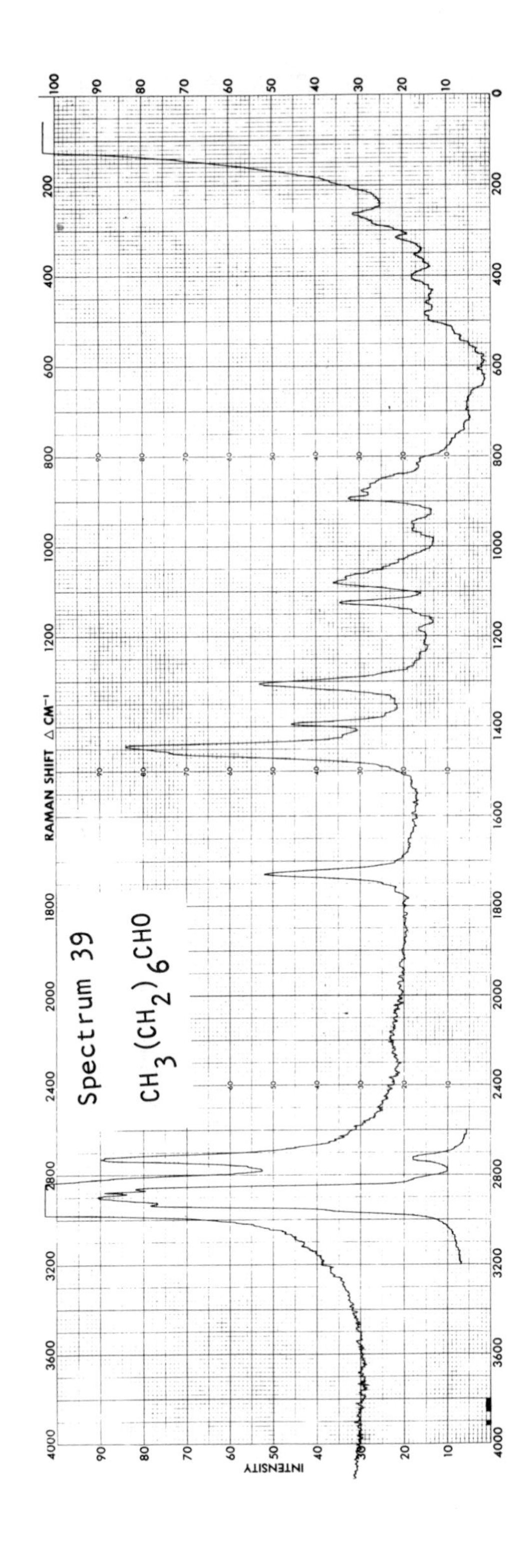
Spectrum 39
CH3(CH2)6CHO
RAMAN SHIFT Δ CM⁻¹
INTENSITY
4000
3600
3200
2800
2400
2000
1800
1600
1400
1200
1000
800
600
400
200
0
100
90
80
70
60
50
40
30
20
10

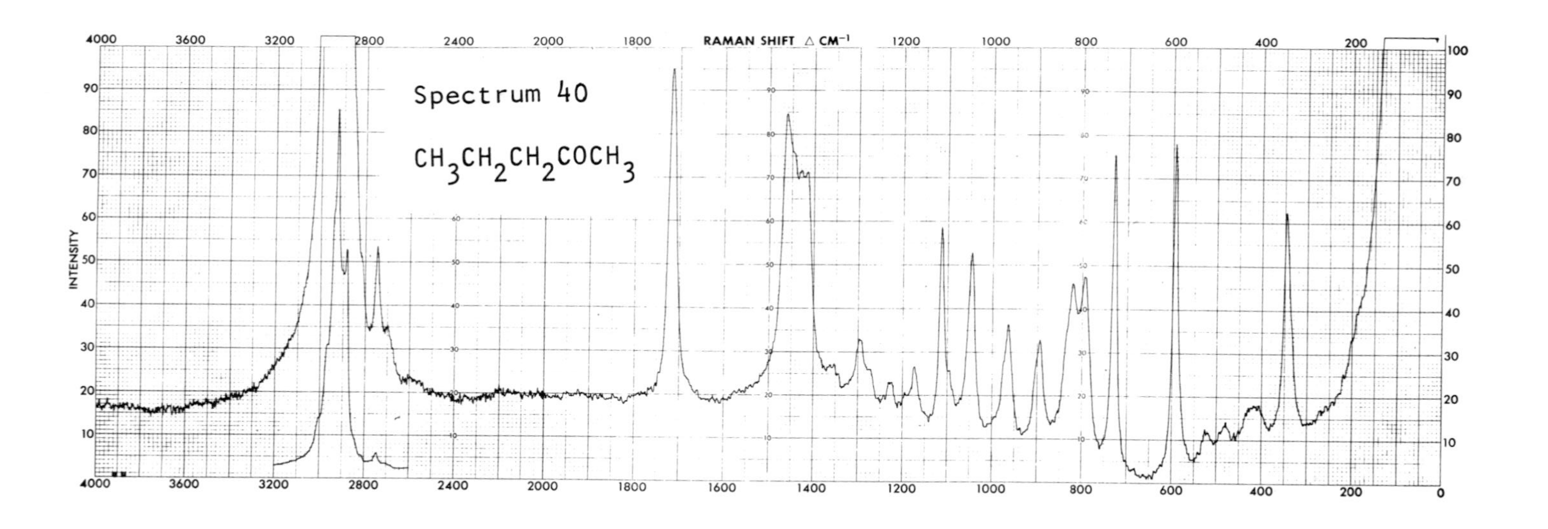
Spectrum 40
$CH_3CH_2CH_2COCH_3$
RAMAN SHIFT Δ CM^{-1}
INTENSITY

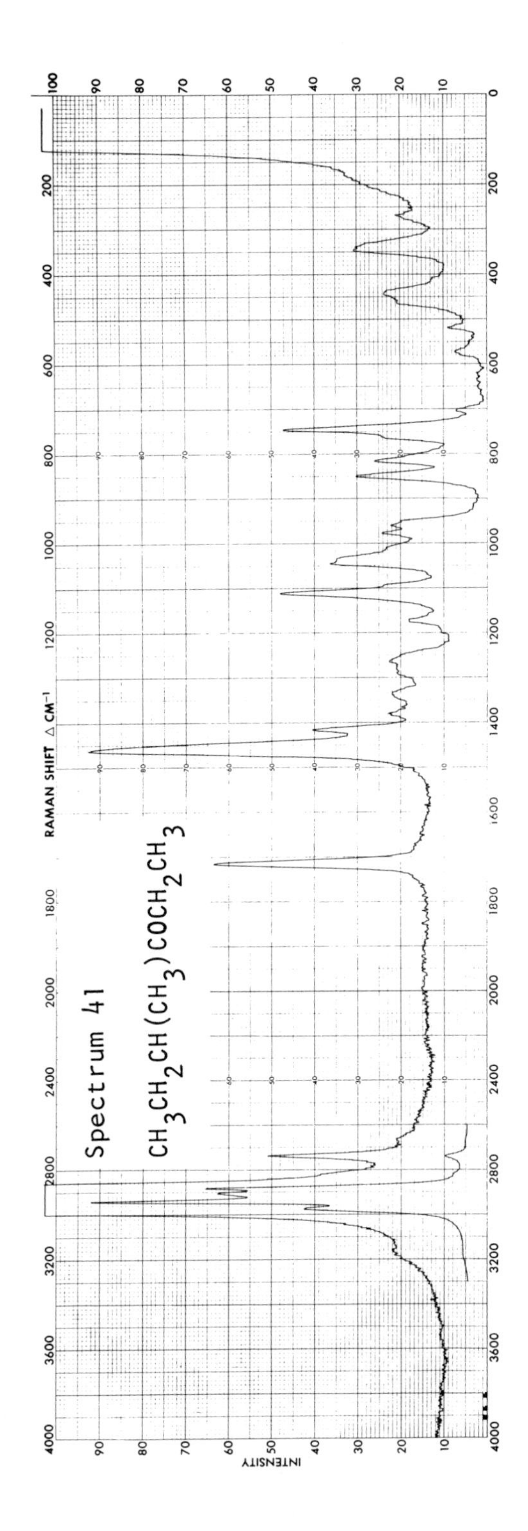
Spectrum 41
CH3CH2CH(CH3)COCH2CH3
RAMAN SHIFT Δ CM⁻¹
INTENSITY

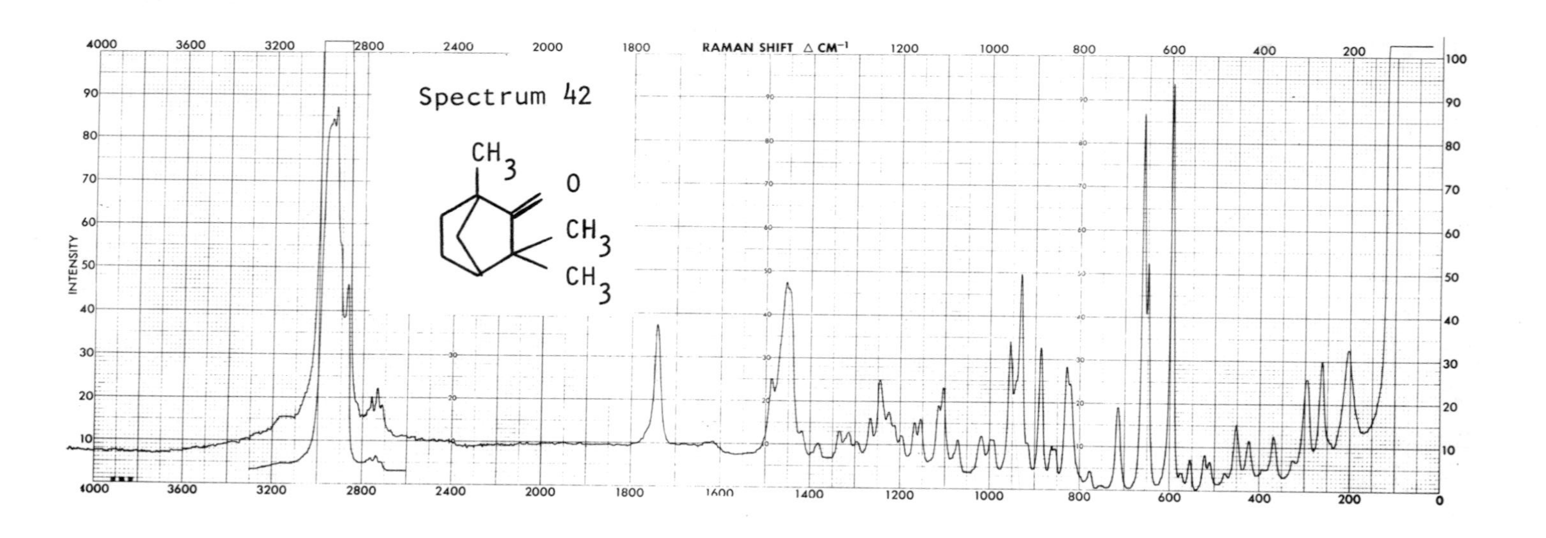

Spectrum 42
CH3
O
CH3
CH3
RAMAN SHIFT △ CM⁻¹
INTENSITY
4000
3600
3200
2800
2400
2000
1800
1600
1400
1200
1000
800
600
400
200
0
10
20
30
40
50
60
70
80
90
100

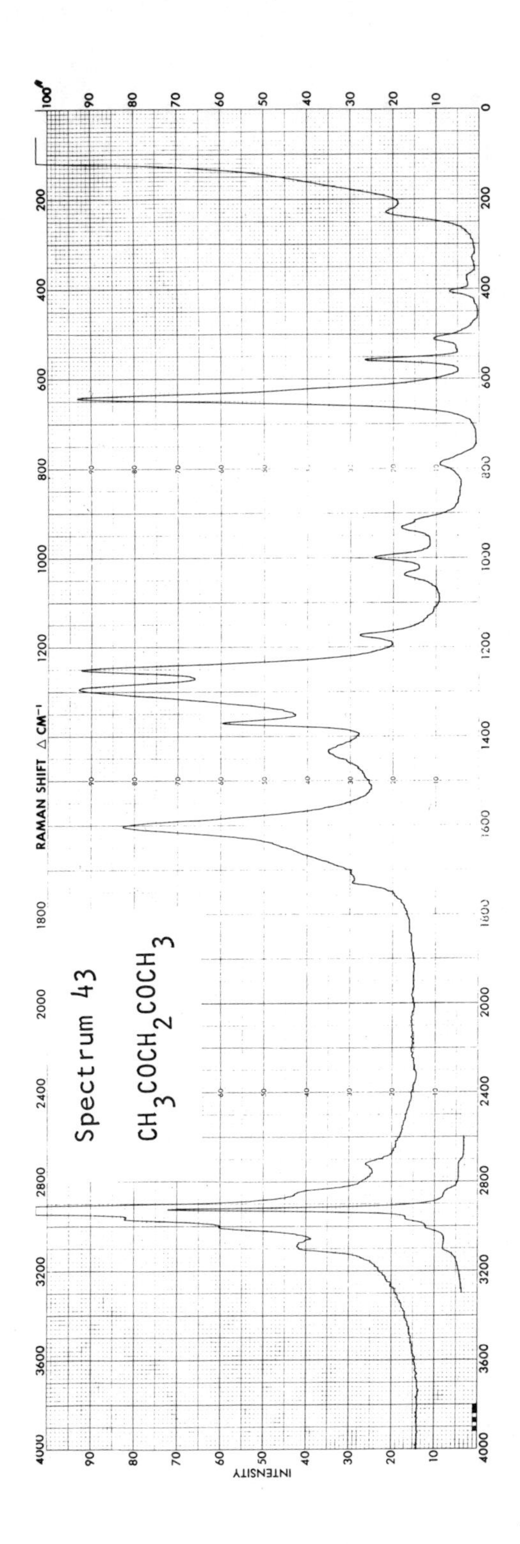
Spectrum 43
CH3COCH2COCH3
RAMAN SHIFT Δ CM⁻¹
INTENSITY

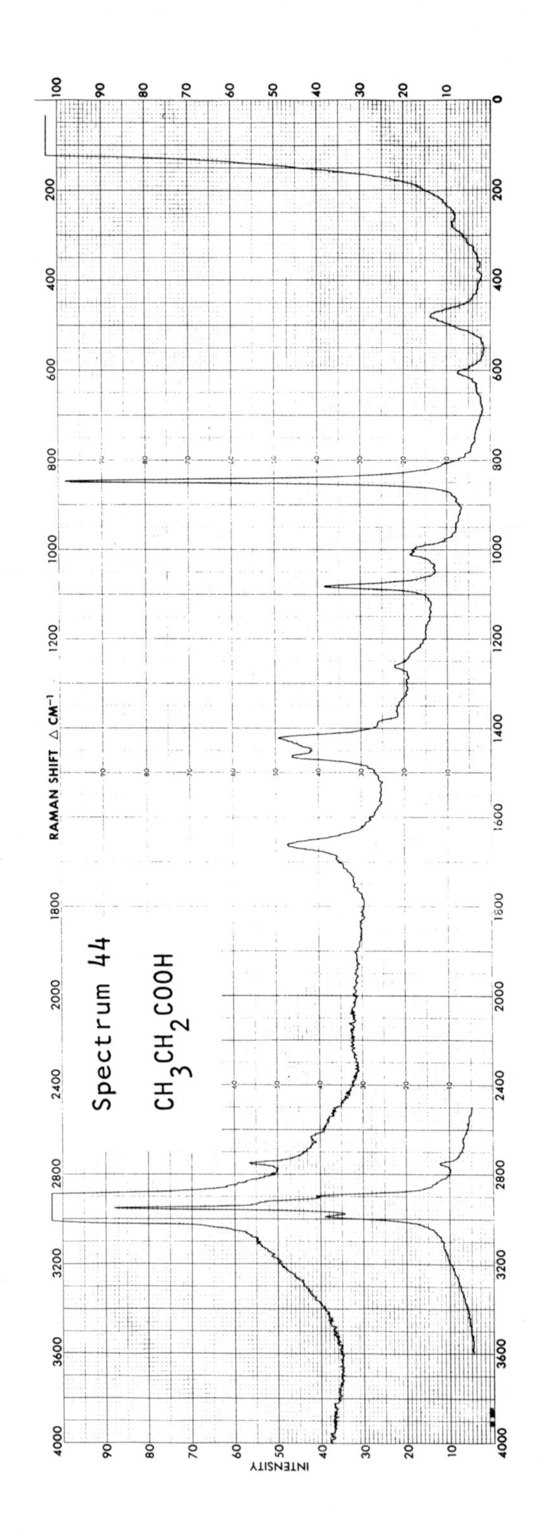

Spectrum 44
CH3CH2COOH
RAMAN SHIFT Δ CM−1
INTENSITY

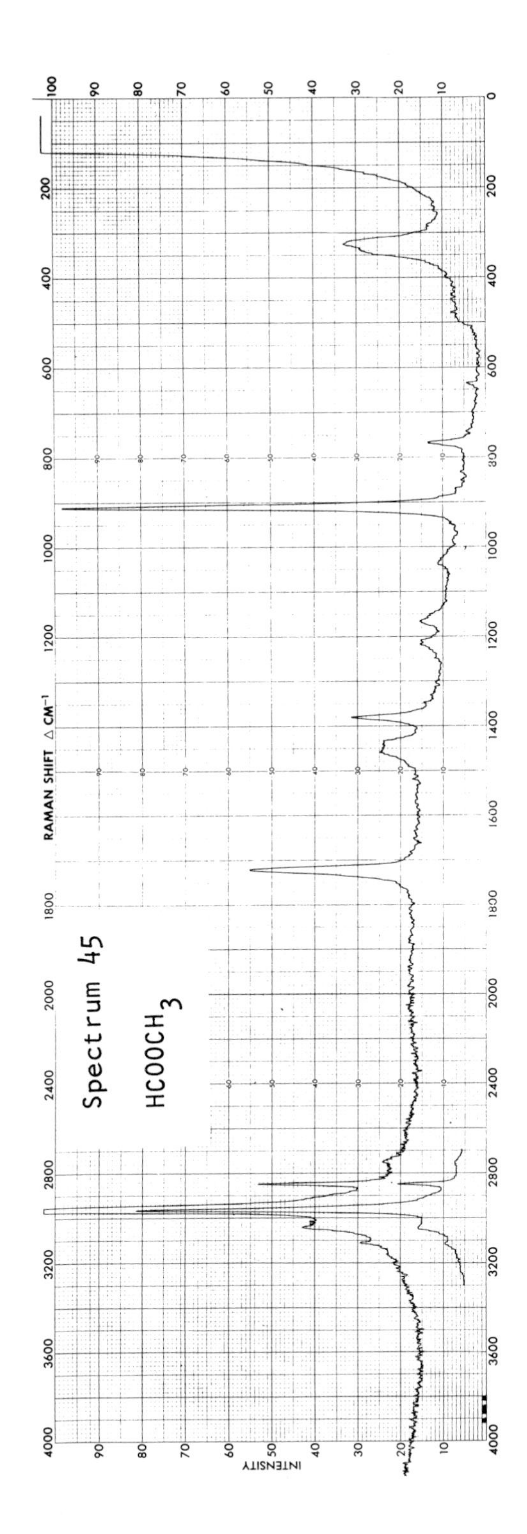
Spectrum 45
HCOOCH$_3$
RAMAN SHIFT Δ CM^{-1}
INTENSITY

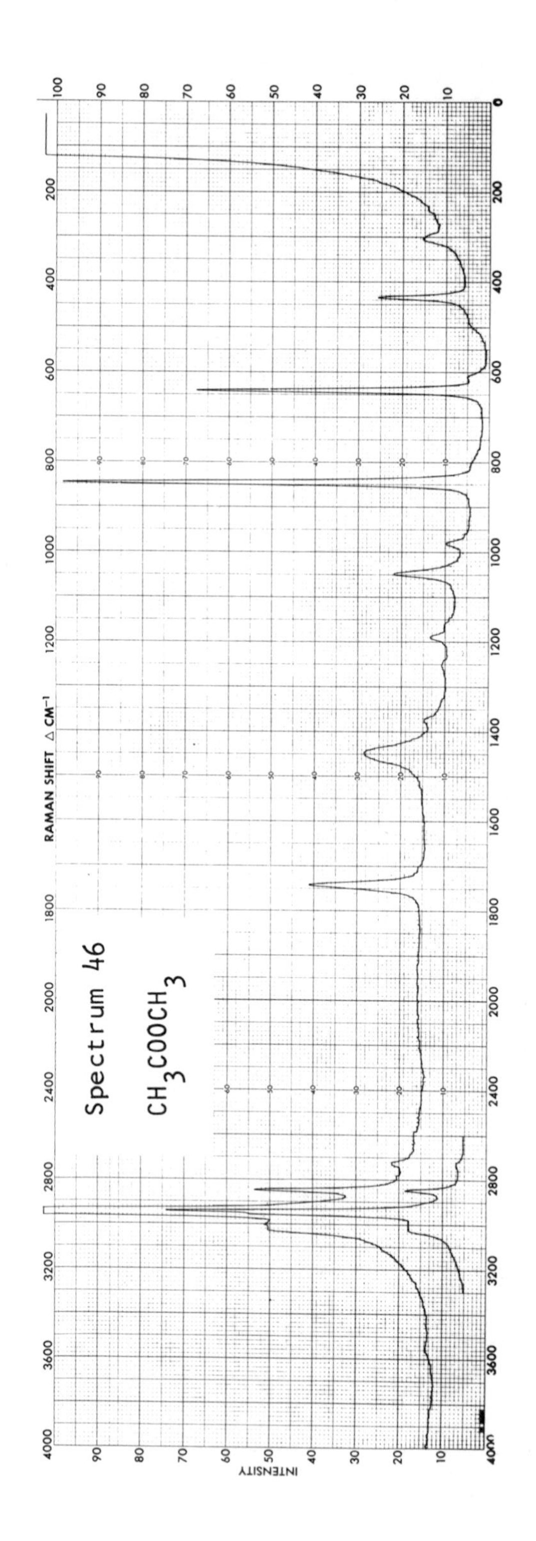
Spectrum 46
CH_3COOCH_3
RAMAN SHIFT △ CM⁻¹
INTENSITY
200
400
600
800
1000
1200
1400
1600
1800
2000
2400
2800
3200
3600
4000
100
90
80
70
60
50
40
30
20
10
0

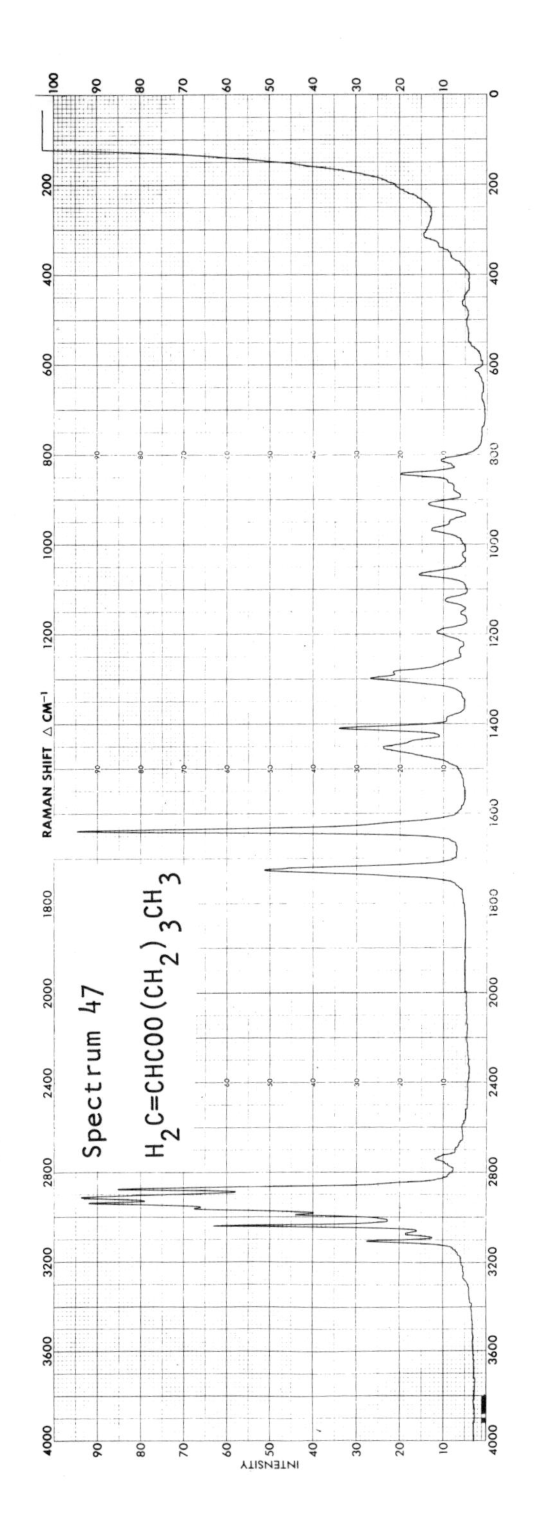

Spectrum 47
$H_2C{=}CHCOO(CH_2)_3CH_3$
RAMAN SHIFT Δ CM^{-1}
INTENSITY

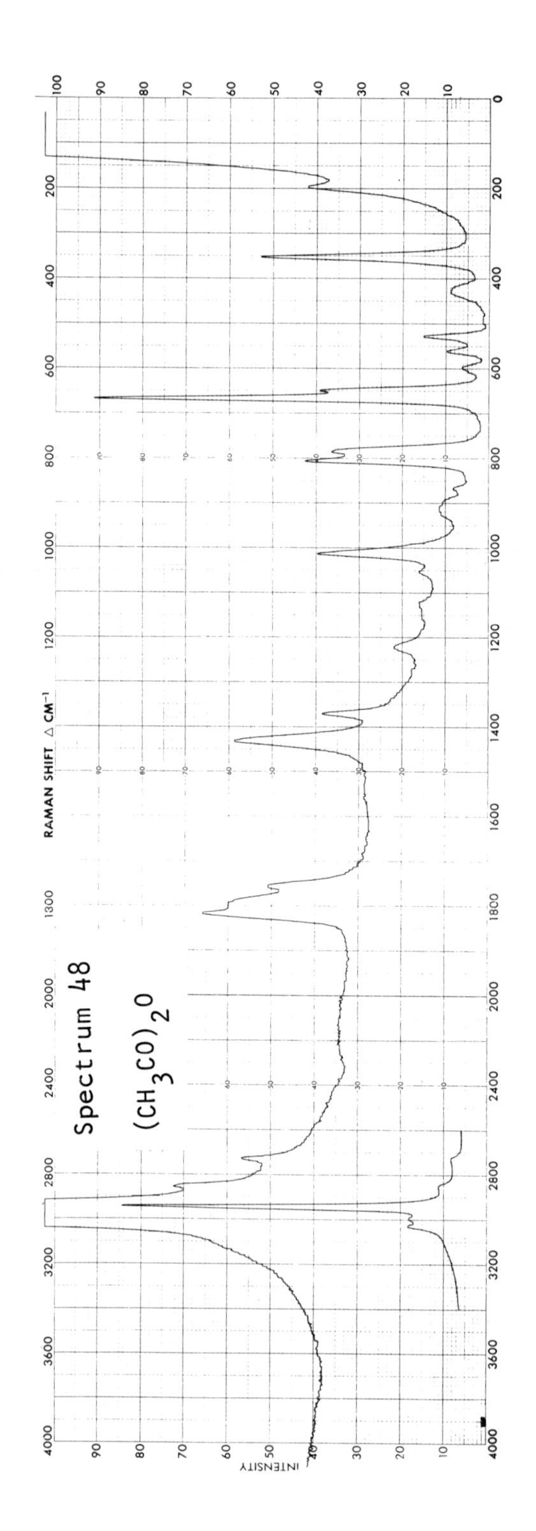

Spectrum 48
$(CH_3CO)_2O$
RAMAN SHIFT △ CM⁻¹
INTENSITY

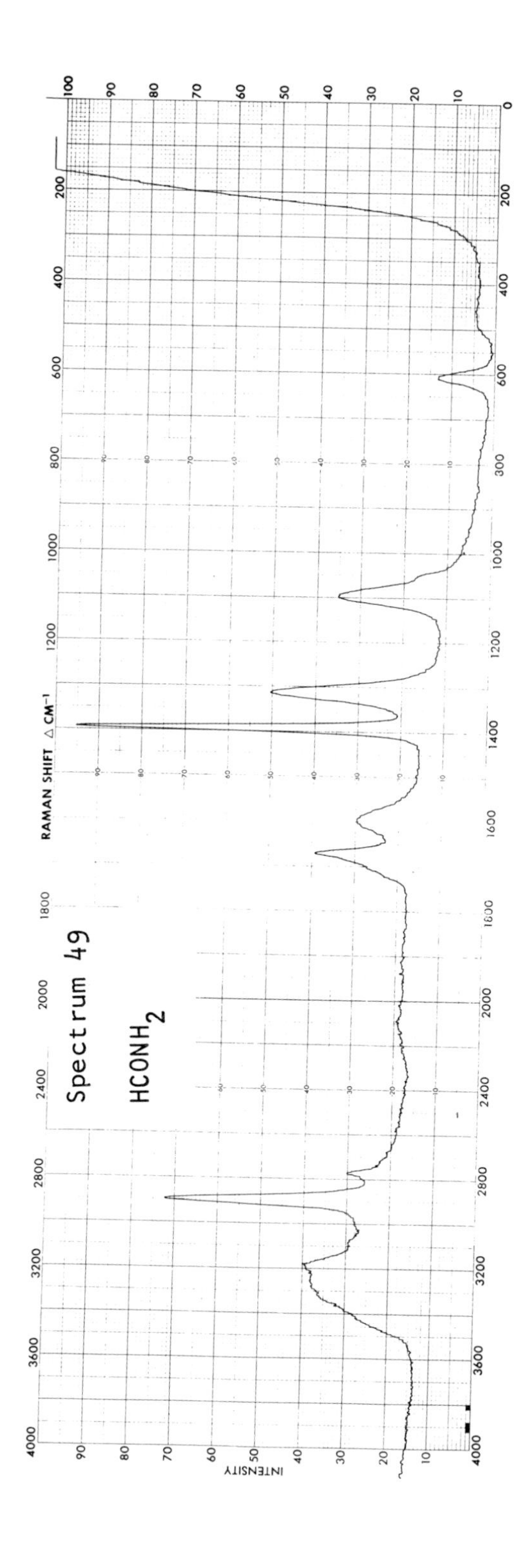
Spectrum 49
HCONH2
RAMAN SHIFT △ CM⁻¹
INTENSITY

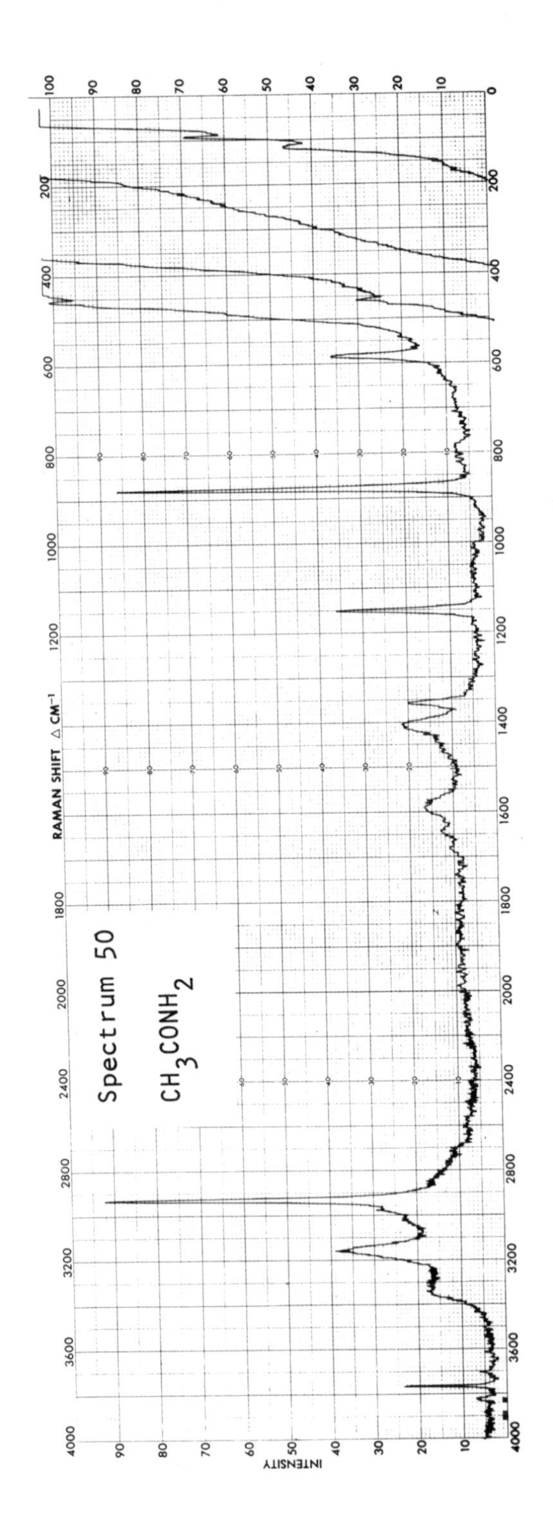
Spectrum 50
CH3CONH2
RAMAN SHIFT △ CM⁻¹
INTENSITY

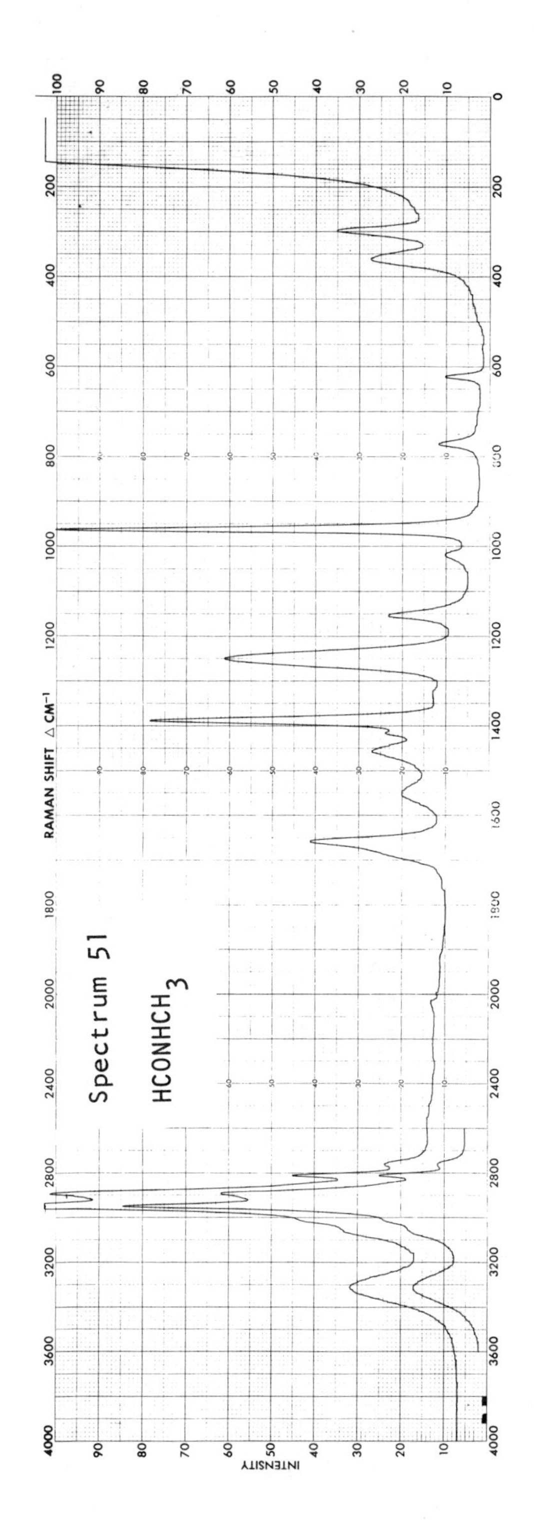
Spectrum 51
HCONHCH$_3$
RAMAN SHIFT Δ CM^{-1}
INTENSITY

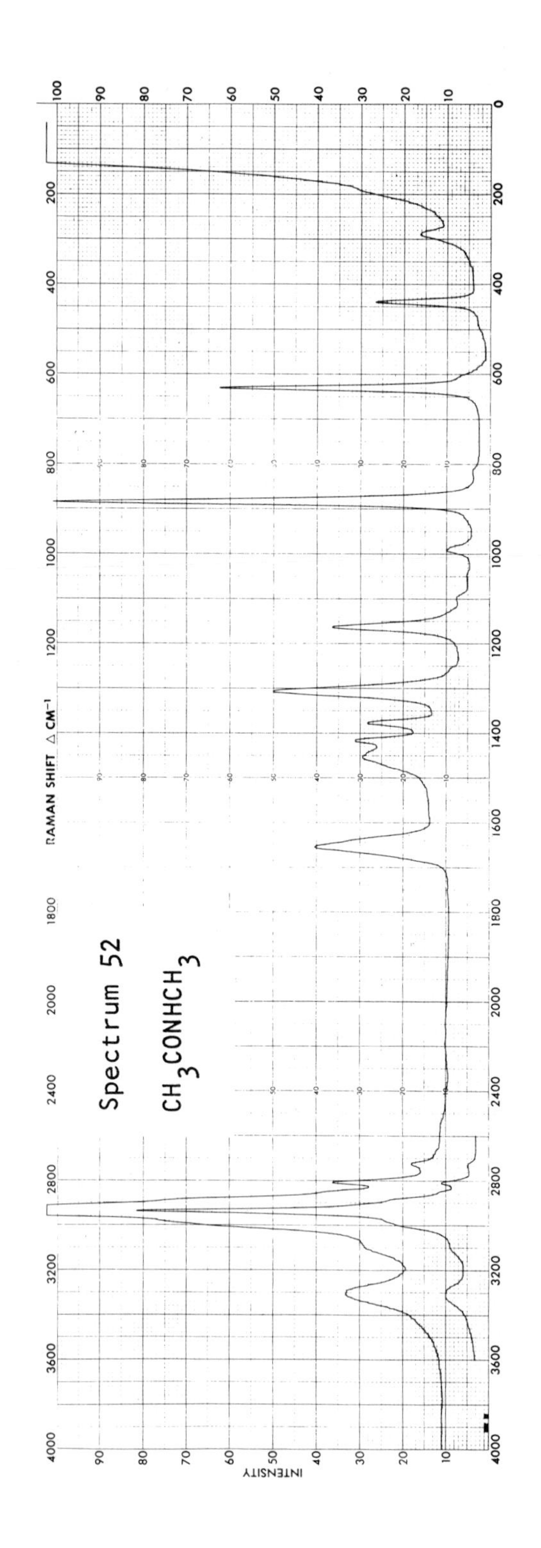
Spectrum 52
$CH_3CONHCH_3$
RAMAN SHIFT Δ CM^{-1}
INTENSITY
4000 3600 3200 2800 2400 2000 1800 1600 1400 1200 1000 800 600 400 200 0
90 80 70 60 50 40 30 20 10

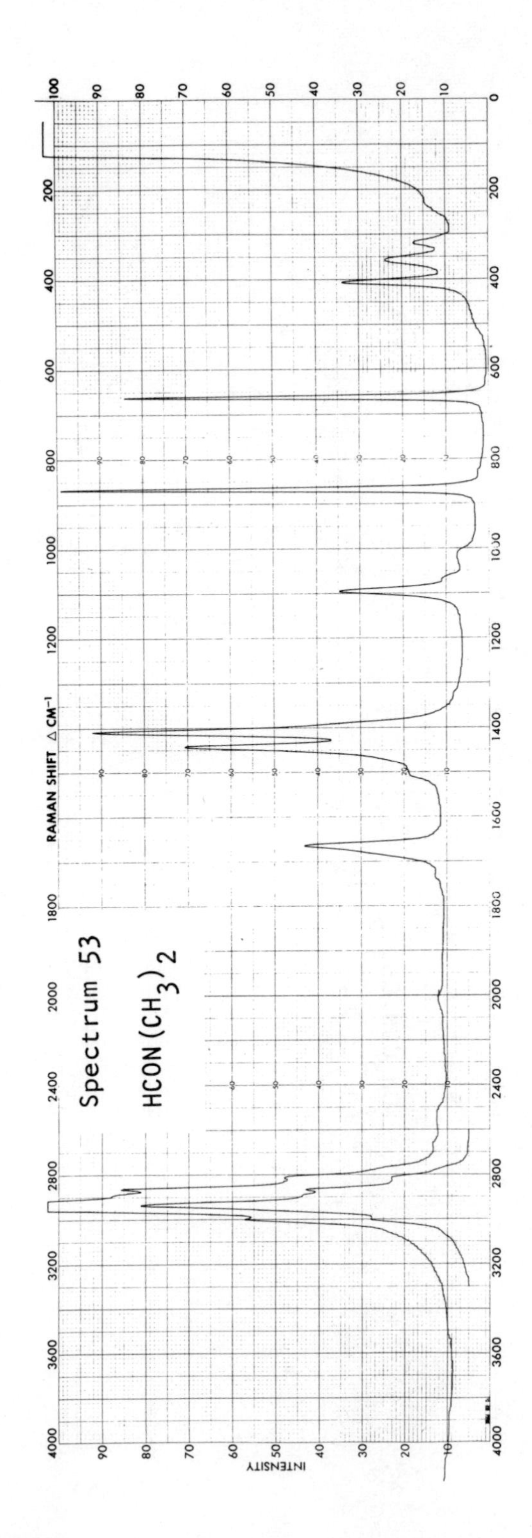

Spectrum 53
HCON(CH3)2
RAMAN SHIFT Δ CM⁻¹
INTENSITY
4000
3600
3200
2800
2400
2000
1800
1600
1400
1200
1000
800
600
400
200
0
100
90
80
70
60
50
40
30
20
10

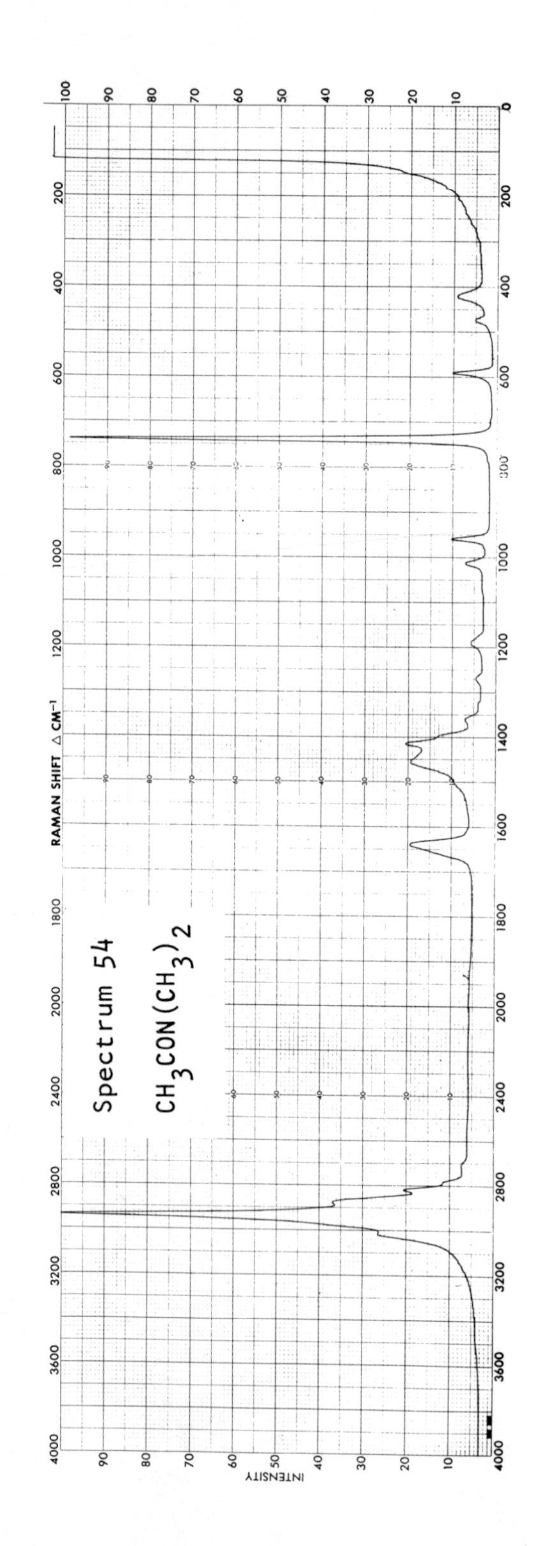

Spectrum 54
$CH_3CON(CH_3)_2$
RAMAN SHIFT Δ CM⁻¹
INTENSITY

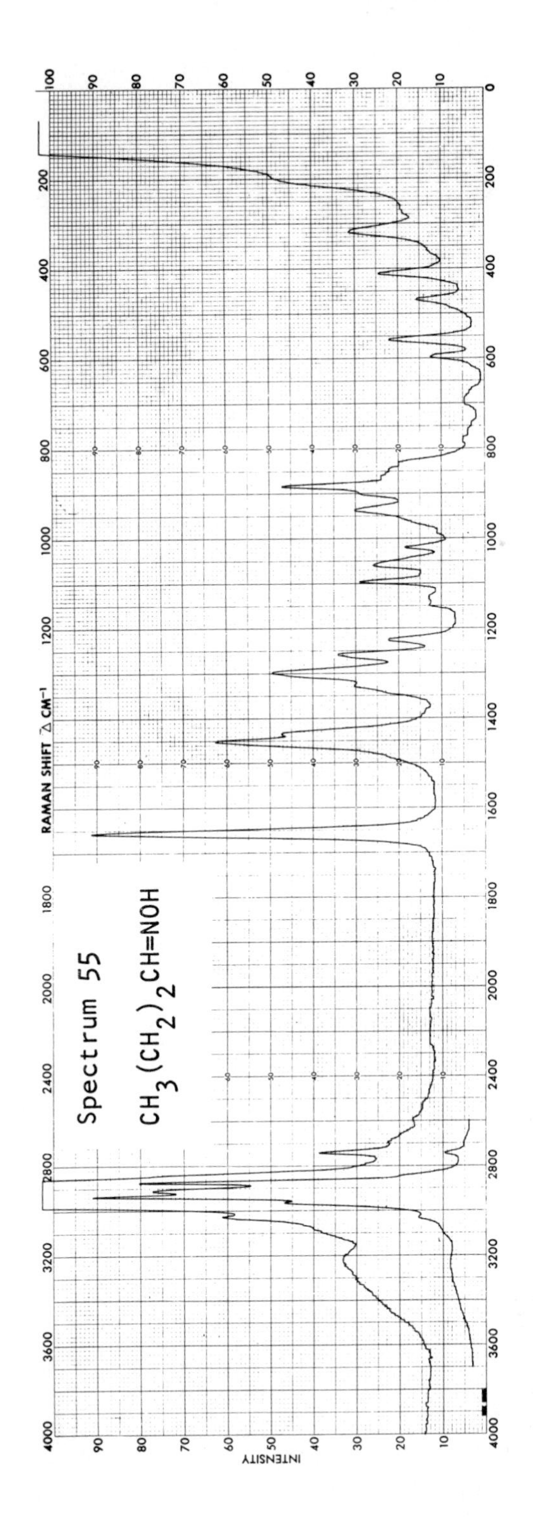

Spectrum 55
$CH_3(CH_2)_2CH{=}NOH$
RAMAN SHIFT Δ CM^{-1}
INTENSITY

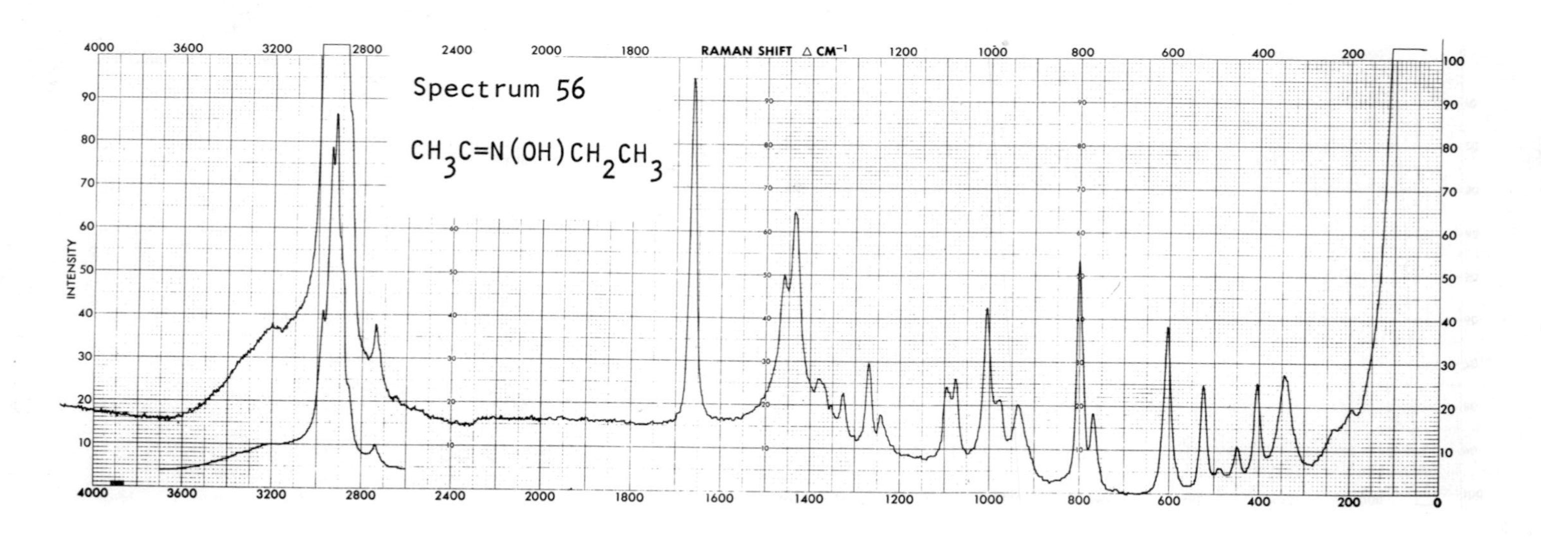

Spectrum 56
$CH_3C{=}N(OH)CH_2CH_3$
RAMAN SHIFT Δ CM⁻¹
INTENSITY

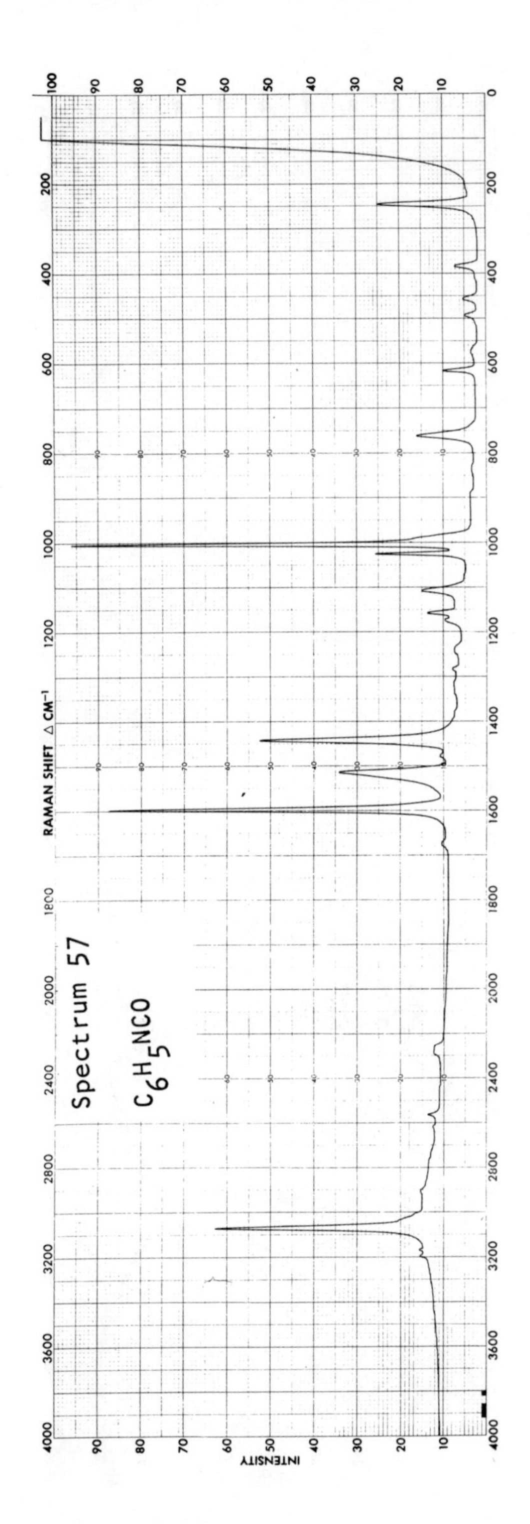
Spectrum 57
C6H5NCO
RAMAN SHIFT Δ CM⁻¹
INTENSITY

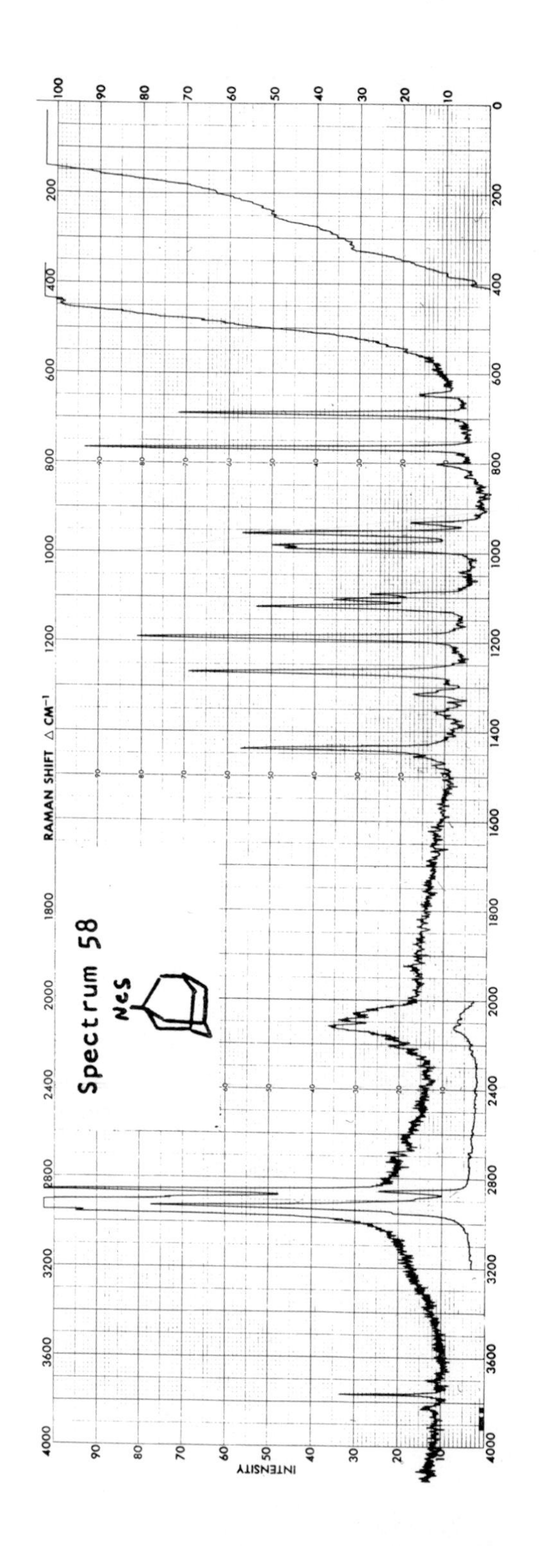
Spectrum 58
RAMAN SHIFT Δ CM⁻¹
INTENSITY
4000
3600
3200
2800
2400
2000
1800
1600
1400
1200
1000
800
600
400
200
0
100
90
80
70
60
50
40
30
20
10

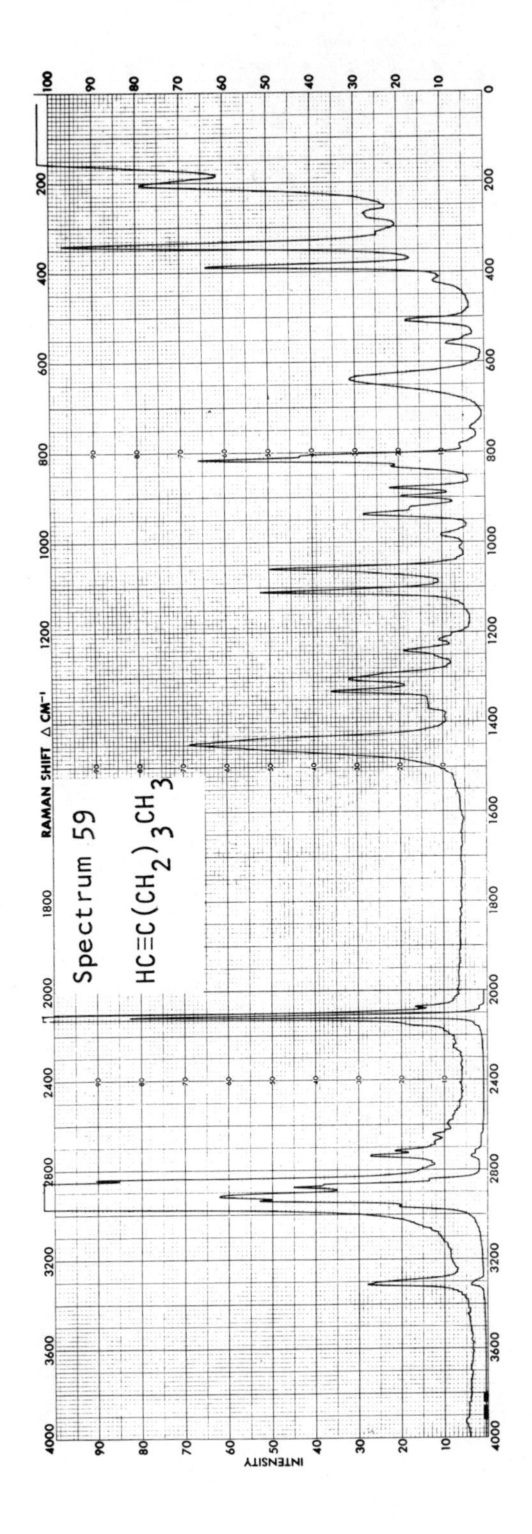
Spectrum 59
HC≡C(CH2)3CH3
RAMAN SHIFT Δ CM⁻¹
INTENSITY
4000
3600
3200
2800
2400
2000
1800
1600
1400
1200
1000
800
600
400
200
100
90
80
70
60
50
40
30
20
10
0

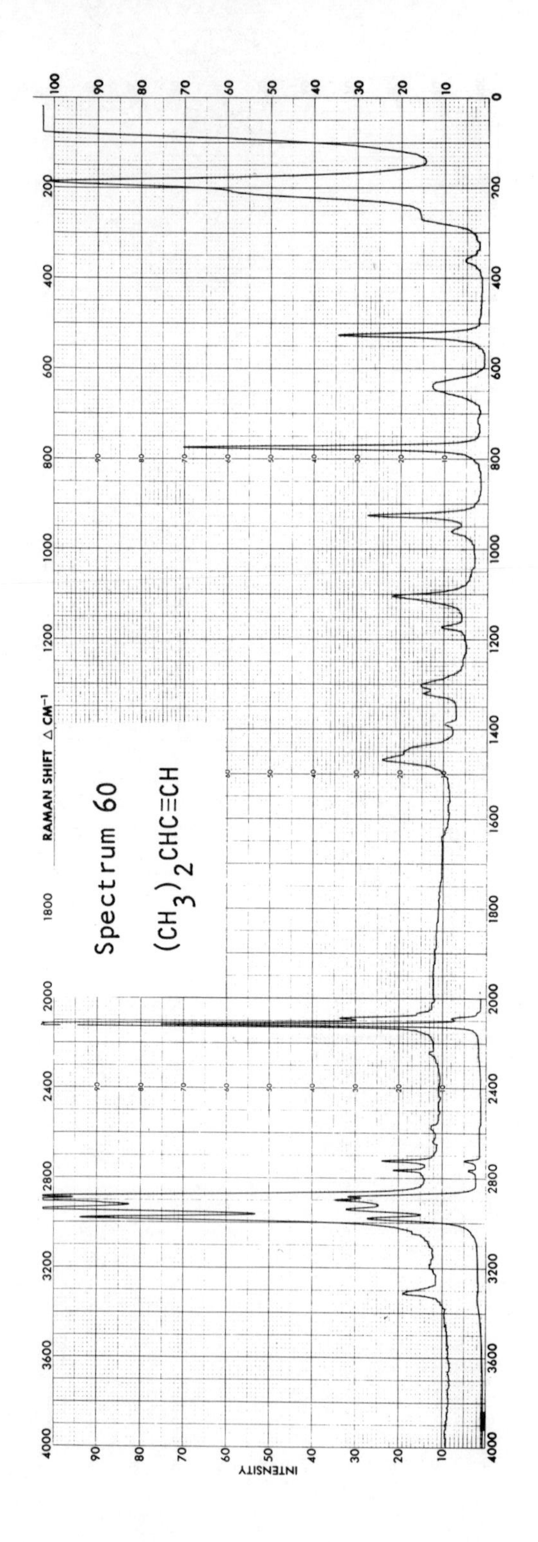

RAMAN SHIFT Δ CM⁻¹
INTENSITY
Spectrum 60
(CH3)2CHC≡CH

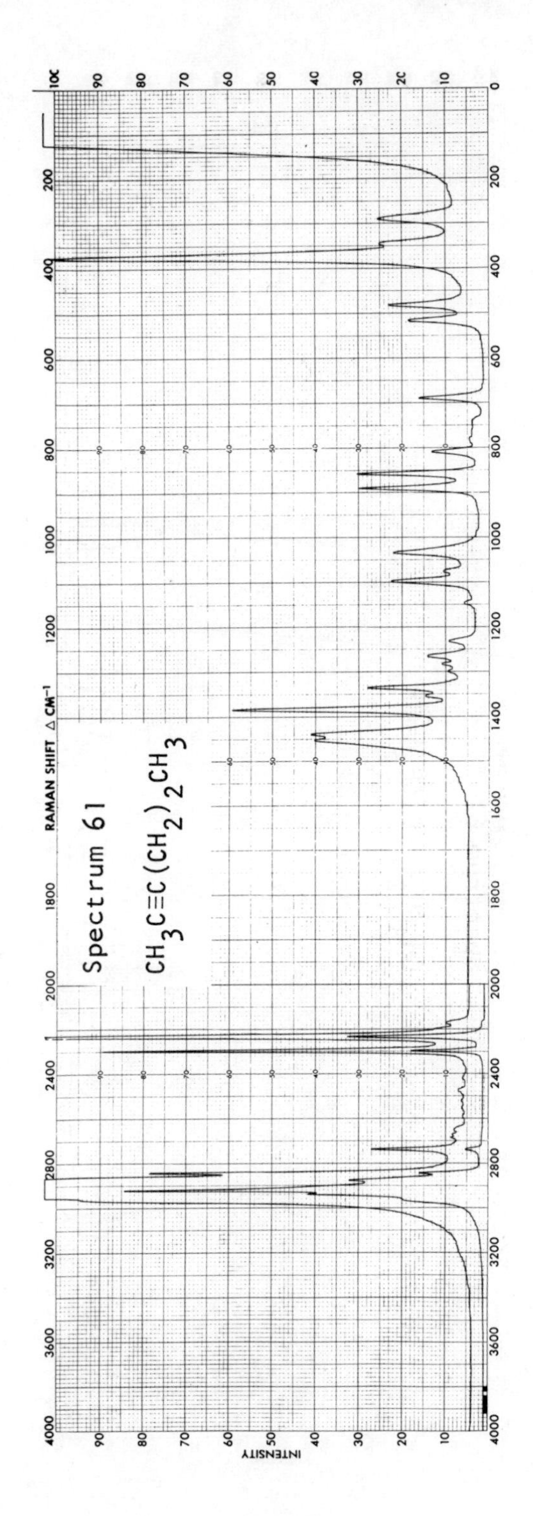

Spectrum 61
CH3C≡C(CH2)2CH3
RAMAN SHIFT Δ CM⁻¹
INTENSITY

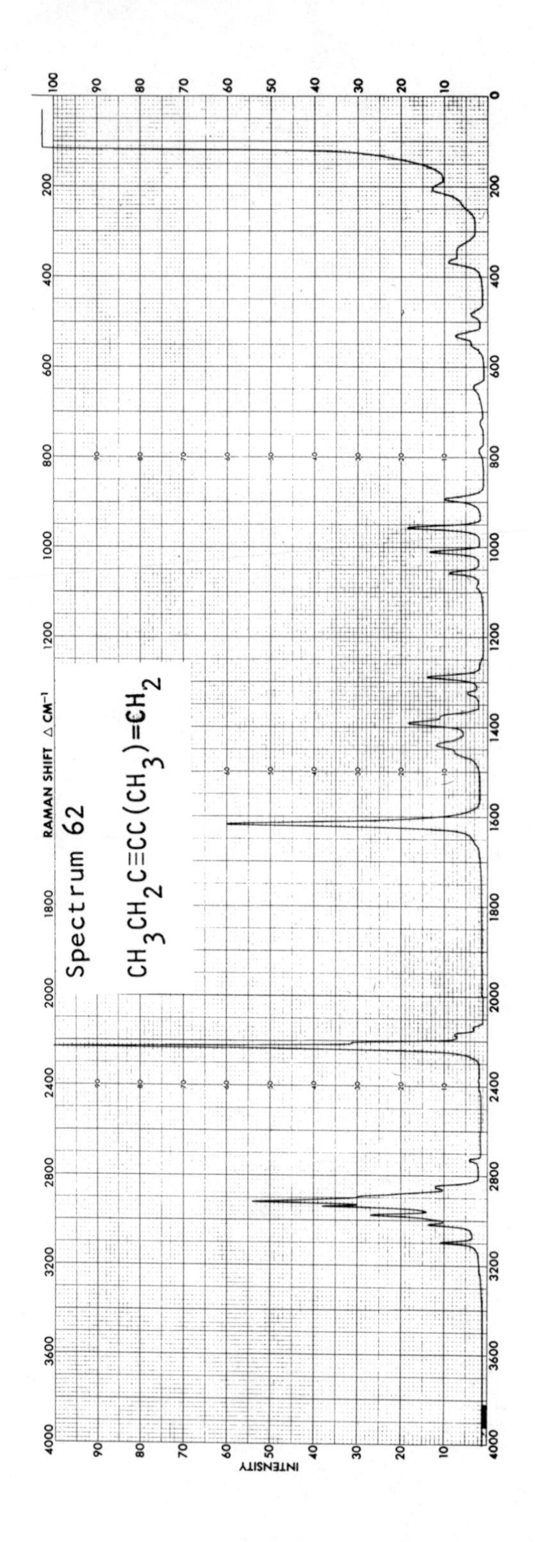
Spectrum 62
$CH_3CH_2C{\equiv}CC(CH_3){=}CH_2$
RAMAN SHIFT Δ CM⁻¹
INTENSITY

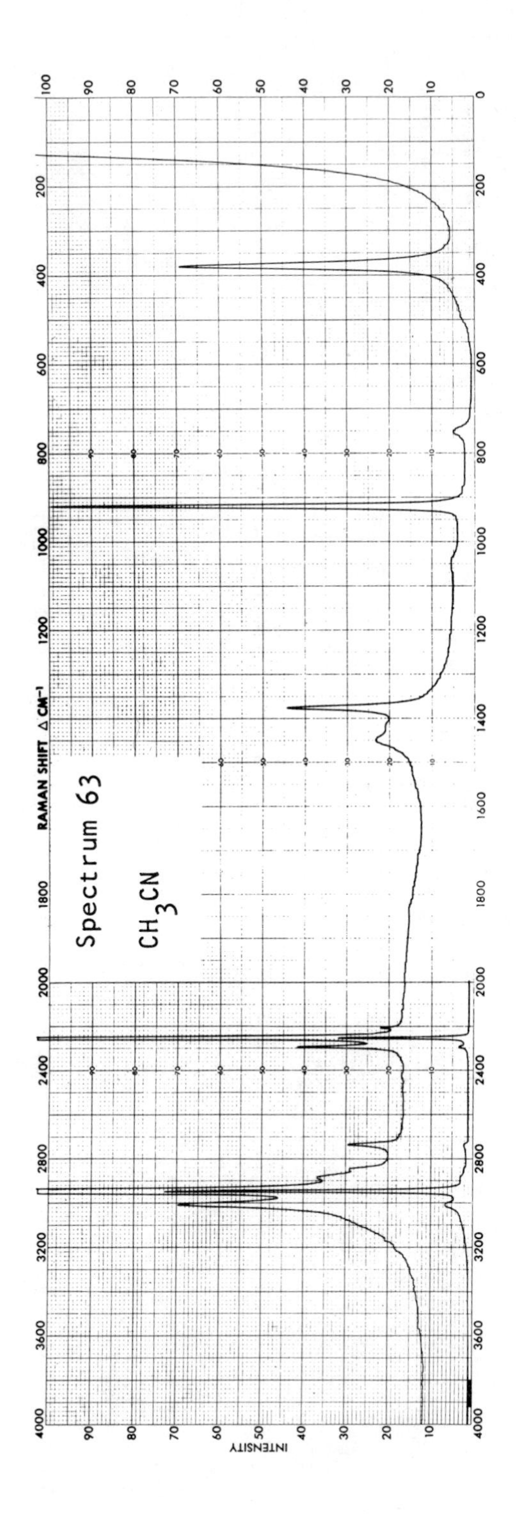

Spectrum 63
CH_3CN
RAMAN SHIFT Δ CM^{-1}
INTENSITY

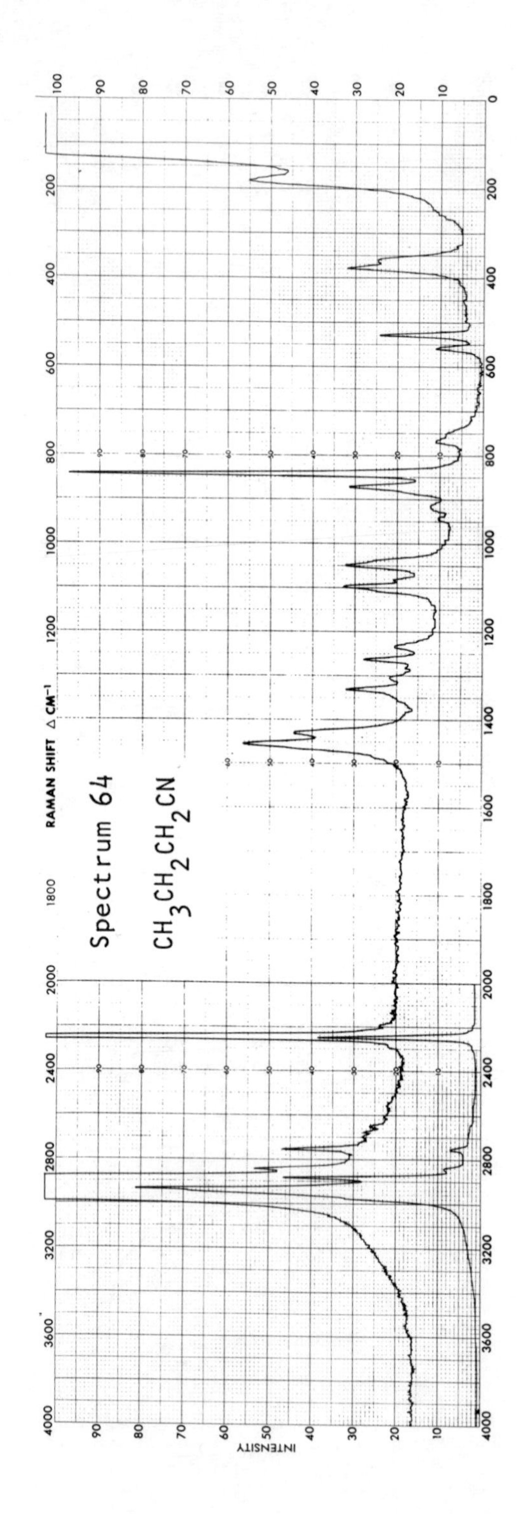

Spectrum 64
CH3CH2CH2CN
RAMAN SHIFT Δ CM⁻¹
INTENSITY
4000
3600
3200
2800
2400
2000
1800
1600
1400
1200
1000
800
600
400
200
0
10
20
30
40
50
60
70
80
90
100

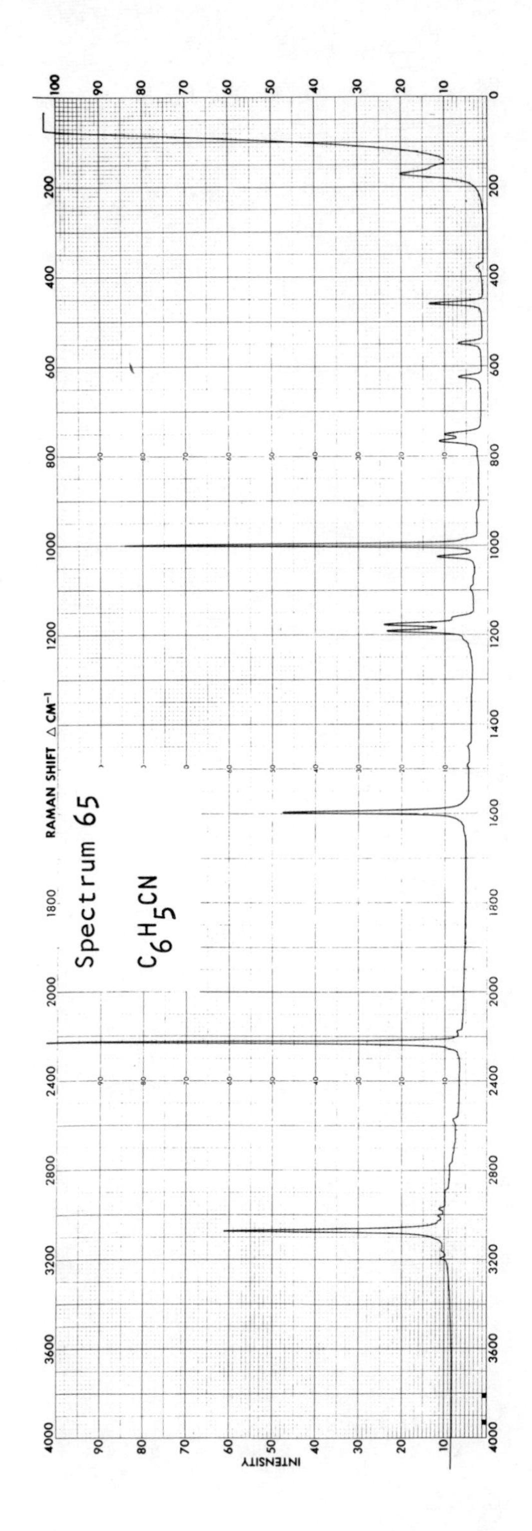
Spectrum 65
C_6H_5CN
RAMAN SHIFT Δ CM^{-1}
INTENSITY
4000 3600 3200 2800 2400 2000 1800 1600 1400 1200 1000 800 600 400 200
100 90 80 70 60 50 40 30 20 10 0

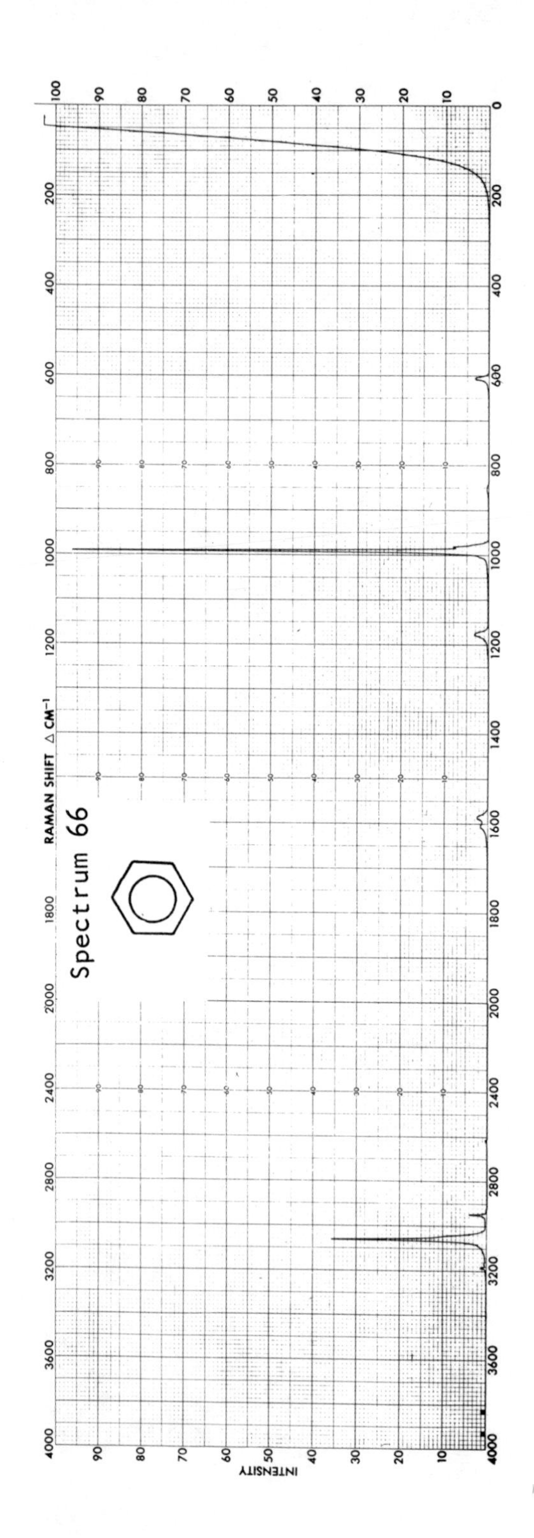
Spectrum 66
RAMAN SHIFT Δ CM⁻¹
INTENSITY

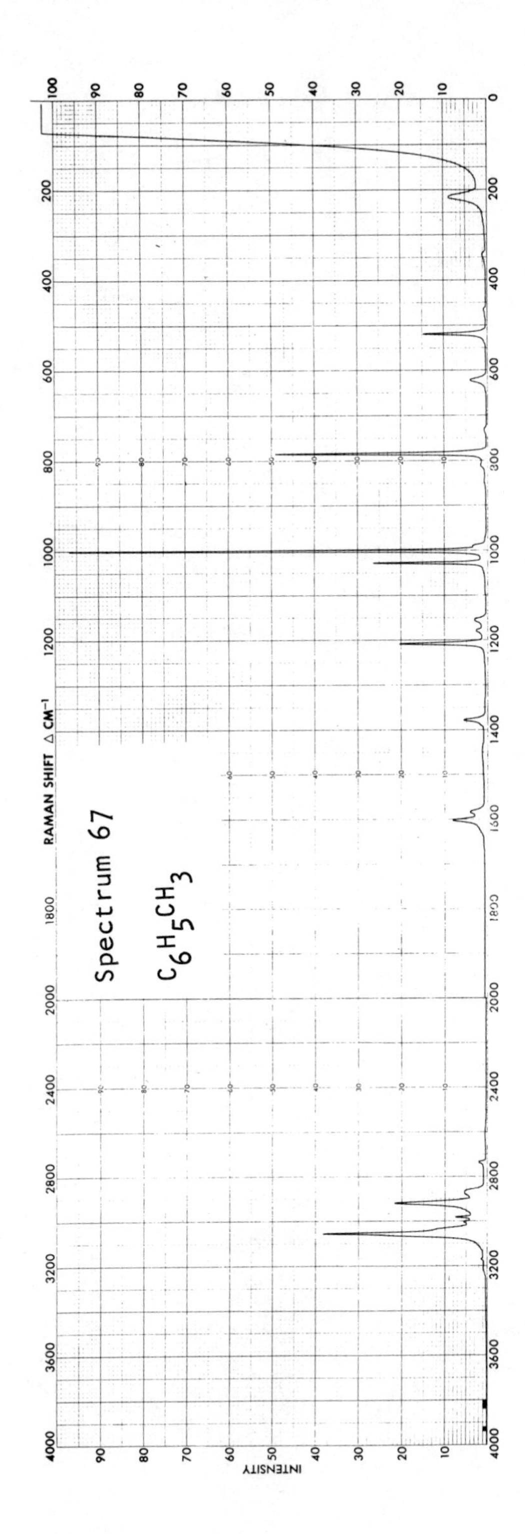
Spectrum 67
$C_6H_5CH_3$
RAMAN SHIFT Δ CM^{-1}
INTENSITY

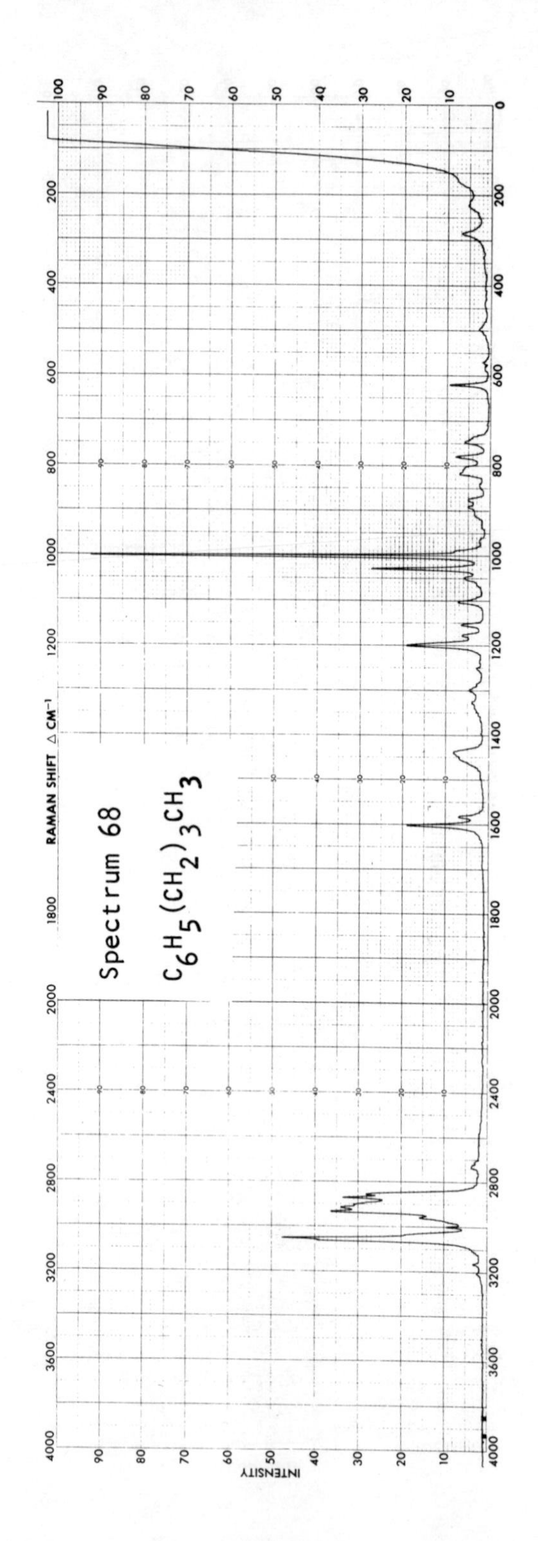

Spectrum 68
$C_6H_5(CH_2)_3CH_3$
RAMAN SHIFT Δ CM^{-1}
INTENSITY
4000 3600 3200 2800 2400 2000 1800 1600 1400 1200 1000 800 600 400 200 0
100 90 80 70 60 50 40 30 20 10

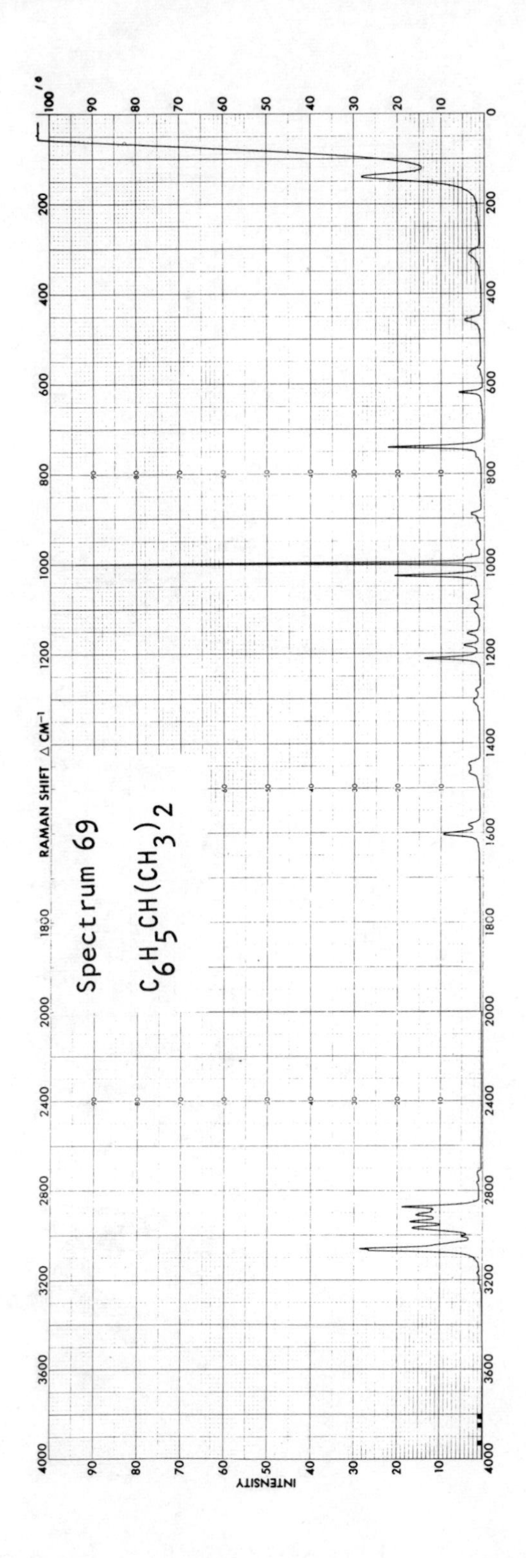
Spectrum 69
$C_6H_5CH(CH_3)_2$
RAMAN SHIFT Δ CM⁻¹
INTENSITY

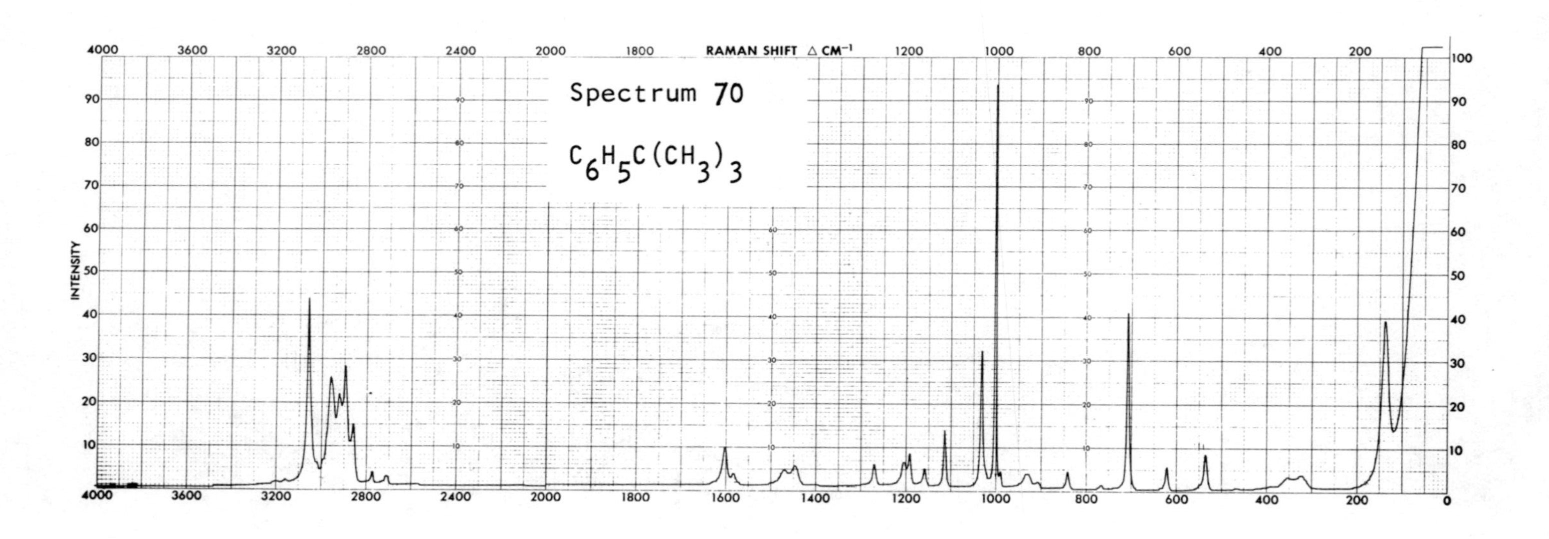
RAMAN SHIFT △ CM⁻¹
INTENSITY
Spectrum 70
C6H5C(CH3)3

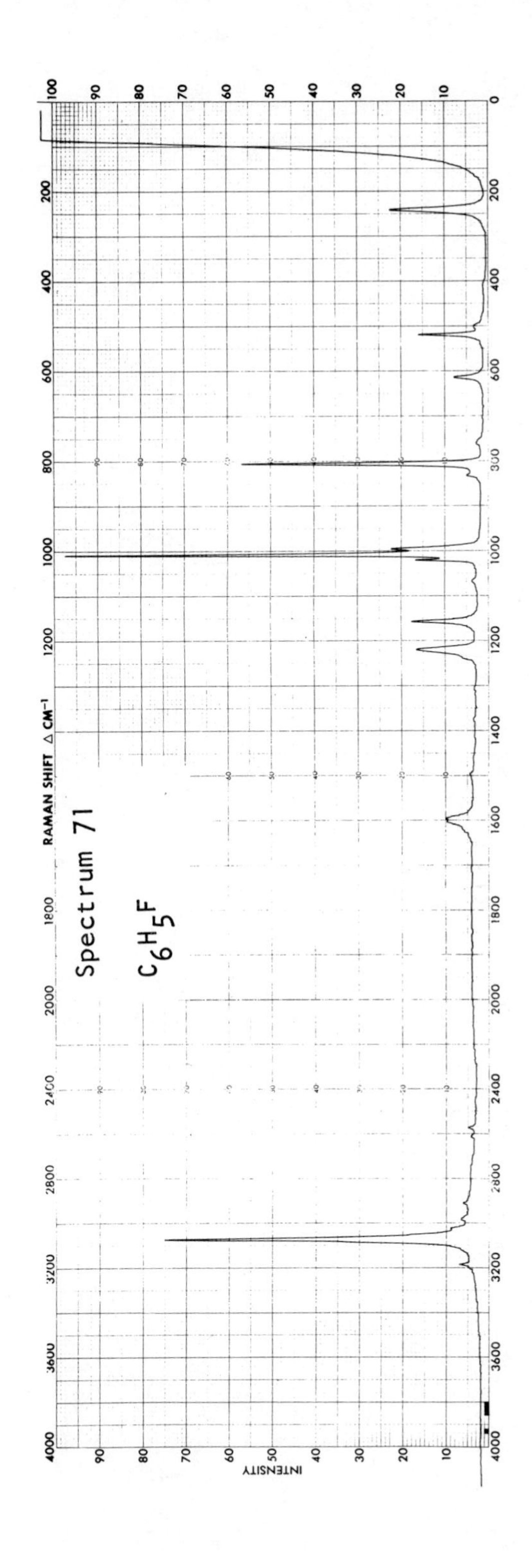

Spectrum 71
C_6H_5F
RAMAN SHIFT Δ CM^{-1}
INTENSITY

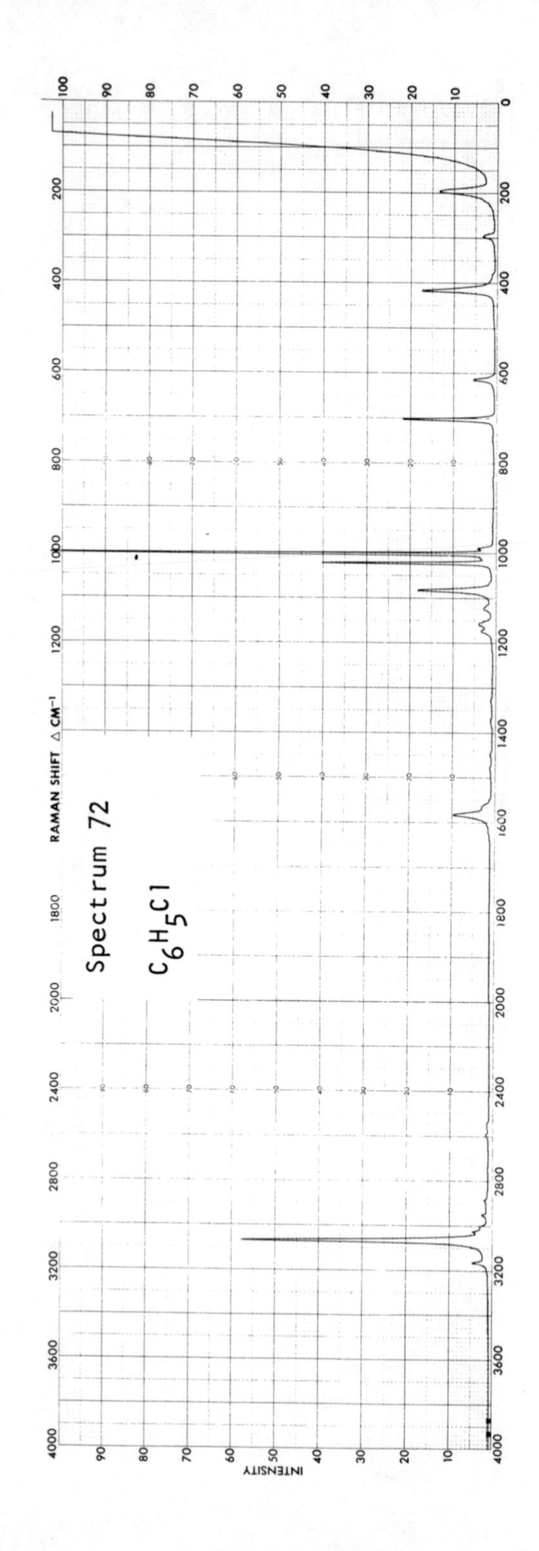
Spectrum 72
C_6H_5Cl
RAMAN SHIFT Δ CM⁻¹
INTENSITY

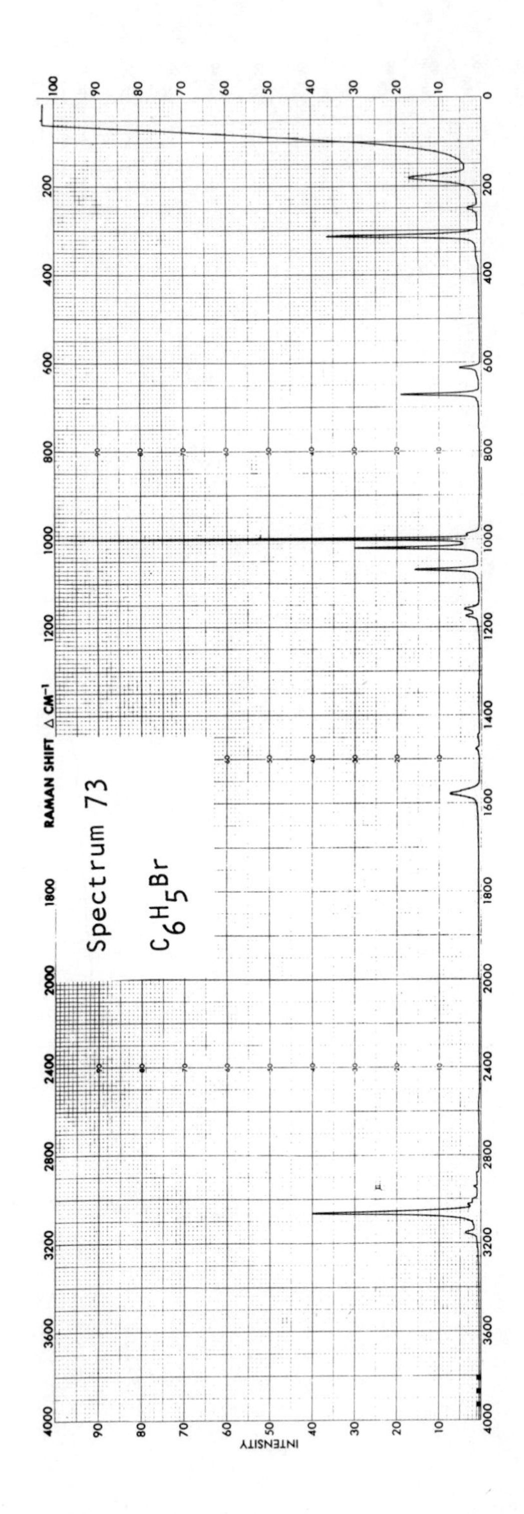

Spectrum 73
C_6H_5Br
RAMAN SHIFT Δ CM⁻¹
INTENSITY

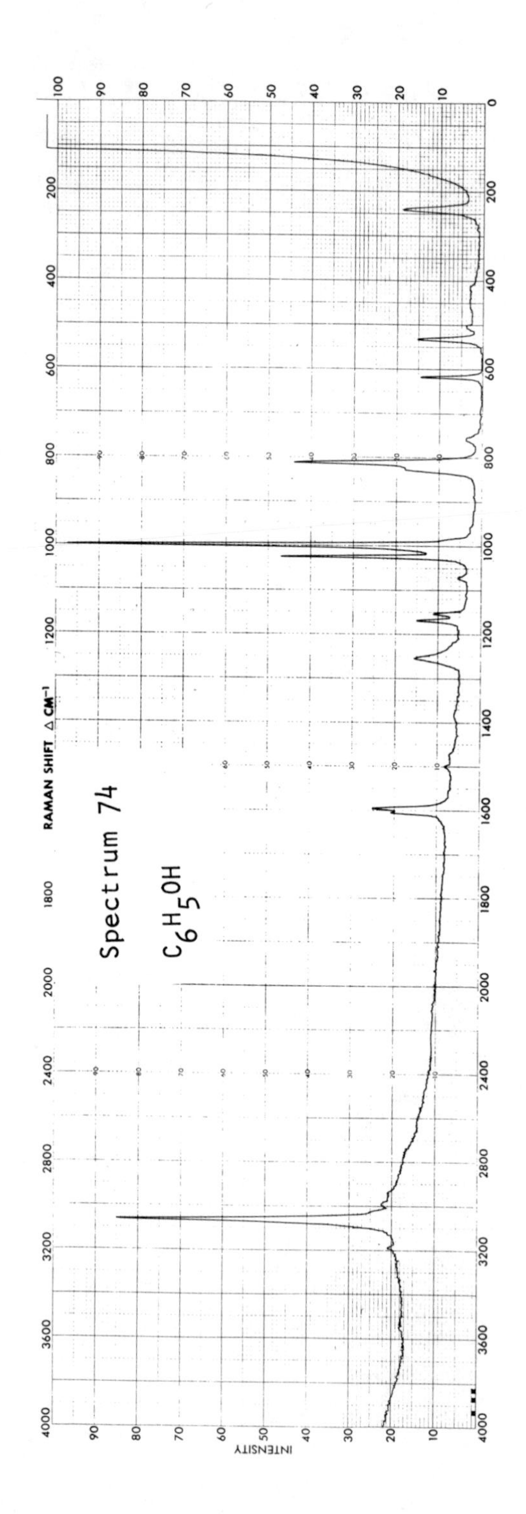

RAMAN SHIFT Δ CM⁻¹
INTENSITY
Spectrum 74
C6H5OH

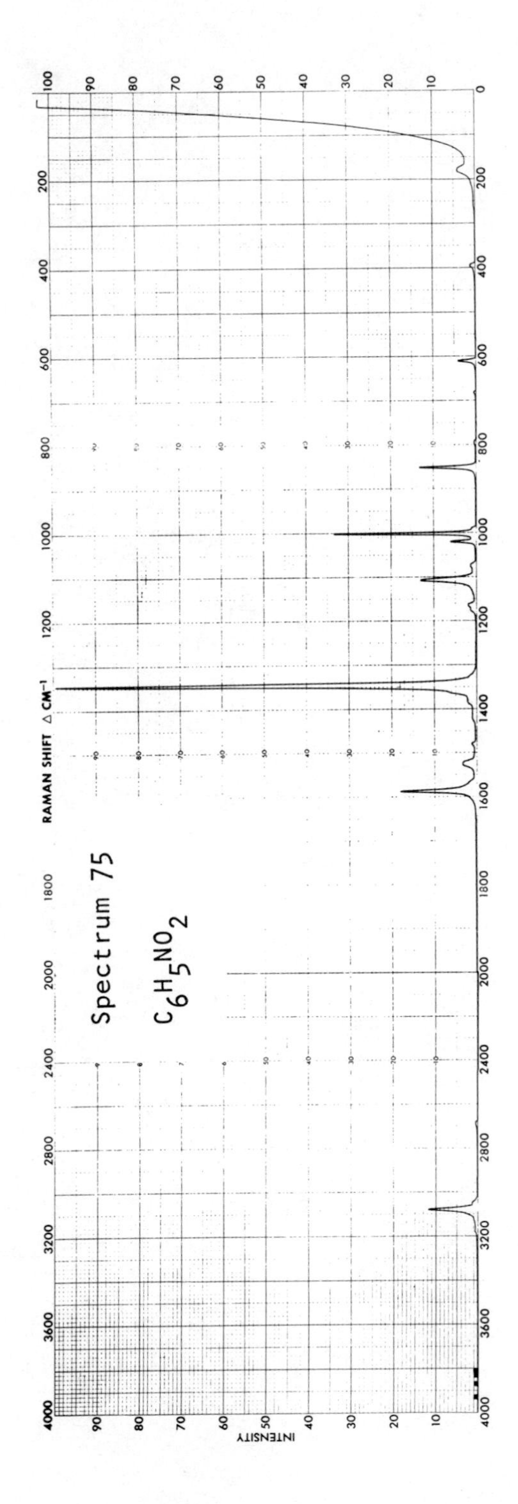
RAMAN SHIFT Δ CM^{-1}
INTENSITY
Spectrum 75
$C_6H_5NO_2$

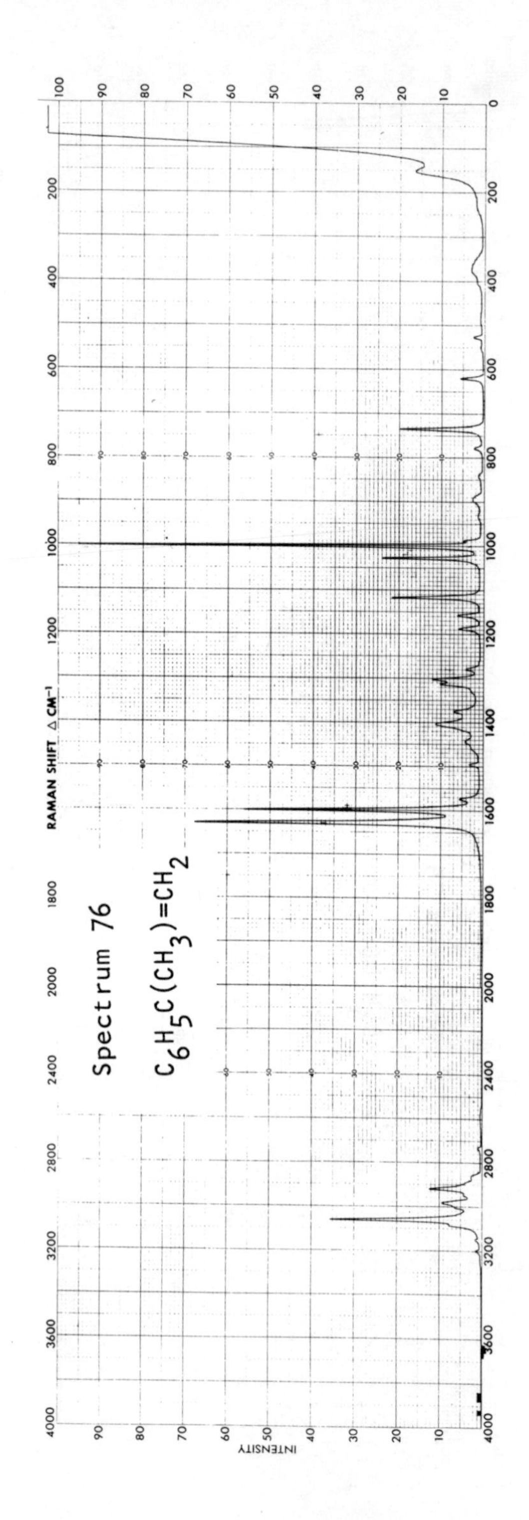
Spectrum 76
C6H5C(CH3)=CH2
RAMAN SHIFT Δ CM⁻¹
INTENSITY

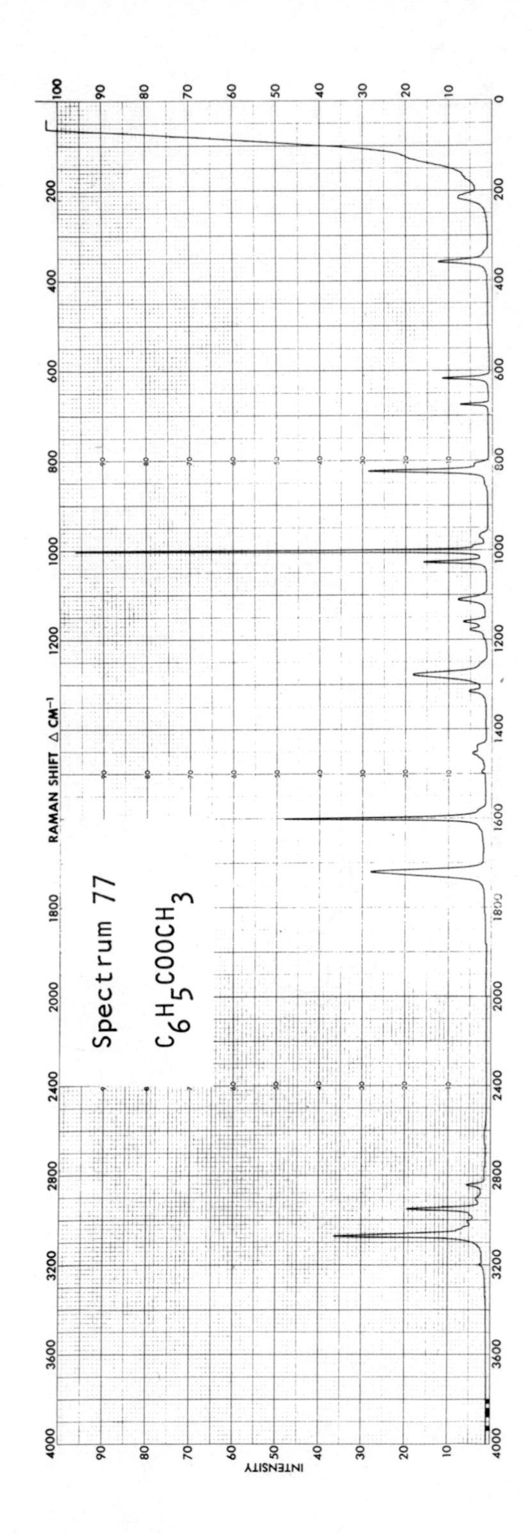
Spectrum 77
$C_6H_5COOCH_3$
RAMAN SHIFT Δ CM⁻¹
INTENSITY

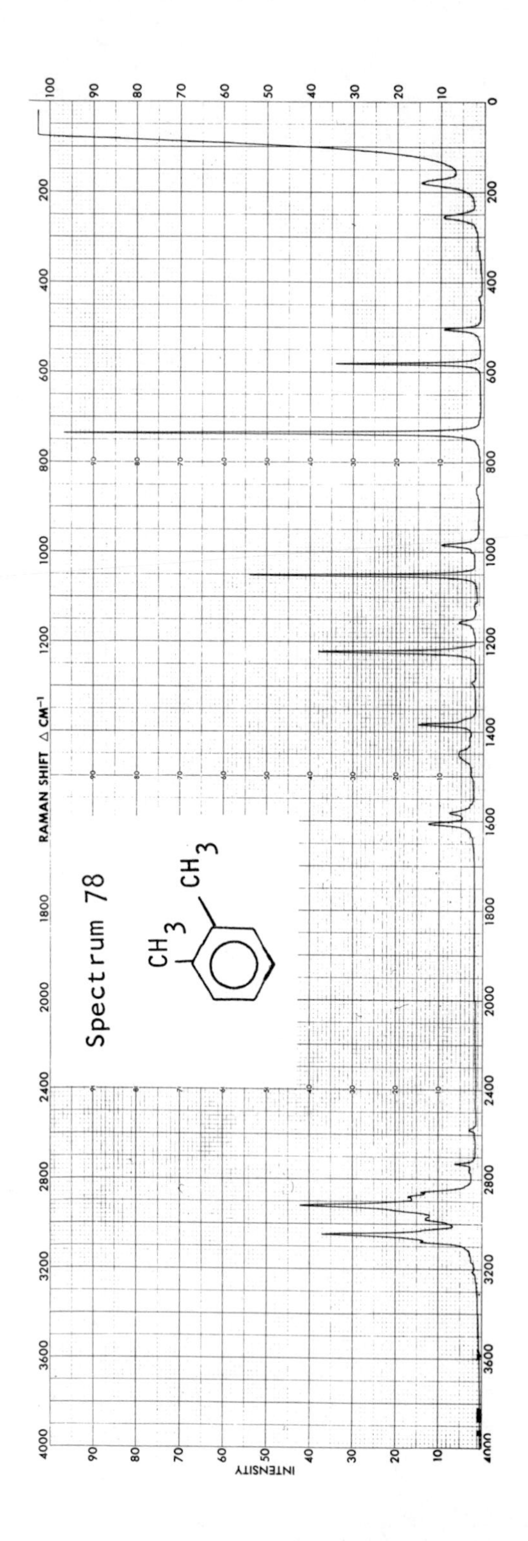
Spectrum 78
CH3
CH3
RAMAN SHIFT Δ CM⁻¹
INTENSITY
4000 3600 3200 2800 2400 2000 1800 1600 1400 1200 1000 800 600 400 200
0 10 20 30 40 50 60 70 80 90 100

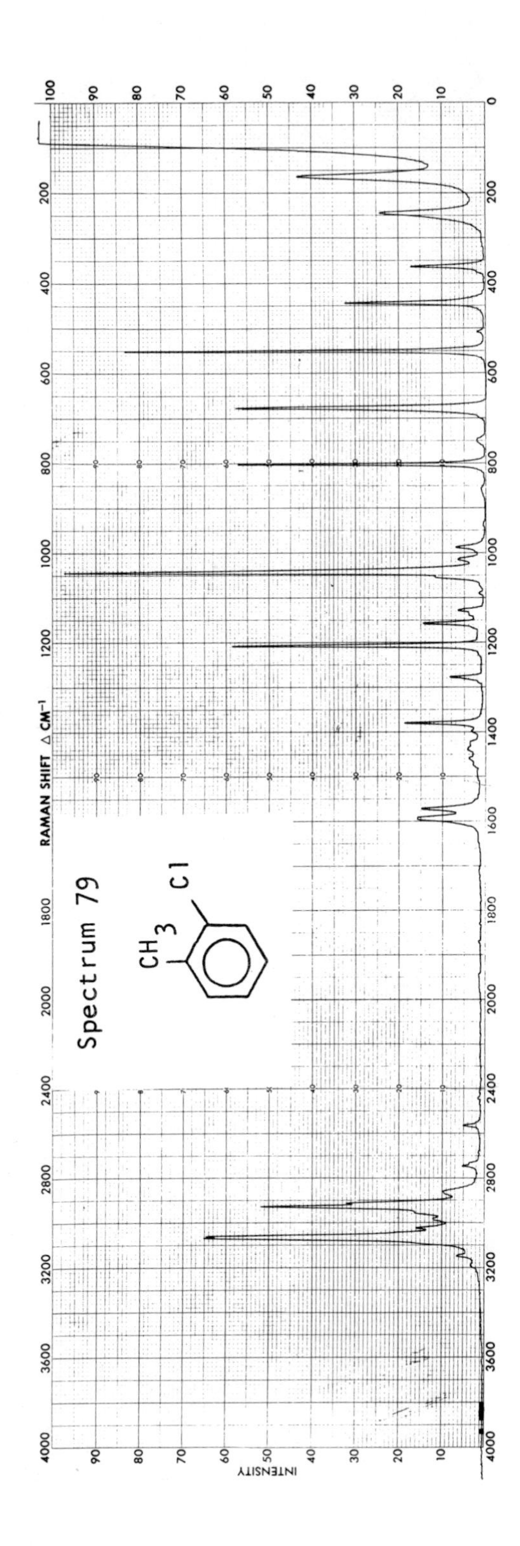
Spectrum 79
CH
3
Cl
RAMAN SHIFT Δ CM⁻¹
INTENSITY
4000 3600 3200 2800 2400 2000 1800 1600 1400 1200 1000 800 600 400 200
0 10 20 30 40 50 60 70 80 90 100

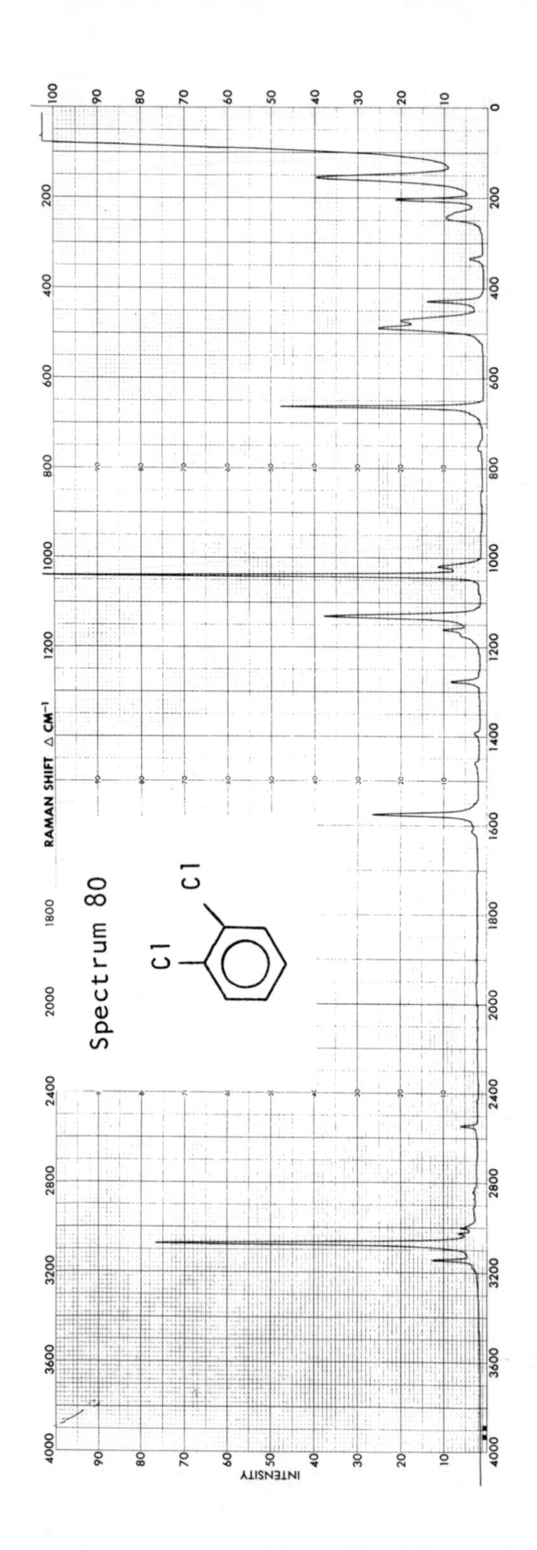
Spectrum 80
Cl
Cl
RAMAN SHIFT Δ CM⁻¹
INTENSITY

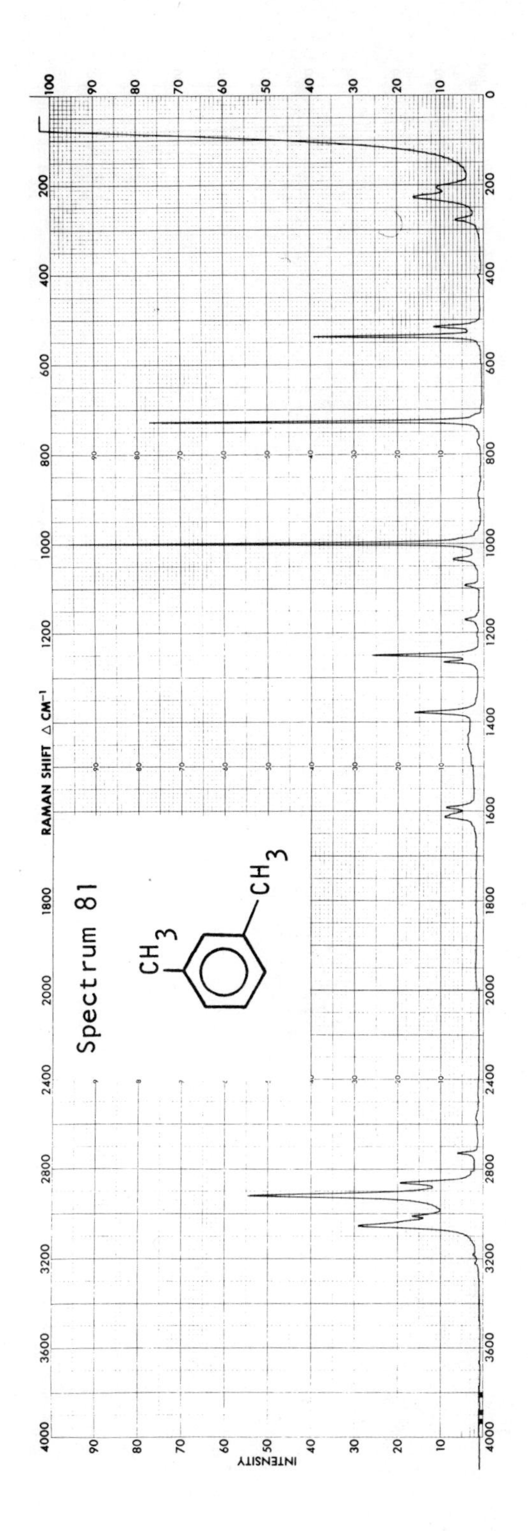

Spectrum 81
CH3
CH3
RAMAN SHIFT Δ CM⁻¹
INTENSITY
0 200 400 600 800 1000 1200 1400 1600 1800 2000 2400 2800 3200 3600 4000
10 20 30 40 50 60 70 80 90 100

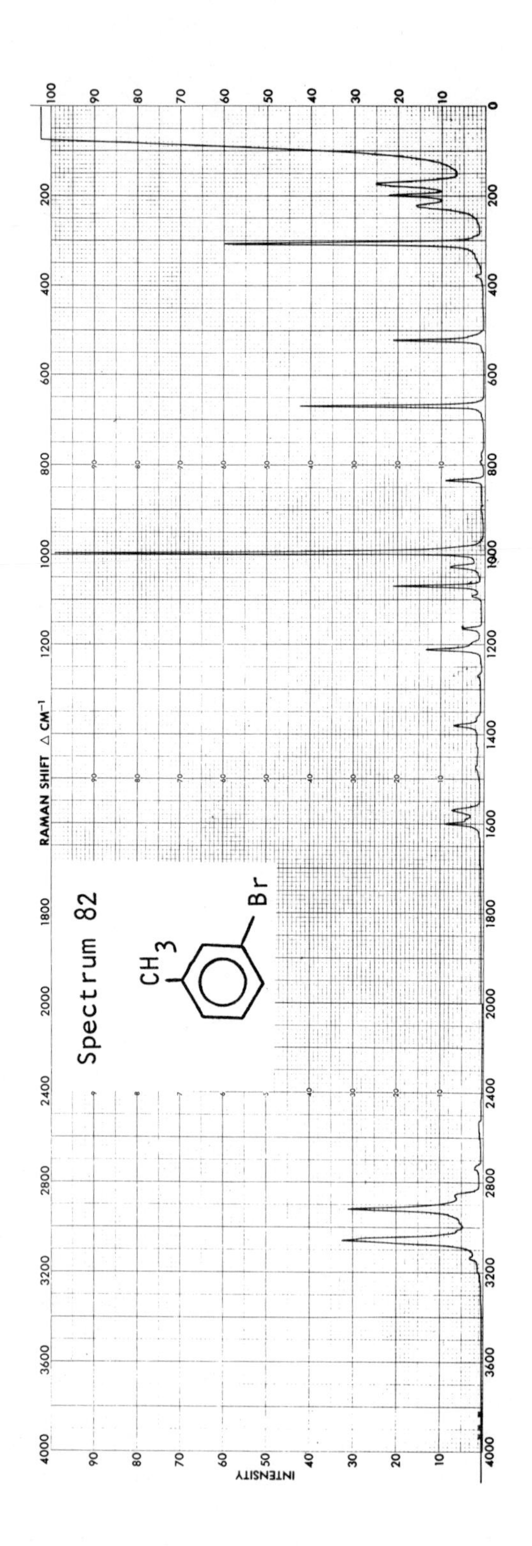

Spectrum 82
CH3
Br
RAMAN SHIFT Δ CM⁻¹
INTENSITY

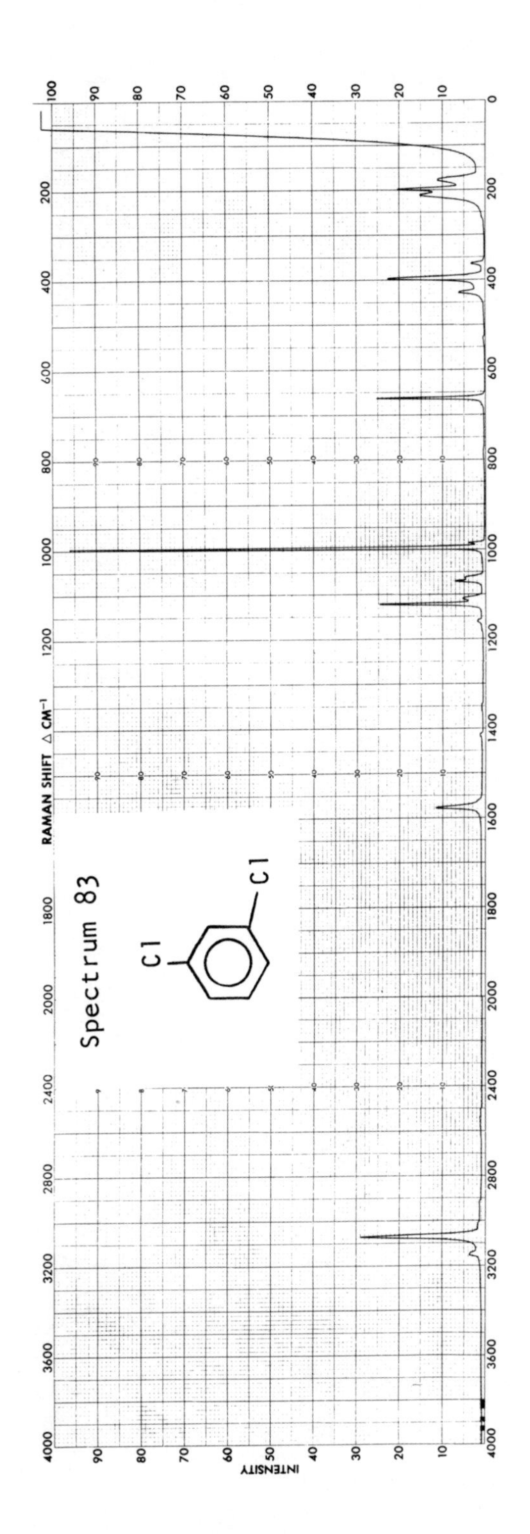
Spectrum 83
Cl
Cl
RAMAN SHIFT Δ CM⁻¹
INTENSITY

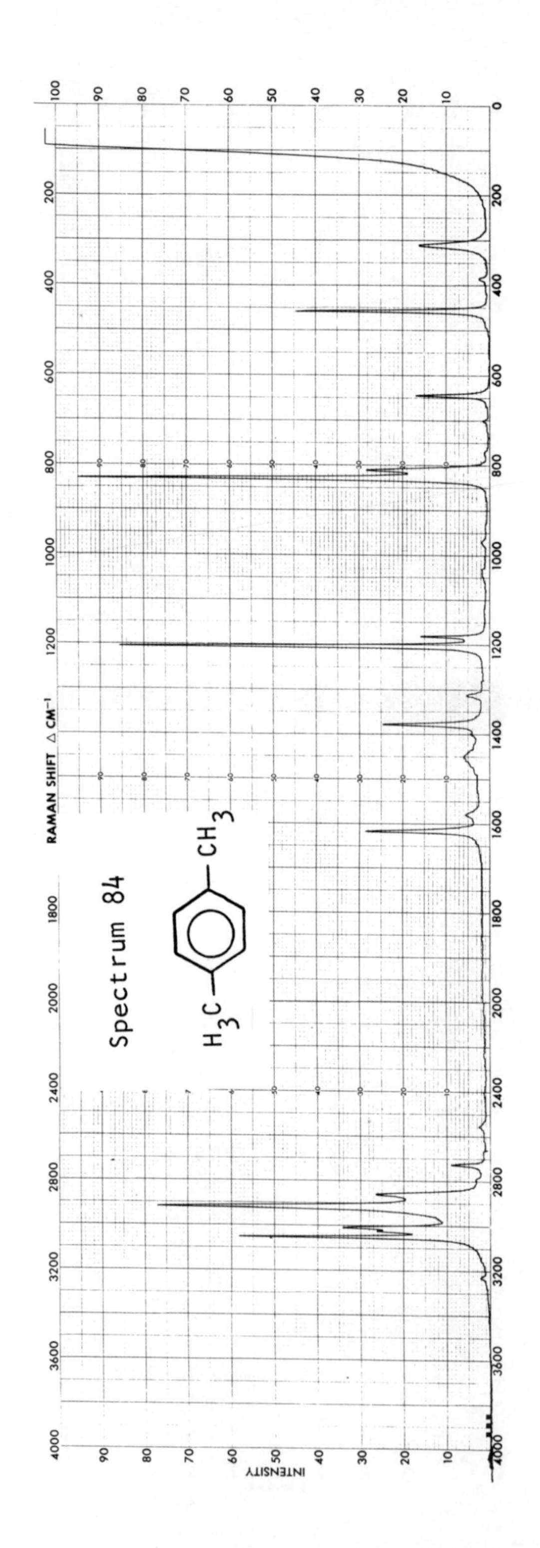
Spectrum 84
H_3C
CH_3
RAMAN SHIFT Δ CM⁻¹
INTENSITY

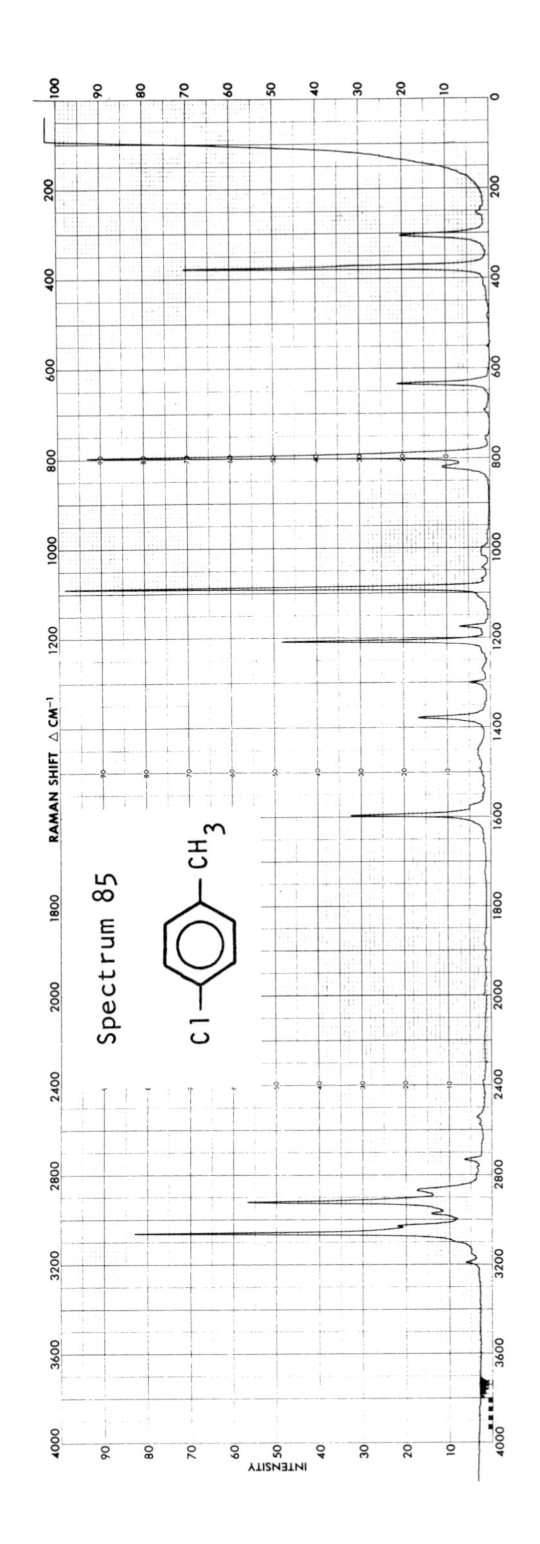
Spectrum 85
Cl
CH3
RAMAN SHIFT Δ CM⁻¹
INTENSITY

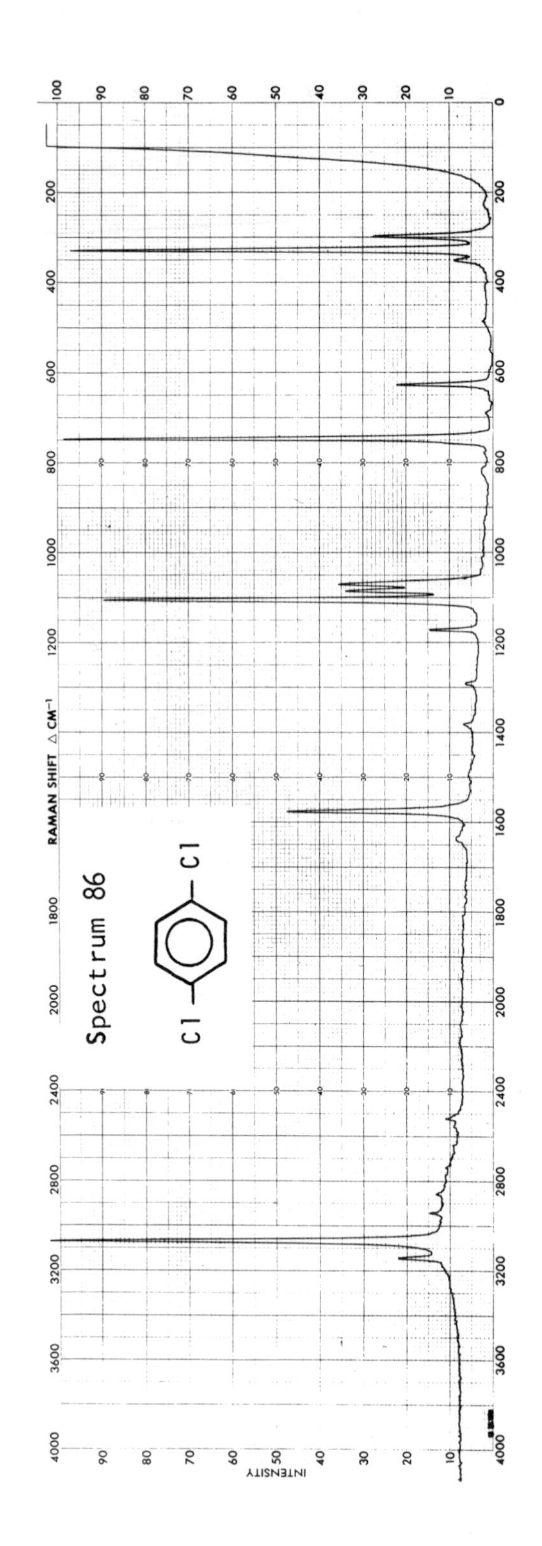
Spectrum 86
Cl
Cl
RAMAN SHIFT △ CM⁻¹
INTENSITY
0
200
400
600
800
1000
1200
1400
1600
1800
2000
2400
2800
3200
3600
4000
10
20
30
40
50
60
70
80
90
100

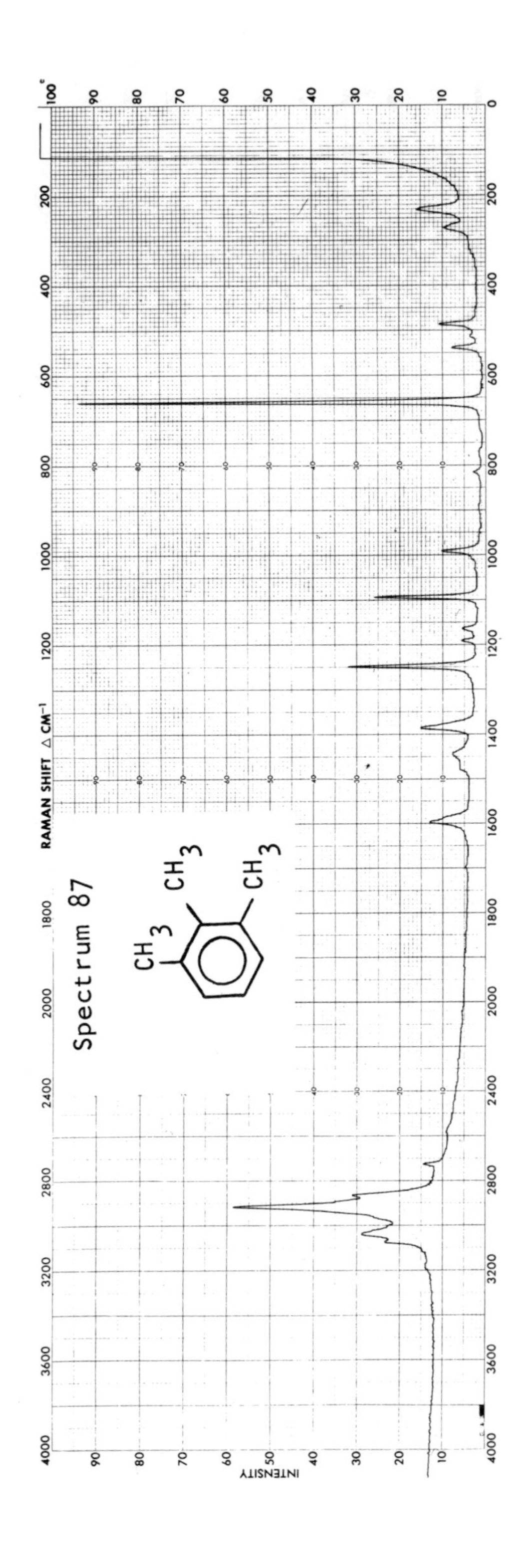
Spectrum 87
CH3
CH3
CH3
RAMAN SHIFT Δ CM⁻¹
INTENSITY

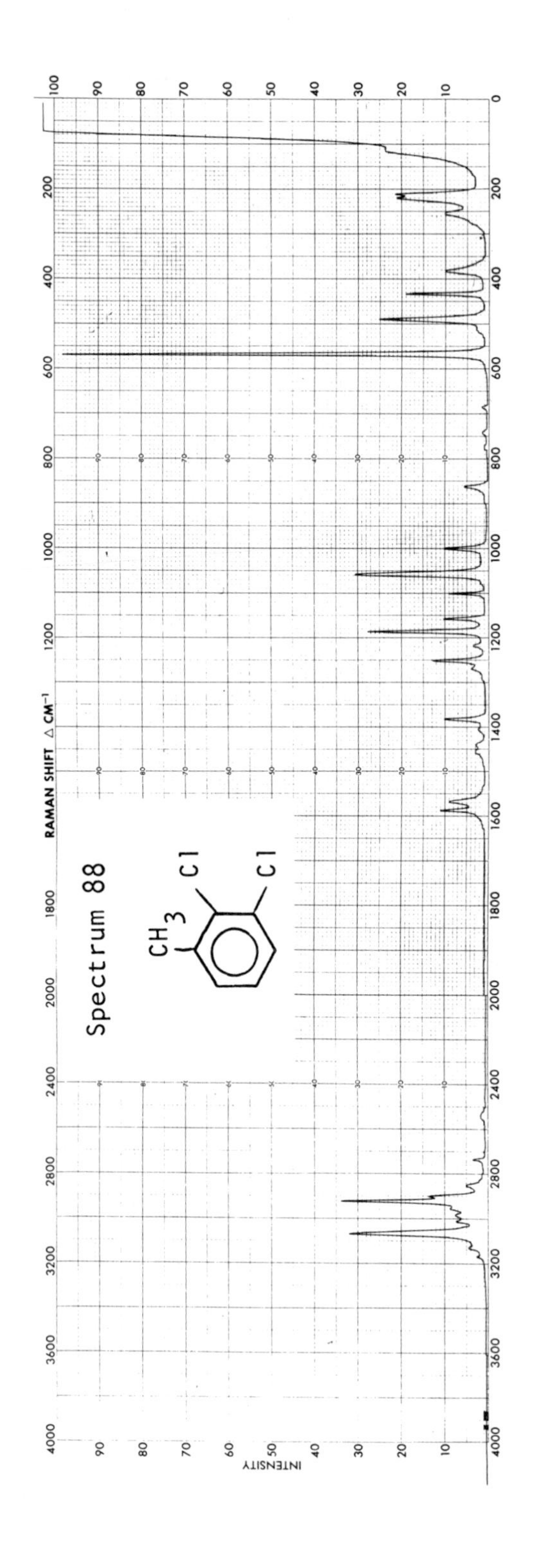
Spectrum 88
CH3
Cl
Cl
RAMAN SHIFT Δ CM⁻¹
INTENSITY

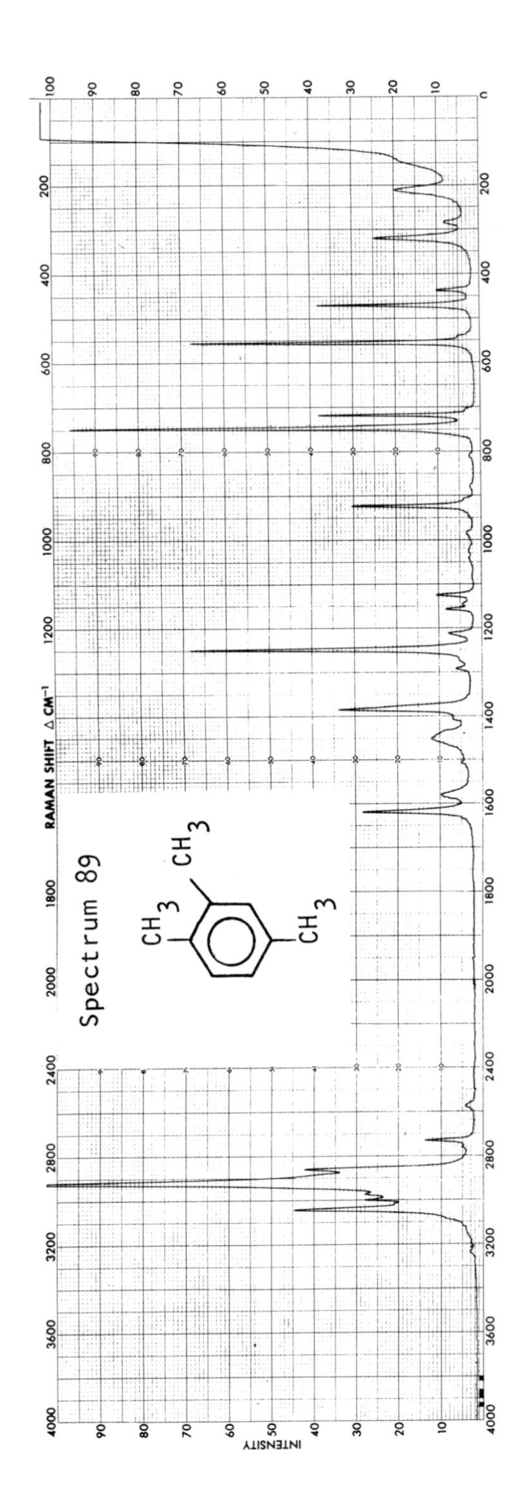

Spectrum 89
RAMAN SHIFT Δ CM⁻¹
INTENSITY
CH3
CH3
CH3

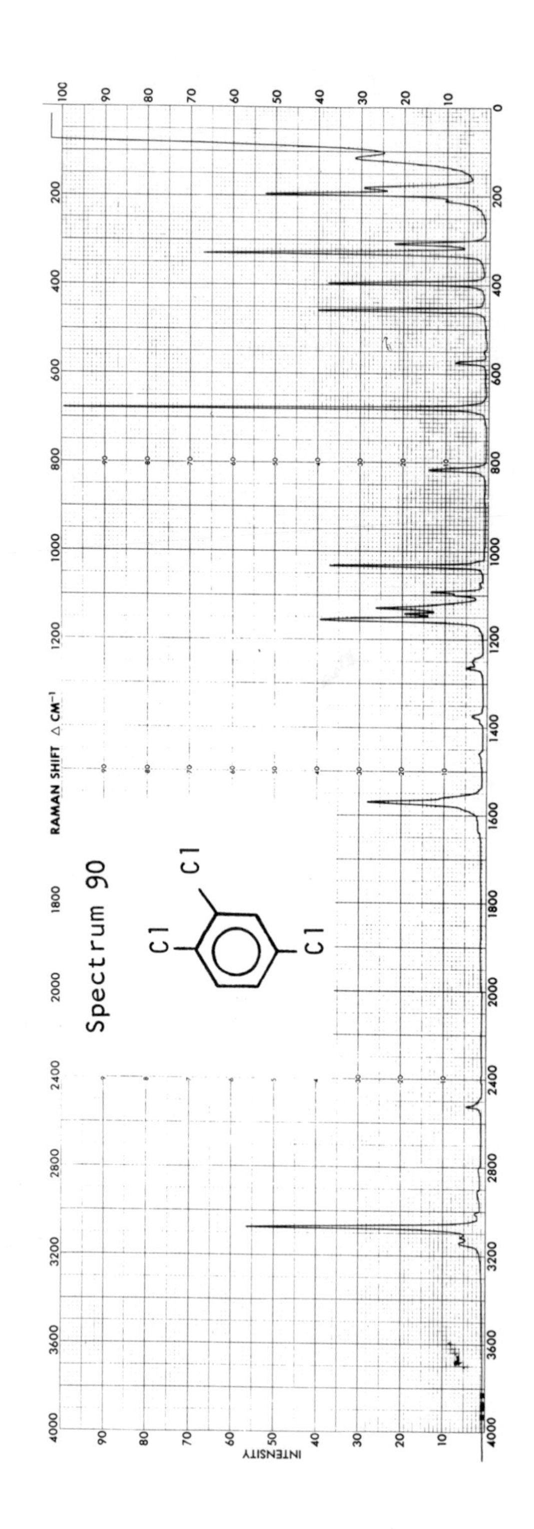
Spectrum 90
Cl
Cl
Cl
RAMAN SHIFT Δ CM⁻¹
INTENSITY

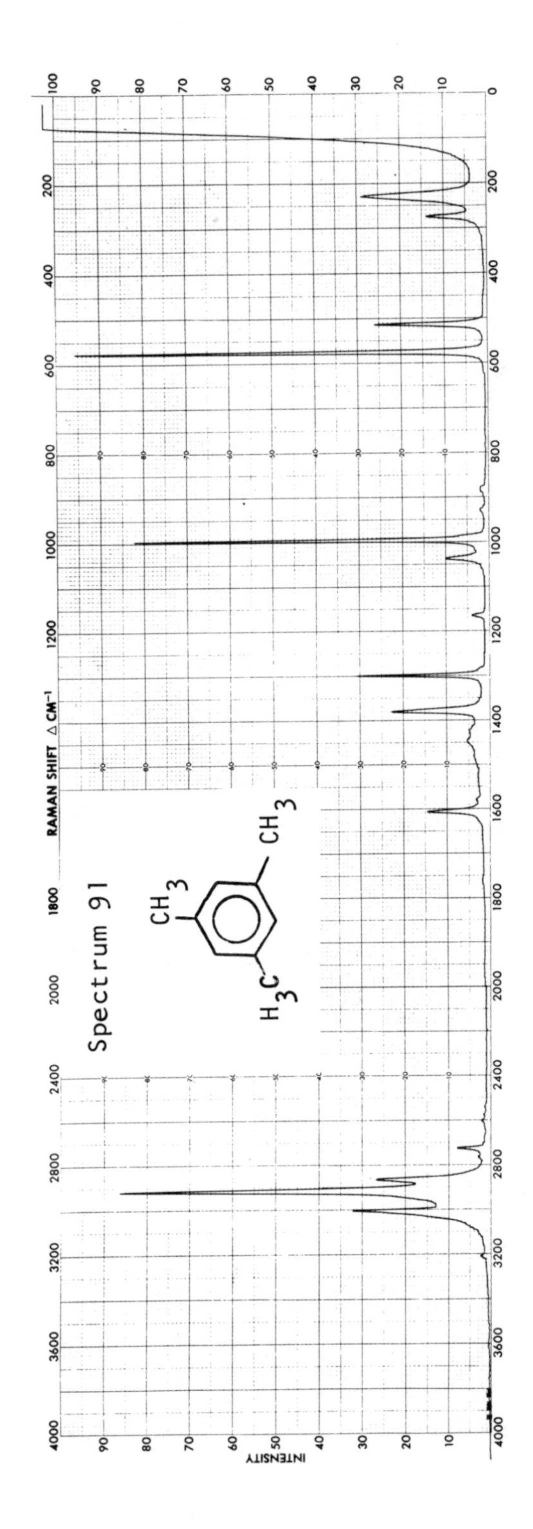
Spectrum 91
RAMAN SHIFT Δ CM⁻¹
INTENSITY
CH3
CH3
H3C

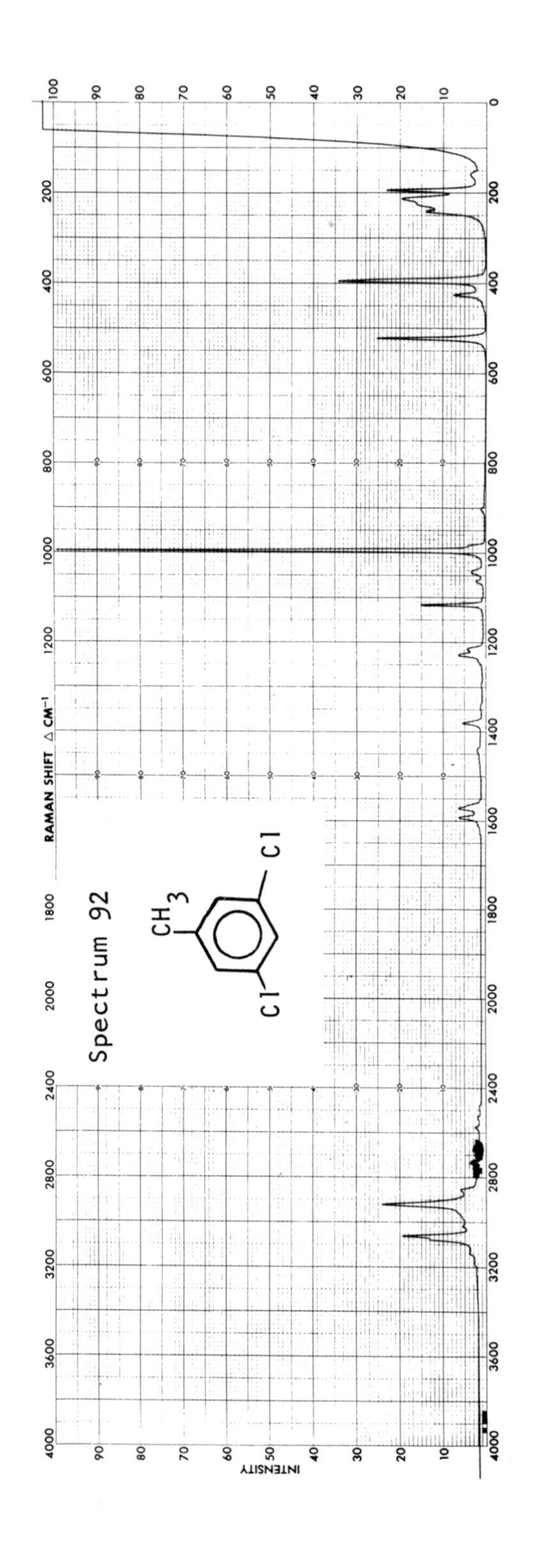
Spectrum 92
RAMAN SHIFT Δ CM⁻¹
INTENSITY
CH3
Cl
Cl

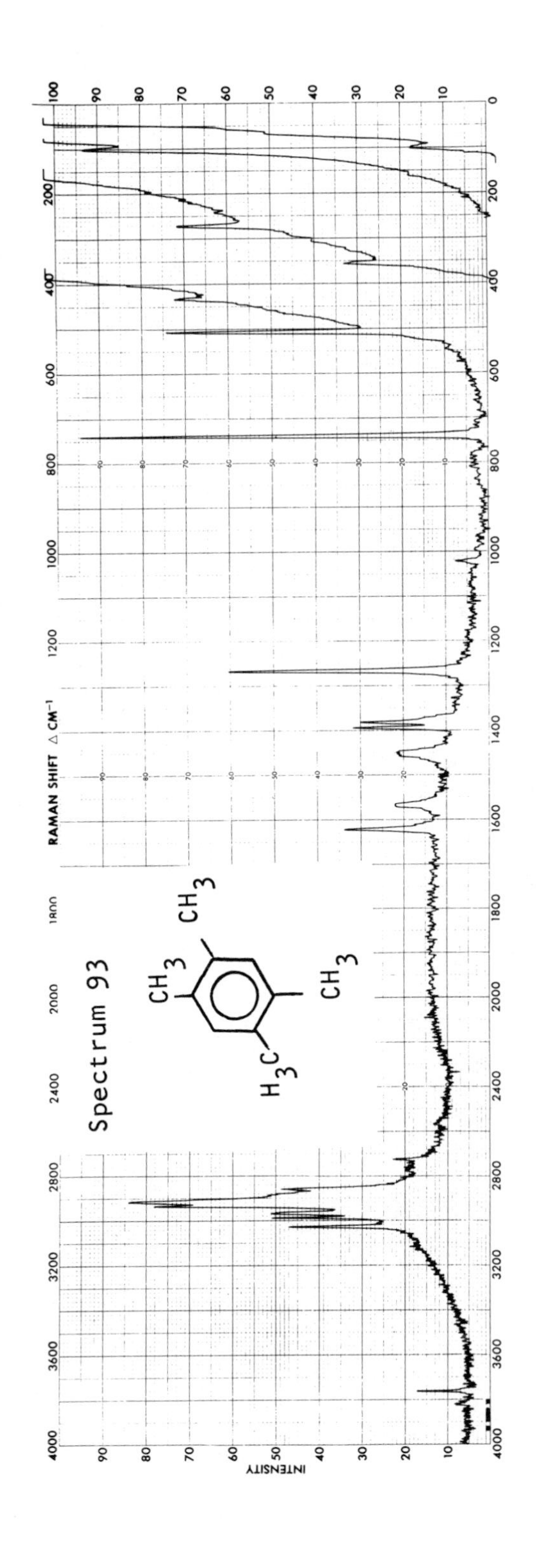
Spectrum 93
RAMAN SHIFT Δ CM⁻¹
INTENSITY
CH3
CH3
CH3
H3C

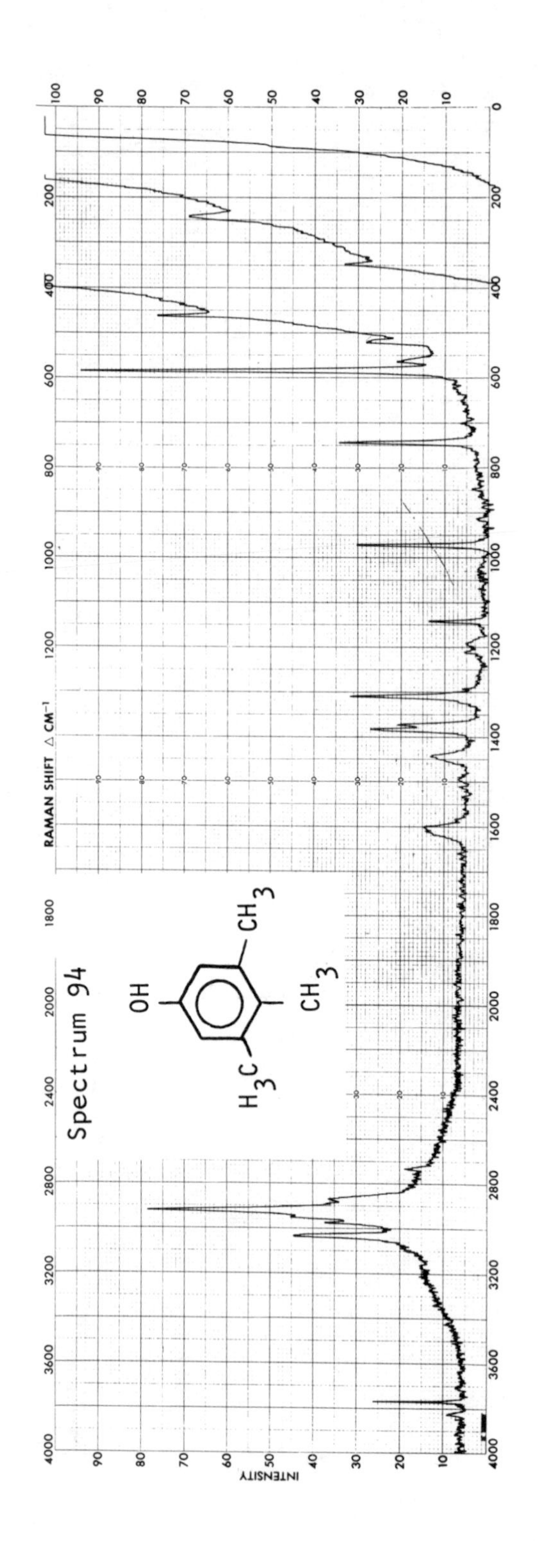
Spectrum 94
OH
H_3C
CH_3
CH_3
RAMAN SHIFT Δ CM⁻¹
INTENSITY

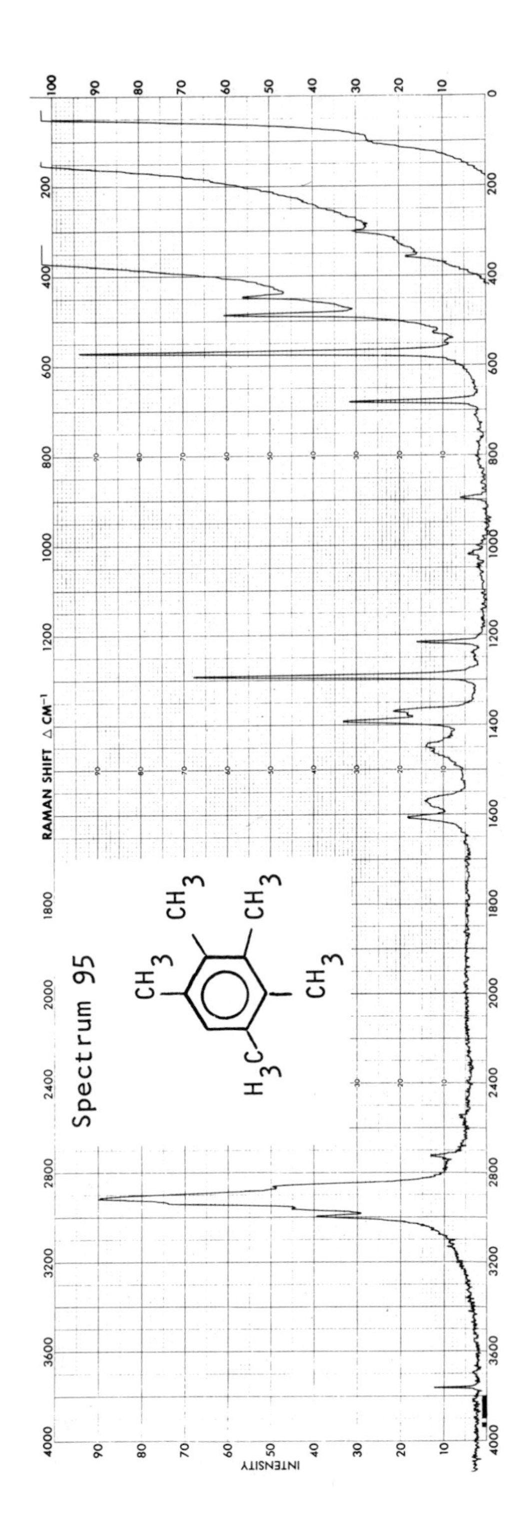
Spectrum 95
RAMAN SHIFT Δ CM⁻¹
INTENSITY
CH3
CH3
CH3
CH3
H3C

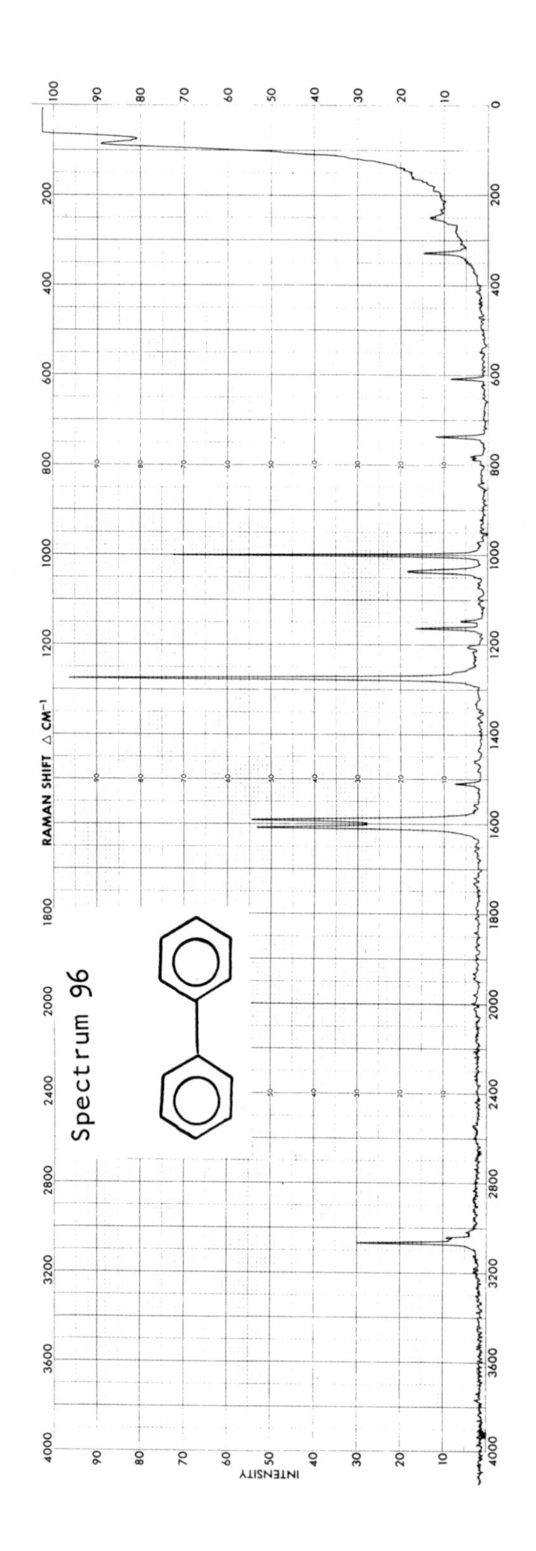
Spectrum 96
RAMAN SHIFT △ CM⁻¹
INTENSITY
0
200
400
600
800
1000
1200
1400
1600
1800
2000
2400
2800
3200
3600
4000
10
20
30
40
50
60
70
80
90
100

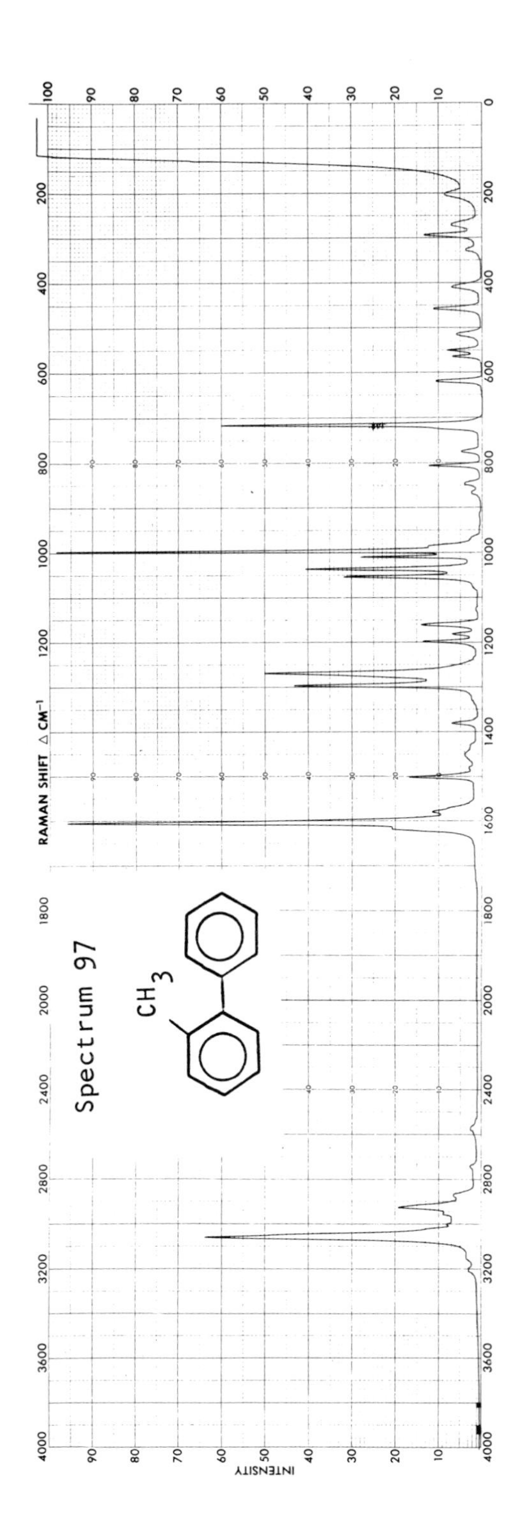
Spectrum 97
CH3
RAMAN SHIFT Δ CM⁻¹
INTENSITY

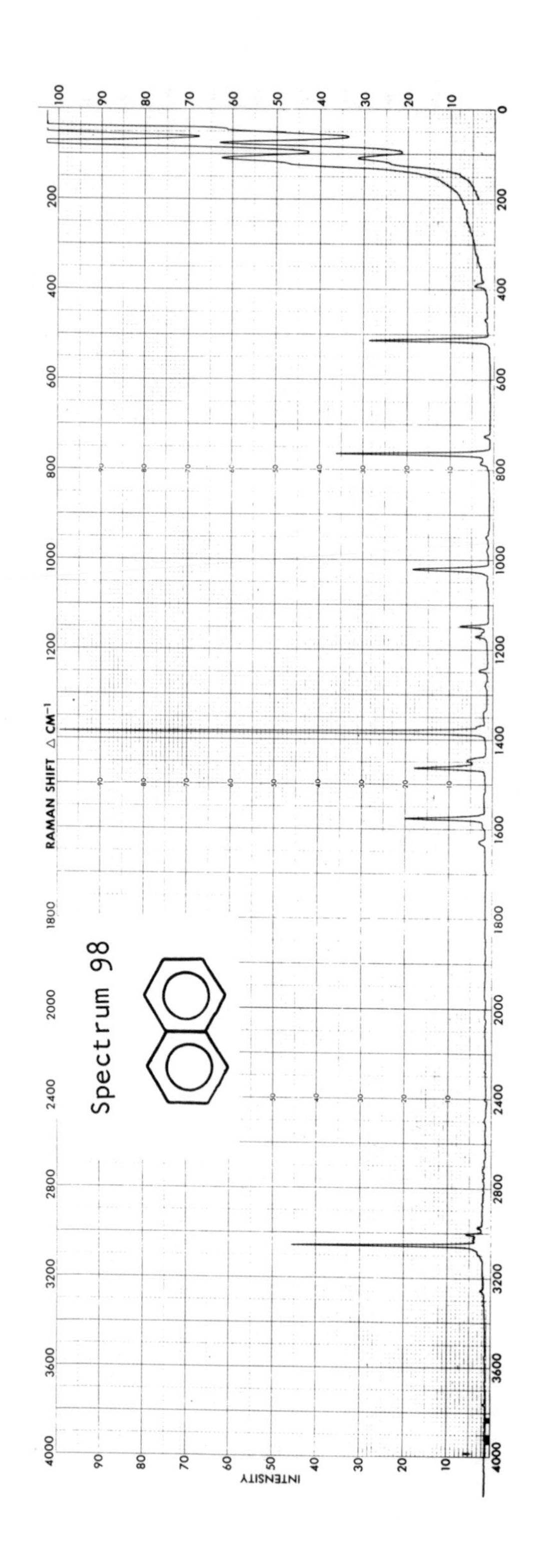
Spectrum 98
RAMAN SHIFT Δ CM⁻¹
INTENSITY

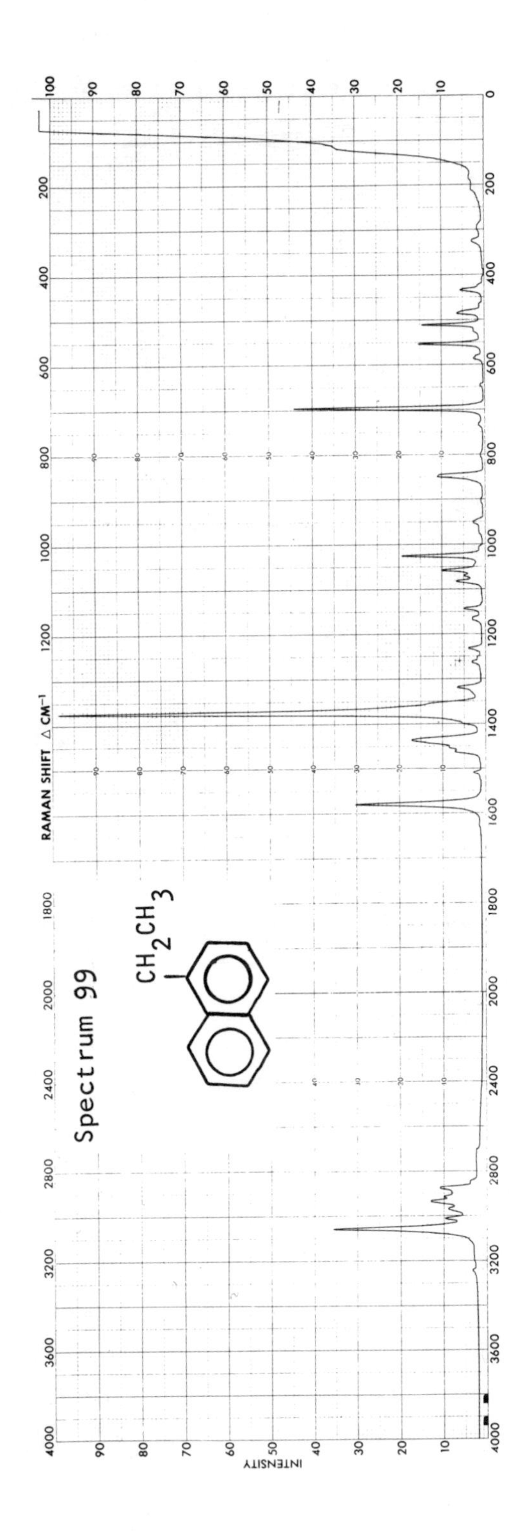
Spectrum 99
CH_2CH_3
RAMAN SHIFT Δ CM^{-1}
INTENSITY
4000
3600
3200
2800
2400
2000
1800
1600
1400
1200
1000
800
600
400
200
0
100
90
80
70
60
50
40
30
20
10

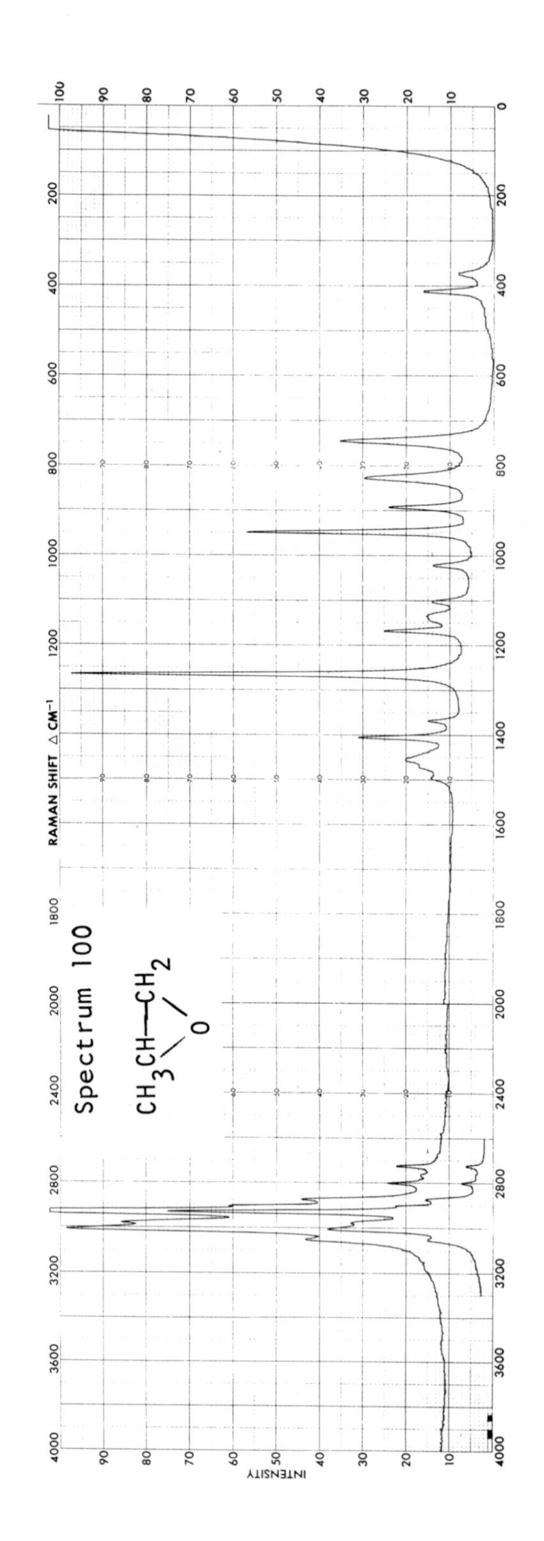
Spectrum 100
$CH_3CH—CH_2$ (with O bridging the two carbons, epoxide ring)
RAMAN SHIFT Δ CM⁻¹
INTENSITY

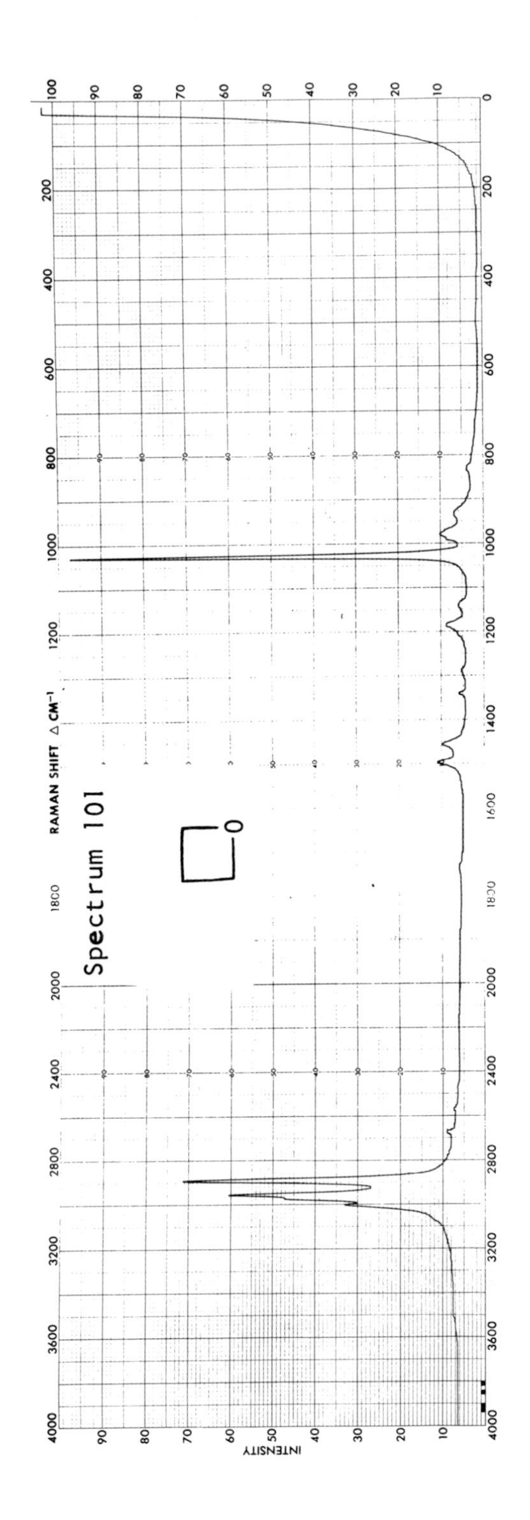
Spectrum 101
RAMAN SHIFT Δ CM⁻¹
INTENSITY

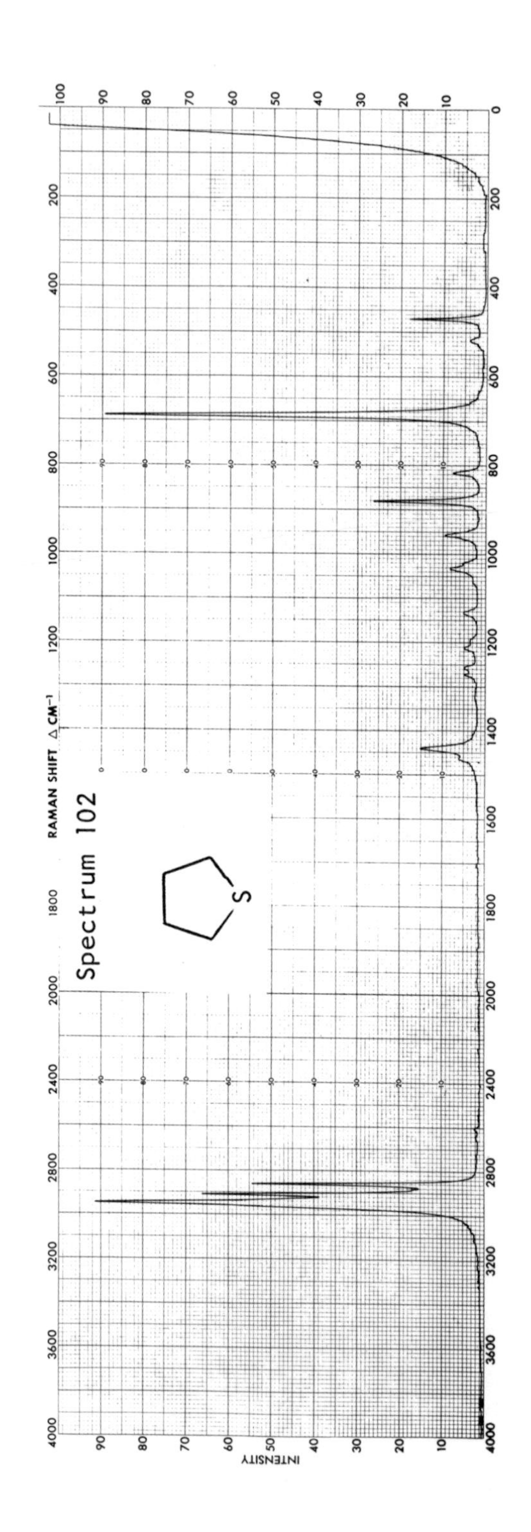
Spectrum 102
S
RAMAN SHIFT Δ CM⁻¹
INTENSITY

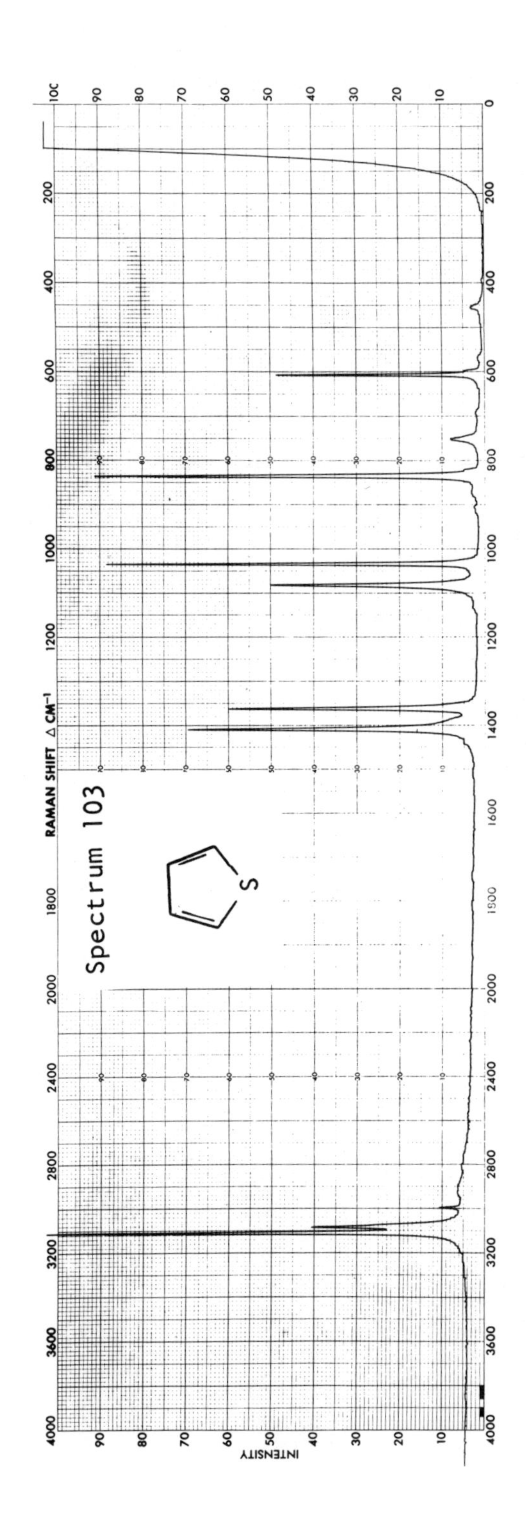
Spectrum 103
RAMAN SHIFT Δ CM⁻¹
INTENSITY
S

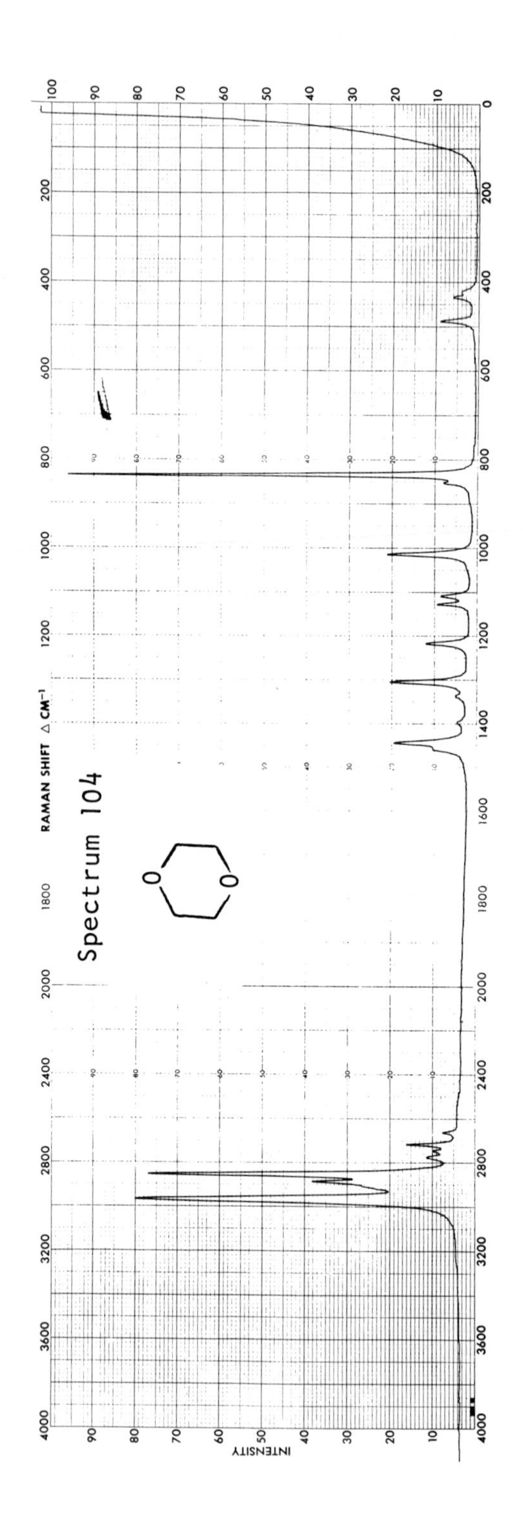
Spectrum 104
RAMAN SHIFT Δ CM⁻¹
INTENSITY

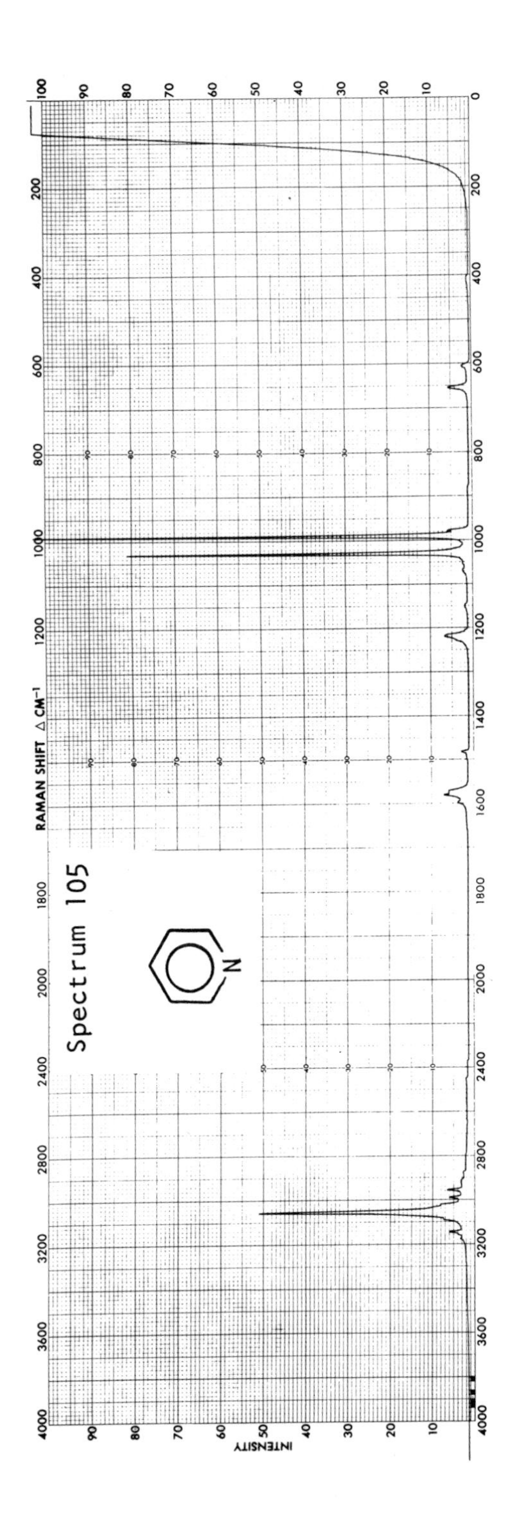
Spectrum 105
N
RAMAN SHIFT △ CM⁻¹
INTENSITY

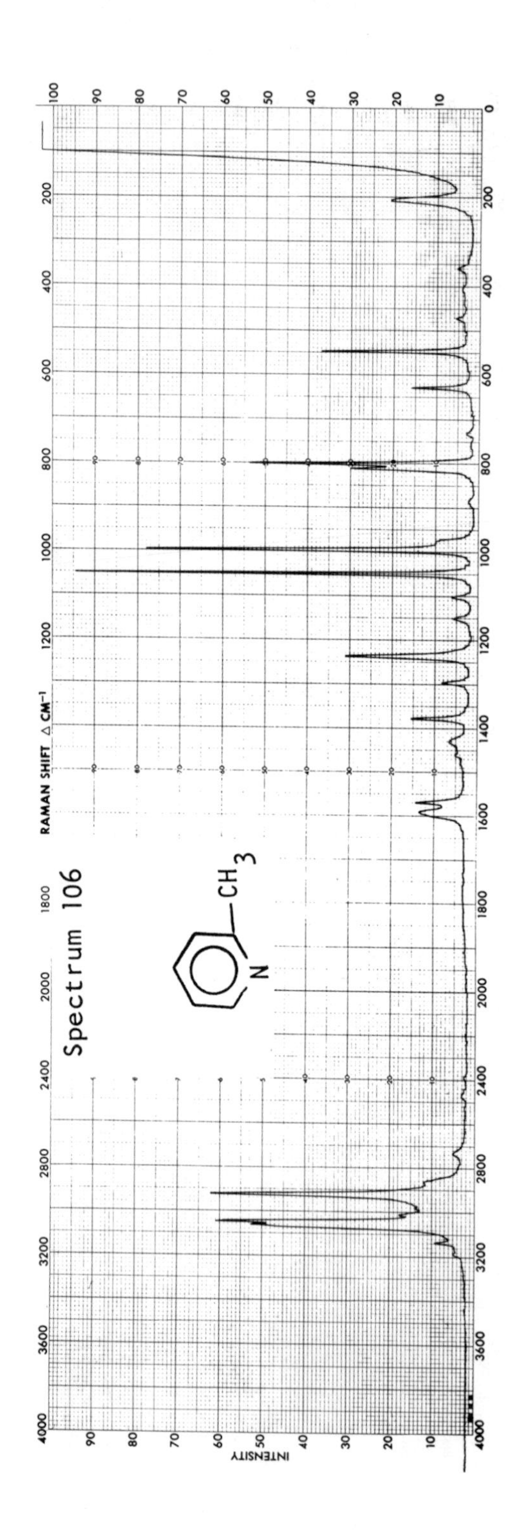
Spectrum 106
RAMAN SHIFT Δ CM⁻¹
INTENSITY
CH3
N

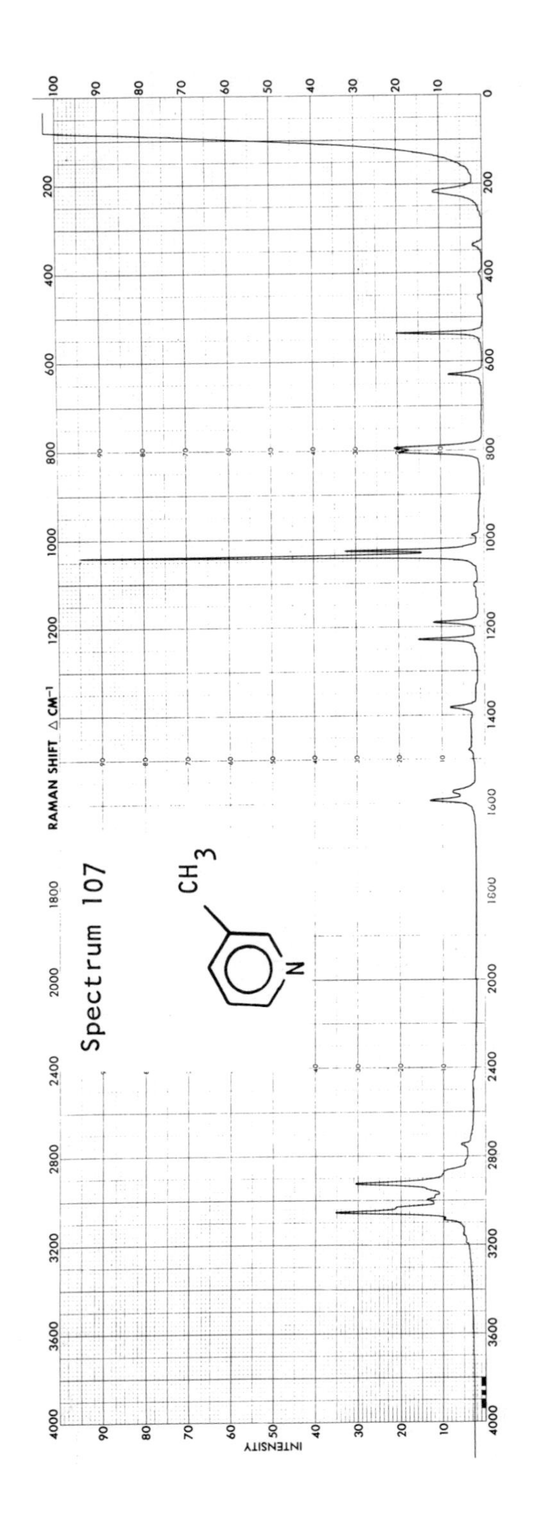
Spectrum 107
RAMAN SHIFT Δ CM⁻¹
INTENSITY
CH3
N

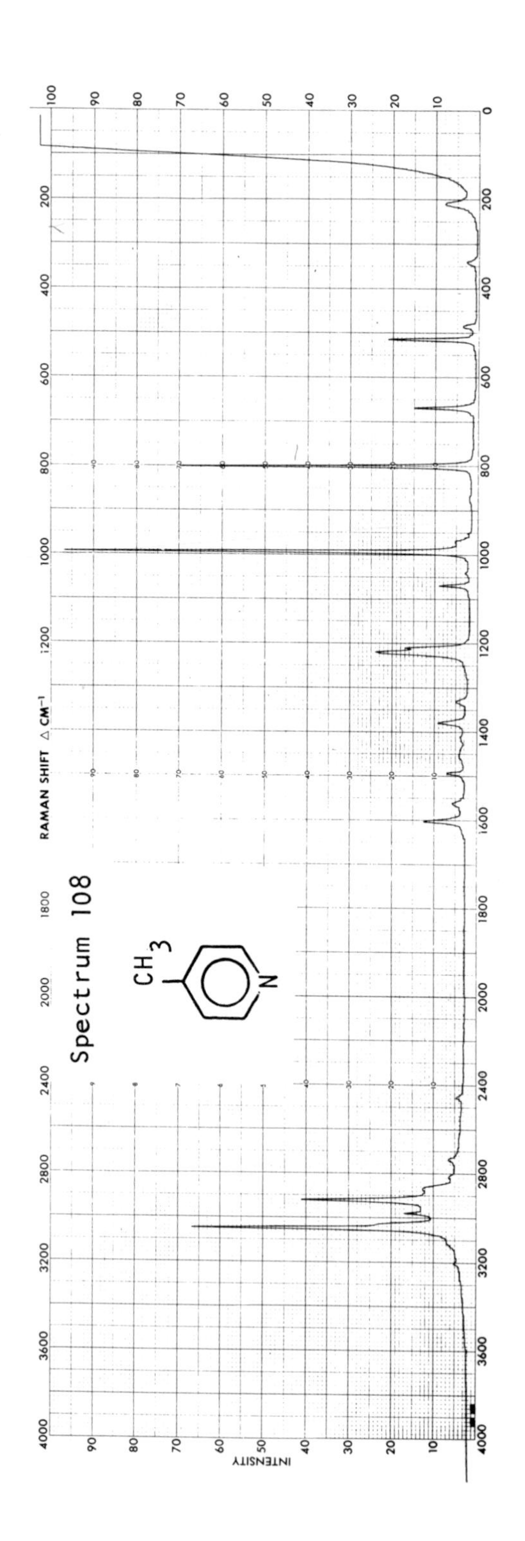
Spectrum 108
CH3
N
RAMAN SHIFT △ CM⁻¹
INTENSITY
4000
3600
3200
2800
2400
2000
1800
1600
1400
1200
1000
800
600
400
200
0
100
90
80
70
60
50
40
30
20
10

POSTSCRIPT

". . . . and so there ain't nothing more to write about, and I am rotten glad of it, because if I'd 'a' knowed what a trouble it was to make a book I wouldn't 'a' tackled it, and ain't a-going to no more."

Mark Twain, *Huckleberry Finn*

AUTHOR INDEX

Numbers in parentheses indicate the citation of an author's work when his name does not appear on the page. Numbers in italics refer to pages on which the complete reference is given.

SUBJECT INDEX

COMPOUND INDEX

A compound is listed if specific Raman frequencies are given in the text and/or reference to the complete spectrum in the literature is provided. Page numbers in italics indicate a representative Raman spectrum in Appendix Two.